CAMBRIDGE LIBRARY COLLECTION

Books of enduring scholarly value

Physical Sciences

From ancient times, humans have tried to understand the workings of the world around them. The roots of modern physical science go back to the very earliest mechanical devices such as levers and rollers, the mixing of paints and dyes, and the importance of the heavenly bodies in early religious observance and navigation. The physical sciences as we know them today began to emerge as independent academic subjects during the early modern period, in the work of Newton and other 'natural philosophers', and numerous sub-disciplines developed during the centuries that followed. This part of the Cambridge Library Collection is devoted to landmark publications in this area which will be of interest to historians of science concerned with individual scientists, particular discoveries, and advances in scientific method, or with the establishment and development of scientific institutions around the world.

The Life of William Thomson, Baron Kelvin of Largs

The mathematician and physicist William Thomson, 1st Baron Kelvin, (1824–1907) was one of Britain's most influential scientists, famous for his work on the first and second laws of thermodynamics and for devising the Kelvin scale of absolute temperature. Silvanus P. Thompson (1851–1916) began this biography with the co-operation of Kelvin in 1906, but the project was interrupted by Kelvin's death the following year. Thompson, himself a respected physics lecturer and scientific writer, decided that a more comprehensive biography would be needed and spent several years reading through Kelvin's papers in order to complete these two volumes, published in 1910. Volume 1 covers Kelvin's life to 1871, including his student days, his election (aged 22) as professor in Glasgow, his ground-breaking theoretical work on thermodynamics, his applied work on telegraphs including the Atlantic cable, and his involvement in a geological controversy about the age of the earth.

Cambridge University Press has long been a pioneer in the reissuing of out-of-print titles from its own backlist, producing digital reprints of books that are still sought after by scholars and students but could not be reprinted economically using traditional technology. The Cambridge Library Collection extends this activity to a wider range of books which are still of importance to researchers and professionals, either for the source material they contain, or as landmarks in the history of their academic discipline.

Drawing from the world-renowned collections in the Cambridge University Library, and guided by the advice of experts in each subject area, Cambridge University Press is using state-of-the-art scanning machines in its own Printing House to capture the content of each book selected for inclusion. The files are processed to give a consistently clear, crisp image, and the books finished to the high quality standard for which the Press is recognised around the world. The latest print-on-demand technology ensures that the books will remain available indefinitely, and that orders for single or multiple copies can quickly be supplied.

The Cambridge Library Collection will bring back to life books of enduring scholarly value (including out-of-copyright works originally issued by other publishers) across a wide range of disciplines in the humanities and social sciences and in science and technology.

The Life of William Thomson, Baron Kelvin of Largs

VOLUME 1

SILVANUS PHILLIPS THOMPSON

CAMBRIDGE UNIVERSITY PRESS

Cambridge, New York, Melbourne, Madrid, Cape Town, Singapore,
São Paolo, Delhi, Dubai, Tokyo, Mexico City

Published in the United States of America by Cambridge University Press, New York

www.cambridge.org
Information on this title: www.cambridge.org/9781108027175

This edition first published 1910
This digitally printed version 2011

ISBN 978-1-108-02717-5 Paperback

LIFE OF LORD KELVIN

MACMILLAN AND CO., LIMITED
LONDON · BOMBAY · CALCUTTA
MELBOURNE

THE MACMILLAN COMPANY
NEW YORK BOSTON · CHICAGO
ATLANTA · SAN FRANCISCO

THE MACMILLAN CO. OF CANADA, LTD.
TORONTO

Kelvin

From a photograph by Annan, Glasgow. 1897

THE LIFE

OF

WILLIAM THOMSON

BARON KELVIN OF LARGS

BY

SILVANUS P. THOMPSON

IN TWO VOLUMES

VOL. I

MACMILLAN AND CO., LIMITED
ST. MARTIN'S STREET, LONDON
1910

Utinam caetera naturae phaenomena ex principiis mechanicis eodem argumentandi genere derivare licet.

NEWTON,
Phil. Nat. Principia Mathematica (Praef.).

PREFACE

THIS Biography was begun in June 1906 with the kind co-operation of Lord Kelvin, who himself furnished a number of personal recollections and data. His death in December 1907 affected the project of the work by necessarily extending its scope to present a much more comprehensive account of his career than the sketch originally planned. The mass of letters, diaries, and other documents which he left became available for filling in the outlines, and the task of arrangement and selection from these greatly extended the period of preparation.

The sympathy which has been so universally felt for Lady Kelvin in her prolonged illness and gradual recovery has manifested itself in many ways; and various friends have lightened for the author the responsibility of dealing with the available materials out of which to frame an authentic record of Lord Kelvin's long and strenuous career.

Thanks are due to many relations and scientific friends of Lord Kelvin, who have generously placed at the author's disposal letters covering every period of Lord Kelvin's life. Amongst the many who have thus aided him, the author ventures to mention in particular Dr. and Mrs. James T. Bottomley, Mr. James Thomson, and Miss Mary Hancock Thomson, the Misses King, Mrs. Ramsay MacDonald, Miss Jessie Crum, Miss May Crum, Mrs. Tait, Miss Andrews, Mrs. FitzGerald, Mrs. Hopkinson, Sir Edward Fry, Sir James Pender, Prof. G. F. Barker, and Miss Jane Barnard. Frau Ellen von Siemens has with great generosity furnished a long series of letters written to her lamented father Excellenz H. von Helmholtz. Madame Mascart has similarly supplied others written to the late M. Mascart. Lord Rayleigh and Sir George Darwin each placed at the author's disposal a very large number of letters, many of them of great scientific interest, and of which a selection only is printed here. Of the long series of letters which passed from 1846 to 1903 between Lord Kelvin and Sir George Stokes, none have been inserted in the present work, save isolated extracts of the year 1896. Sir Joseph Larmor, who edited for publication the two volumes of Stokes's *Memoirs and Scientific Correspondence*, has prepared these letters for publication in a separate volume which it is now proposed to amplify by including selections from Lord Kelvin's other scientific correspondence, along with excerpts from his diaries and unpublished manuscripts. Hence the author has deliberately omitted many letters of great scientific value, giving rather such as seemed to possess a more general interest.

With grateful thanks the author acknowledges his indebtedness for advice and help during the writing and printing of the book to Mr. James Thomson, Miss Mary Hancock Thomson, Dr. and Mrs. Bottomley, the Misses King, and Mr. J. D. Hamilton Dickson, all of whom have assisted either in criticism or in proof-reading. The last-named in particular, as an old pupil of Lord Kelvin and a Fellow of Peterhouse, possesses a unique fund of knowledge, which he has unstintingly placed at the author's disposal, correcting innumerable points of detail.

Four veteran contemporaries of Lord Kelvin in his Cambridge days—Professor Frederick Fuller, Professor Hugh Blackburn, the Rev. Canon Grenside, and the Rev. J. A. L. Airey—were so good as to furnish reminiscences of that time. Alas! while these sheets have been passing through the press the Rev. J. A. L. Airey. Professor Blackburn, and Professor Fuller have all passed away.

An intimate family narrative written by Lord Kelvin's eldest sister, Mrs. David King, who died in 1896, now edited by her daughters, has just been published. It gives a picture of the life, from childhood to adolescence, of Lord Kelvin as a member of a singularly gifted and harmonious family. The author of the present work has purposely abstained from trenching on that narrative, possessing, as it does, an intrinsic value of its own, quite apart from the information it affords of Lord Kelvin's early years.

It has been the author's desire to let documents and letters speak as far as possible for themselves; and if he has not always been able to avoid letting his own views tinge these pages, he has at least endeavoured to avoid attributing to others that which is only his own. Doubtless there are many of Lord Kelvin's former pupils who will find gaps in the presentation of his life and character, as must needs be when the author can himself claim no nearer association than that of disciple. But the disciple of one who was himself conspicuously faithful in little things, must at least try to be faithful. The peculiar and affectionate admiration, amounting in some almost to worship, which characterizes those who had the high privilege of that more intimate association, spreads far beyond their circle to the disciple. Let it be hoped that the affectionate admiration which he too shares may not have warped his judgment.

The late Professor Ayrton kindly gave the author permission to appropriate extracts from his article on "Kelvin in the 'Sixties," in which he narrated his own experiences when a member of Lord Kelvin's enthusiastic volunteer laboratory corps.

In dealing with Lord Kelvin's contributions to Geology, to Mathematics, and some other departments of knowledge, the author has had to rely greatly upon the judgment of others. In this particular connexion he gratefully acknowledges help given by Professor J. W. Gregory, Professor A. E. H. Love, Professor George Forbes, and Professor J. A. Ewing. Professor Andrew Gray, formerly pupil, then assistant, lastly successor of Lord Kelvin in the Chair of Natural Philosophy at Glasgow, has very kindly permitted the author to appropriate the extracts on pp. 651-653 which relate to Lord Kelvin's lectures to his students; and he has helped the author in various other ways in relation to Lord Kelvin's work in the University.

Miss Agnes G. King has kindly furnished the portrait-photograph reproduced in Plate XIII; Professor J. D. Cormack the original photographs for Plates VIII, X, and XV; and Professor Edgar Crookshank that for Plate XIV.

To the proprietors of *Punch* the author acknowledges the special permission given to reprint the extracts from poems given on pp. 576 and 610. To the proprietors of the *Daily Graphic* similar thanks are due for the sketch-portrait of p. 899. The author gladly acknowledges the services of his assistant Mr. Ernest W. Moss in the preparation of the Bibliography and the verification of references.

CONTENTS

CHAPTER I

CHILDHOOD, AND UPBRINGING AT GLASGOW

CHAPTER II

CAMBRIDGE

CHAPTER III

POST-GRADUATE STUDIES AT PARIS AND PETERHOUSE

CHAPTER IV

THE GLASGOW CHAIR

CHAPTER V

THE YOUNG PROFESSOR

CHAPTER VI

THERMODYNAMICS

CHAPTER VII

THE LABORATORY

CHAPTER VIII

THE ATLANTIC TELEGRAPH: FAILURE

CHAPTER IX

STRENUOUS YEARS

CHAPTER X

THE EPOCH-MAKING TREATISE

CHAPTER XI

THE ATLANTIC TELEGRAPH: SUCCESS

CHAPTER XII

LABOUR AND SORROW

CHAPTER XIII

THE GEOLOGICAL CONTROVERSY

CHAPTER XIV

LATER TELEGRAPHIC WORK: THE SIPHON RECORDER

LIST OF PLATES

VOLUME I

VOLUME II

CHAPTER I

CHILDHOOD, AND UPBRINGING AT GLASGOW

WILLIAM THOMSON, Baron Kelvin of Largs, was born in Belfast on the 26th of June 1824. The family was of Scottish origin. Three brothers, named respectively James, John, and Robert Thomson, migrated from the Lowlands of Scotland about the year 1641 in the troublous times of the civil wars. From papers in the possession of the family it appears that John Thomson settled in County Down at Ballymaglave (or Ballymaglymph), and for nearly two hundred years his descendants continued to occupy a farm called Annaghmore, near Spa Well, Ballynahinch. On his house, on a quoin of a building now used as a barn, James Thomson, grandson of John Thomson, cut his name, with the date 1707. This James Thomson had three sons, two of whom (John and Robin) emigrated about 1755 to Buffalo Valley, New York State, and set up as millers. The second son, James, the grandfather of Lord Kelvin, born about 1738, remained at Ballynahinch. On 29th September 1768 he married Agnes Nesbitt, who bore him three sons, also named Robert, John, and James, and three daughters. At this date the Thomsons owned about one-quarter of the township of Ballymaglave. According to tradition they nearly all bore the character of being "religious, moral, patriotic, honest, large, athletic, handsome men."

James Thomson, the father of Lord Kelvin, was born at Annaghmore on the 13th of November 1786. He was a man of remarkable abilities and strong character. Brought up on the land as a farm labourer, and receiving from his father the rudiments of education, he studied for himself, without either skilled teachers or good text-books, the art of dialling, making for himself a sun-dial, and also a night-dial to tell the time by the position of one of the stars of *Ursa Major*. The following story is told of him:—

> It was when he was about eleven or twelve years old, that one day the boy was observed to be working with a slate and a bit of stone for a pencil. In the evening he was again working by the light of a handful of shavings he had brought in to make a blaze until the candle should be lighted. After a little he exclaimed to his eldest brother Robert, who was thirteen years his senior, "Robert, I have made a discovery. I have found out how to make dials for any latitude." "Can you show me?" said the brother. "Yes," said he; and he showed him so clearly that his brother quite understood the method.

Three of James Thomson's dials are now in the possession of his grandson, James Thomson, of Newcastle-on-Tyne. On them his name is spelt Thompson, in the fashion more common in England.

Indeed the name is thus spelt throughout in the old family Bible belonging to his father, and in other documents. It is believed that James Thomson changed the spelling when he found that in Scotland the name was usually written without the letter *p*.

In view of the intellectual abilities displayed by James Thomson, his father allowed him to go as a pupil to a small school[1] kept by Dr. Samuel Edgar (minister of the "Secession" Presbyterian Church at Ballynahinch) at Ballykine, near his native place, to learn classics and mathematics; and his abilities were such that he was soon promoted to be assistant teacher. It was his intention to become a Presbyterian minister. Nothing shows more clearly the force of character of the youth than the determined way in which he strove for self-improvement. While still teaching at Ballynahinch during the summers to gain his livelihood, he for four consecutive years, from 1810 to 1814, spent the six winter months studying at the University of Glasgow, the session of which lasted from November to May. He graduated M.A. in 1812. Nearly eighty years afterwards, Lord Kelvin, on the occasion of his installation as Chancellor of the University, related the story of his father's experience as follows:—

"There were no steamers, nor railways, nor motor cars in those days. Can the young persons of the present time imagine life to be possible

[1] See a small book, *Three Ballynahinch Boys*, by Rev. Wm. J. Patton, Belfast, 1880.

under such conditions? My father and his comrade students, chiefly aspirants for the ministry of the Presbyterian Synod of Ulster, and for the medical profession in the north of Ireland, had to cross the Channel twice a year in whatever sailing craft they could find to take them. Once my father was fortunate enough to get a passage in a revenue cutter, which took him from Belfast to Greenock in ten hours. Another of his crossings was in an old smack whose regular duty was to carry lime, not students, from Ireland to Scotland. The passage took three or four days, in the course of which the little vessel, becalmed, was carried three times round Ailsa Craig by flow and ebb of the tide.

"At the beginning of his fourth and last University session, 1813-1814, my father and a party of fellow-students, after landing at Greenock, walked thence to Glasgow. On their way they saw a prodigy—a black chimney moving rapidly beyond a field on the left side of the road. They jumped the fence, ran across the field, and saw to their astonishment Henry Bell's *Comet*—then not a year old—travelling on the river Clyde between Glasgow and Greenock. Their successors, five years later, found in David Napier's steamer *Rob Roy* (which in 1818 commenced plying regularly between Belfast and Glasgow) an easier, if a less picturesque and adventurous, way between the College of Glasgow and their homes in Ireland."

James Thomson's persistency in his studies met with reward: on the completion of his course in

Glasgow in 1814 he received the appointment of teacher of Mathematics at the Royal Belfast Academical Institution, at first in the school department, being the first person to hold that post. His duties comprised the teaching of geography as well as arithmetic and book-keeping. In 1815 he was made Professor of Mathematics in the College department. In the summer of 1817 he was married to Miss Margaret Gardner, daughter of a Glasgow merchant, who at the time of the war of American Independence had gone as a volunteer to fight on the British side.

James and Margaret Thomson had seven children: Elizabeth, born in 1819, married the Rev. David King, LL.D., and died in 1896; Anna, born in 1820, married William Bottomley, and died in 1857; James (LL.D., F.R.S. and Professor of Engineering, first in Belfast, afterwards in Glasgow), born in 1822 and died in 1892; William (Lord Kelvin), born in 1824; John, born in 1826 and died in 1847; Margaret, born in 1827 and died in 1831; and Robert, born in 1829 and died in Australia in 1905.

The Thomsons lived in College Square East, Belfast, in a house still standing, which was built by Professor Thomson. Here all his children, except the eldest daughter, were born. On the flags in front of the house the future Lord Kelvin and his brother James used to whip their tops, and doubtless became familiar with the phenomenon of the precession of a spinning body.

"One of my earliest memories," said Lord Kelvin, "of those old Belfast days, is of 1829, when the joyful intelligence came that the Senate of the University of Glasgow had conferred on my father the honorary degree of Doctor of Laws." But the joy of the family was overshadowed by a sad event. Margaret Thomson died in 1830, when her eldest daughter was but twelve years old, and her youngest boy only twelve months. The future Lord Kelvin was but six, and his brother James eight. Their father devoted himself to his children, taking the two boys to sleep in his bedroom, and teaching them himself, save that James and William both went for a few months to the writing-school in Belfast. He taught them in particular the use of the globes, and began Latin with them on the Hamiltonian system of teaching. The elder daughter Elizabeth compiled in later years a deeply interesting narrative of the family life, giving many details.

In 1832 the chair of Mathematics at Glasgow became vacant by the retirement of Professor James Millar, who had held it for thirty-six years; and it was offered to James Thomson, who migrated with his young family to Glasgow in that year. He still kept the education of his sons in his hands. He was indeed a gifted person—a good scholar, capable on emergency of teaching the University classes in classics; and that his mathematical knowledge was sound is attested by the text-books he produced—including one on Differential and Integral Calculus—books[1] which, though now superseded,

[1] James Thomson's books cover a considerable range. In 1819 he published in Belfast *A Treatise on Arithmetic in Theory and Practice*, a small

long held their own for clear exposition. He also made several original contributions to mathematics. James Thomson was known as a successful teacher. It was his practice to catechise his class at the beginning of each lecture on the work of the preceding day, *viva voce* questions being passed with energy and enthusiasm from bench to bench, a practice which his distinguished son was wont at times to pursue. The following anecdote is narrated by Sir William Ramsay, whose father was at one time a member of Thomson's class.

One day Professor Thomson asked a certain Highland student, "Mr. M'Tavish, what do you understand by a point?" The answer was, "It's just a dab!" Again, in the course of construction of a diagram, the question came, "What should I do, Mr. M'Tavish?" "Tak' a chalk in your hand." "And what next?" "Draw a line." Professor Thomson complied, and, pausing, said, "How far shall I produce the line?" "*Ad infinitum*," was the astonishing reply.

The boys James and William were allowed to attend informally their father's lectures at the University, and also those of some of the other

duodecimo volume, which had a very large sale. The seventy-second edition of this work, revised by his two sons and edited by Sir William Thomson, was published by Messrs. Longmans in 1880. In 1827 he produced two books, an *Introduction to Modern Geography* and *The Romance of the Heavens*. In 1830, while still in Belfast, he issued the *Elements of Plane and Spherical Trigonometry*, with a chapter on the "First Principles of Analytical Geometry," of which a fourth edition was published in London in 1844. In 1834 he edited an edition of *Euclid's Elements of Geometry*, and wrote an excellent *Algebra*. He was the first systematically to apply Horner's method of solving algebraic equations to the arithmetical extraction of cube roots and roots of higher powers. In 1831 appeared his *Introduction to the Differential and Integral Calculus*, of which a second edition was printed in 1848.

professors. They often repeated at home in a juvenile way the demonstrations they attended. In the year 1834 or 1835 they made themselves electrical machines and Leyden jars, and administered electric shocks to their friends, and later they constructed voltaic batteries.

In October 1834 both James and William Thomson matriculated in the University of Glasgow, James being then twelve years of age, and William ten years and three months. The Matriculation Album for the session 1834-35 bears the entry :—

Gulielmus Thomson, filius natus secundus Jacobi, Math. Prof. in Academia Glasguensi.

The signature is in William Thomson's own handwriting; the remaining words in that of William Ramsay, Professor of Humanity, in whose Class he and his elder brother were duly enrolled. The University classes in those days consisted largely of raw Highland lads, sent from the farm to train as theological students, of all ages from fourteen to twenty-four, with others intending to follow law or medicine. The following excerpt by Dr. H. S. Carslaw from *The Book of the Jubilee*, 1901, gives a picture of interest respecting young William Thomson :—

"It is somewhat difficult to picture the classes of the time. It is equally surprising to find that at the end of his first winter's work he carried off two prizes in the Humanity Class; this before he was eleven. In the next session we follow him to the

QUADRANGLE OF THE OLD GLASGOW COLLEGE.

The rooms used as Laboratory of Natural Philosophy are in the dark corner on the right.

classes of Natural History and Greek—we wonder what the present occupants of these chairs would say to a stripling under twelve who presented himself at their lectures—and his name figures in both prize-lists. Sympathy is not lacking for the hard-worked schoolboy of to-day; but what would the child of twelve think of the holiday task of translating Lucian's *Dialogues of the Gods*, with full parsing of the first three dialogues! This is the piece of work for which William Thomson, Glasgow College, receives a prize in May 1836. Next session we find the two brothers together in the Junior Mathematical Class, of the Junior Division of which they are first and second prizemen. They appear again at the head of the list for the Monthly Voluntary Examinations on the work of the class and its applications. Proceeding to the Senior Mathematical Class in 1837-38, they again stand at the top, nor have they failed to present themselves for the Voluntary Examinations. William is not satisfied with this class, but in addition receives the second prize in the Junior Division of Prof. Robert Buchanan's Logic Class, having as a near rival John Caird, Greenock, the name of our late revered Principal now appearing in the lists." At the close of the session of 1838-39 William and James Thomson took the first and second places as prizemen in Natural Philosophy, and in that of 1839-40 William gained the class prize in Astronomy, and was awarded a University medal for an essay *On the Figure of the Earth*, the manuscript[1] of which is still extant. In 1840-41 his name appears once more in the prize-lists, being this time fifth prizeman in the Senior Humanity Class under Professor Lushington. Lord Kelvin loved to recur to his student days, and to his teachers of that time — Ramsay, Lushington, Thomas Thomson, Meikleham, and J. P. Nichol. In 1907, at the annual dinner of the London "Glasgow University Club," he spoke of the fine

[1] It is a carefully-written bound volume of eighty-five pages, undated. On the title-page are two quotations :—

> . . . Mount where science guides ;
> Go, *measure earth*, weigh air, and state the tides ;
> Instruct the planets in what orbs to run,
> Correct old time, and regulate the sun.

> Principio terram, ne non aequalis ab omni
> Parte foret, magni speciem glomeravit in orbis.

A brief preface states that the writer has consulted Airy's *Tracts* and his *Encyclopædia Metropolitana* article, and the works of Poisson, Pentécoulant, Pratt, and Laplace. He claims some originality, but fears that more extended reading may show that he has been anticipated. The contents are grouped under four heads :—Physical Theory, Disturbance in the Moon's Action, Geodetic Measures, and Pendulum Observations. In the last part a demonstration is given of Clairaut's Theorem. The mathematical handling throughout is marvellous. The manuscript bears three notes of later dates ; one added December 16, 1844 ; one dated "*Gt. Eastern* at sea, Sep. 13/66" ; a third signed "K. Oct. 21, 1907." After fifty-seven years, and only two months before his death, Lord Kelvin had returned to the study of his boyhood ! Prof. A. E. H. Love, who has kindly examined the text of the Essay, writes : "It is a truly astonishing performance for a boy of sixteen. It has many affinities with Airy's *Tract*, but in the arrangement of the matter, and still more in the general tone, it is quite different from Airy's *Tract*. Airy's writing was meant to be a textbook for the use of students ; Thomson writes like a scientific investigator. Besides this, his work is more complete. For example, he includes the ellipticity deduced from the constant of precession combined with Laplace's hypothetical law of density in the interior of the Earth, and he includes the perturbation of the Moon's motion in longitude. These things are omitted by Airy. Even Pratt in his *Treatise* omits the perturbation of the moon's motion in longitude. I don't wonder that Lord Kelvin took the Essay about with him, because it had everything in it in a small compass. But the methods which he used in it are not those which he adopted afterwards in Thomson and Tait. Evidently he learned two things about the subject at a later date—the use of the potential function and the use of the method of harmonic analysis. He had a large share in developing these more powerful methods, and it seems clear that when he came to the task of printing an account of the theory he preferred them to the methods which he had used in his youth."

all-round education afforded by his University in the good old days, and praised its width. "A boy," he said, "should have learned by the age of twelve to write his own language with accuracy and some elegance; he should have a reading knowledge of French, should be able to translate Latin and easy Greek authors, and should have some acquaintance with German. Having learned thus the meaning of words, a boy should study Logic." And then he went off in praise of the advantages of some knowledge of Greek. "I never found that the small amount of Greek I learned was a hindrance to my acquiring some knowledge of Natural Philosophy." Assuredly not in his case. Yet he confessed one day that if he could only find his old note-book with the notes of Lushington's lectures on the Greek play in his last year of study at Glasgow, its pages would show that his mind was often wandering away to matters of Natural Philosophy! He retained a very lively memory of his early University days, and delighted to recall them. Well did he remember "the little tinkling bell in the top of the college tower, calling college servants and workmen to work at six in the morning; the majestic tolling of the great bell wakening at seven the professors (and students, too, in the olden times when students lived in the college); then, again, the lively little tinkling bell calling the professors and students of Moral Philosophy and Senior Greek and Junior Latin at half-past seven to work in their class-rooms.

"Woe to the student of Latin who reached the door ten seconds after the quick little bell's last stroke. He was shut out by the doorkeeper, unfailingly ruthless, by inexorable order, and had to wend his way through the darkness to his lodging, sorrowfully losing the happy hour's reading of Virgil or Horace or Livy with his comrades, under their bright young Professor, William Ramsay, and knowing that he had got an indelible black mark against his name."

The then Professor of Natural Philosophy, William Meikleham, had held the chair since 1803. Though he can scarcely be accounted a distinguished man, he yet had a sound knowledge of the older branches of his science, and certainly succeeded in arousing in his students an interest in physical phenomena. He made them read the *Mécanique analytique* of Lagrange and the *Mécanique céleste* of Laplace, a task that would indeed have been hard but for the excellent mathematical training of Professor James Thomson. In the session of 1838-39 Meikleham broke down in health, and for the remainder of that academic year his lectures were shared between Professor Thomas Thomson (Professor of Chemistry) and Professor John Pringle Nichol (Professor of Astronomy). In the session of 1839-40 Nichol gave all the Natural Philosophy lectures after the first three weeks, and young Wm. Thomson took the Senior Course of Natural Philosophy under him. His note-book of the lectures is still preserved. Nichol was a most accomplished man, of quick parts, with a keen eye for recent advances in science, and a poetical imagination. He fitted up his newly-built observatory with numerous pieces of apparatus of his own possession,[1] particularly optical apparatus. He showed his students the phenomena of diffraction and the spectrum of the sun's light. He also procured Daguerreotype apparatus, and in 1839 initiated the brothers James and William Thomson into the mystery of taking Daguerreotype photographs. He taught William to take transits of the sun and stars with the transit instrument in the old Macfarlane Observatory. The summer of 1839 was in later life described by Lord Kelvin as "a white era, an era of brightness in my memory." Such was the inspiring influence of the teachers[2] under whom he drank in knowledge. Nichol had recently got hold of a new book—a pamphlet of some eighty pages—on Couples, and made his students write Christmas essays on the Theory of Couples. It was Nichol, too, who in 1840 brought to the notice of his eager young student the *Théorie analytique de la chaleur*, of Fourier, which was destined to influence his whole

[1] In the summer of 1840 he travelled to Munich on purpose to procure some new instruments for his observatory. During part of this tour he and Mrs. Nichol and their son (afterwards Professor John Nichol) were with the Thomsons at Frankfort, as narrated at the end of this chapter.

[2] The following extract from Lord Kelvin's inaugural address as Chancellor in 1904 gives a grateful reference to his early teachers :—"My predecessor in the Natural Philosophy chair, Dr. Meikleham, taught his students reverence for the great French mathematicians Legendre, Lagrange, and Laplace. His immediate successor, Dr. Nichol, added Fresnel and Fourier to this list of scientific nobles ; and by his own inspiring enthusiasm for the great French school of mathematical physics, continually manifested in his experimental and theoretical teaching of the wave theory of light and of practical astronomy, he largely promoted scientific study and thorough appreciation of science in the University of Glasgow."

career. Lord Kelvin himself gave me the following account of the matter :—

"The origin of my devotion to these problems is that after I had attended in 1839 Nichol's Senior Natural Philosophy Class, I had become filled with the utmost admiration for the splendour and poetry of Fourier. Nichol was not a mathematician, and did not profess to have really read Fourier, but he was capable of perceiving his greatness and of understanding what he was driving at, and of making us appreciate it. I asked Nichol if he thought I could read Fourier. He replied 'perhaps.' He thought the book a work of most transcendent merit. So on the 1st of May [1840], the very day when the prizes were given, I took Fourier out of the University Library; and in a fortnight I had mastered it—gone right through it."

Fourier's *Théorie analytique de la chaleur* had appeared in Paris in 1822. In this work he set himself to establish on a thorough basis of mathematical analysis the theory of the movement of heat in bodies and between bodies. It is characterised by the same extreme elegance of exposition which distinguishes the writings of Laplace, Lagrange, and Poisson in their treatment of other branches of mathematical physics; while its spacious verbiage and refinement of style is such as to cause Clerk Maxwell to pronounce it a great mathematical poem. At the date of its appearance the application of the methods of analysis to Mechanics and Astronomy was a comparative novelty; and certainly no one before Fourier had had the hardihood to apply analysis to the movement of heat. The success with which he built up, by patient insight, the differential equations for the movement of heat, in the several cases considered, was equalled by his success in discovering the processes for integrating them; leading him not only to establish the famous "Fourier Series" for the expression of periodic quantities, but also to formulate several new integrals of great importance in mathematical physics generally. Fourier's memoirs had attracted but little attention in England, and his work passed almost unrecognised until the events now to be narrated.

William Thomson was already familiar with the French language. He and his three brothers had been taken in the summer of 1839 to London to see the sights of the great city, and then on to Paris, where they were left for about two months to learn French, while their father and their elder sisters went on for a tour round Switzerland and South Germany. At Paris he frequented the Bibliothèque Royale in order to read Laplace's *Mécanique céleste*, in preparation for his University essay on the Figure of the Earth. But for this training in French he would scarcely have been able in a fortnight to go through Fourier's work. It was a part of his father's plan of educating his family that they should acquire a mastery of German also. Accordingly he determined to take his children for a summer residence in Germany—no light undertaking in those days, when the facilities for travel were extremely restricted. For two months the whole family took lessons in German conversation. On May 21, 1840, the father with his six children started from Glasgow. The eldest daughter Elizabeth was almost 22, James 18, William 16, and the youngest boy Robert only 11 years old. They travelled by steamer to Liverpool, thence by train to London. On the 25th they went to see the Queen drive to Buckingham Palace. On the 26th the party visited the Polytechnic to view the latest wonders, and the same night left by steamer for Rotterdam. A note in young Thomson's diary runs:—

> Reached the bar at the mouth of the Maas, near Brill, at about $4\frac{1}{2}$ o'clock in the morning, where we had to lie till 10. The vessel rolled greatly from side to side, but the rolling was intermittent, as every two or three minutes it calmed down and then rose again with perfect regularity. This probably arose from two sets of waves of slightly different lengths coming in in the same direction from two different sources.

On the 28th they visited the Hague; and the diary adds a visit to the Museum to see a stuffed mermaid! Also a visit to a windmill at Delft, where they criticised the primitive machinery. Then they took a river steamer to Emmerich, and thence by Düsseldorf to Bonn, reading *Peter Simple* on the deck, conversing with some acquaintances on painting and animal magnetism, and landing at Cologne to see the cathedral and purchase some of J. M. Farina's *eau véritable.* They reached

Frankfort-on-the-Main on June 16, and put up first at the Würtemburgerhof. On June 19 they moved into a house on the Promenade, near the Eschenheimer Thor, which house they furnished. They remained here until August 2, when they left for Baden. From there the two brothers James and William went by themselves for a walking tour, lasting some days, in the Black Forest. The whole family returned to Glasgow on September 2. If this astonishing expedition reveals the unique personality of the elder Thomson, and the thoroughness of his educational methods with his children, the fact remains to be told that, so far as young William Thomson was concerned, its principal object turned out a failure. In his later life he used to tell with whimsical glee how it was that he never became a good German scholar.

"Going that summer," he said, "to Germany with my father and my brothers and sisters, I took Fourier with me. My father took us to Germany, and insisted that all work should be left behind, so that the whole of our time should be given to learning German. We went to Frankfort, where my father took a house for two months. The Nichols had lodgings adjacent, and came in to meals with us nearly every day. Now, just two days before leaving Glasgow I had got Kelland's book (*Theory of Heat*, 1837), and was shocked to be told that Fourier was mostly wrong. So I put Fourier into my box, and used in Frankfort to go down to the cellar surreptitiously every day to read a bit of Fourier. When my father discovered it he was not very severe upon me."

Kelland, in fact, had been misled by not comprehending that a Fourier series may be expressed either in a double series of sine and cosine terms, or in a single series of either sines or cosines, by appropriate assignment of epochs to the various terms of the series. He had, therefore, hastily concluded that, since many of the expansions of functions given by Fourier are in series of sines or cosines alone, they were "nearly all erroneous." Thomson discovered, while at Frankfort, the cause of the misunderstanding, and wrote thereupon an article "On Fourier's Expansions of Functions in Trigonometrical Series," giving a new demonstration of the expansion, and pointing out the explanation of the apparent discrepancy noticed by Kelland. This article was subsequently published over the pseudonymous signature "P. Q. R." in the short-lived *Cambridge Mathematical Journal*, vol. ii., May 1841, and is reprinted as the first article in vol. i. of Lord Kelvin's *Mathematical and Physical Papers*. Lord Kelvin gave me, in 1906, the following account of it :—"I was filled with indignation at a statement by Kelland that almost everything in Fourier was wrong. When I wrote my paper—my first published original paper—for the *Cambridge Mathematical Journal*, my father sent it to Gregory. Gregory had been beaten recently by Kelland in the competition for the Edinburgh chair of Mathematics. Gregory thought the paper rather controversial, and sent it to Kelland. This was a graceful act on Gregory's part, that he would not put it into the *Journal* without referring it first to Kelland. Kelland wrote back rather tartly, as if piqued. Then my father and I went over the paper and smoothed down a few passages that might have offended Kelland's feelings. Kelland wrote[1] back that he was charmed with the paper, and was quite amiable. So then it was printed." As it appeared, it was dated "Frankfort, July 1840, and Glasgow, April 1841."

In the circle of University acquaintances in Glasgow was one David Thomson, a cousin of the great Faraday. David Thomson (B.A. 1839), of Trinity College, Cambridge, took over the duties of Professor Meikleham's chair from 1842 to 1845, during the latter's illness. He subsequently held the chair of Natural Philosophy at Aberdeen. He wrote the article on "Acoustics" for the eighth edition of the *Encyclopædia Britannica*, and died in 1880. By him William Thomson was, as he himself expressed it, "inoculated with Faraday fire." He indoctrinated the youthful student into Faraday's then heterodox notions of electric action in a medium. Hitherto the doctrines taught him respecting electricity and magnetism had been on the then accepted lines of Newtonian forces acting at a distance, with all the weight of Poisson and Laplace to support the analytical theory. Of the

[1] The letters which passed, in February and March 1841, between Gregory, Kelland, and James Thomson, were mostly preserved by him, and were found amongst Lord Kelvin's correspondence.

Boscovichian theory of atoms as centres of force acting at a distance he had learned from Nichol. But now David Thomson inculcated the Faraday conception of electric and magnetic forces acting along curved lines in the medium, and the further possibility of the screening of electric forces by the interposition of a conducting sheet. At first William Thomson rejected these notions, thinking them incompatible with first principles, and argued eagerly against Faraday's views. Ultimately he was convinced, and ever afterwards retained the most sincere admiration for Faraday and his work.

And so with the advent of April 1841 came to an end William Thomson's sixth and last session as a student in the University of Glasgow. He left the University[1] without even taking a degree!

APPENDIX TO CHAPTER I.

THE VISIT TO FRANKFORT.

In the Memoir of John Nichol (Professor of English Literature in Glasgow, 1862 to 1889), son of Professor J. P. Nichol, there are some autobiographical notes, written in 1861, which throw an interesting light upon the Thomson family, and, in particular, upon the episode of the trip to Frankfort. John Nichol was then seven years old. From these notes the following passages are extracted :—

---

[1] Nevertheless he sat for the degree examinations at Glasgow. A certificate, still preserved, reads as follows :—"COLLEGE OF GLASGOW, *April* 22, 1839. William Thomson. Examined and approved for the Degree of A.B. by us, Robert Buchanan, William Fleming, William Ramsay, E. L. Lushington." At that date Thomson had not completed his fifteenth year. He purposely abstained from applying for the formal conferment of the degree, in order that he might not be prejudiced in entering as an undergraduate at Cambridge.

The day came when we started for Germany,—my father, my mother, and myself. . . . We went, I think, from Edinburgh to Glasgow, and then to Liverpool, and then to London. . . . I have no memory of our embarkation. Light breaks upon one next at Ostend. . . . We went to the Continent alone—we three—but our friends, the Thomsons, had arranged to meet us on the way; they spent some considerable time with us on the Rhine, so I had better explain who they were. Had I more leisure and a clearer memory, I think I could write something about the Old College Court. The dingy old place has for me some pleasant associations. . . . When we first lived there, Hill had not begun to send forth his platitudes from the chair, . . . nor the most illustrious of the Thomsons to make new discoveries in electricity. . . . Members of that great *gens* literally filled one-half of the chairs in the University. I will not venture to say how many I have known. There was Tommy Thomson the chemist; William Thomson of Materia Medica; Allen Thomson of Anatomy, brother of the last; Dr. James Thomson of Mathematics; William, his son, etc., etc. Old Dr. James was one of the best of Irishmen, a good mathematician, an enthusiastic and successful teacher, the author of several valuable school-books, a friend of my father's, and himself the father of a large family, the members of which have been prosperous in the world. They lived near us in the court, and we made a pretty close acquaintanceship with them all. Mrs. Thomson had died before her husband came to Glasgow; but there were two daughters, both clever, good talkers and sketchers, one of them very pretty; and four sons, in their order, James, William, John, and Robert, a pleasant and happy group now scattered far and wide. Dr. James came originally from the North of Ireland, and, to some extent, combined the qualities of the two races who are in that district fused together. He was laborious and precise and acute, destitute of the inventive, but largely endowed with appreciative faculties. Good-hearted, he was shrewdly alive to his interest without being selfish, and would put himself to some trouble, and even expense, to assist his friends. He was a stern disciplinarian, and did not relax his discipline when he applied it to his children, and yet the aim of his life was their advancement. He was impressionable, if not impressible, like the most of Irishmen, and was more tenacious of his impressions than most. He was uniformly kind to me, and I owe him nothing but gratitude.

Of the sons I liked James the best. He was crotchety, and apt to be sulky with those who would not enter into his crotchets; here, as far as I know, his faults end. He was steadfast, straightforward, independent, quiet, unobtrusive, with more Scotch than Irish blood in his veins, and yet it ran warmly enough for his friends, and at a later period I had the honour to be one of them. His passion was engineering; he was always on the eve of inventing something that was going to revolutionise trade. He used to show me lots of models, and often when we were in Arran together he would walk out to try his boats or his wheels on the streams, as a chemist goes to make an experiment that will test the worth or worthlessness of years of toil, or the astronomer goes to look for the star whose place he has predicted with the help of a million figures. I believe some of those inventions were excellent, but there was always some practical obstacle which prevented their bringing to the inventor either the fame or the fortune they merited. James was an idealist in his way.

John was an assiduous and successful student of medicine, and died of a fever caught during his attendance on the hospital. . . .

We stayed some little time at Bonn. We lodged near the verge of the town, where we met the Thomsons, and the younger boys and I used to make little paper boats, and let them sail far away over the roofs of the houses. . . From Bonn, too, my father, with James and Willy Thomson, went to walk for three days among the craters of the district, and came home with their pockets full of specimens, which James still preserves in his cabinet.

"It was upon a trancèd summer night" that we sailed round the corner of the Rhine which reveals the Siebengebirge, and came gliding in to the island of Nonnenwerth. Clear and calm and fair the memory of that night comes back to me from over all the years. One by one the peaks appeared, and stood grandly above the quiet stream, in the grey light which soon faded away beyond their purpling crests. The moon stood out, a glorious crescent on the ridge of Rolandseck, and a bright star led the host of heaven over the brow of Drachenfels. . .

We were on our way to Frankfort when this happened, and there we spent the most considerable part of our time. I remember our getting settled down somewhere into comfortable lodgings up one or two stairs, and our meeting the Thomsons again. . . .

My father went alone to Vienna by Ratisbon and Passau, returning by Innsbruck and the Tyrol and Munich. My mother and I stayed three months at Frankfort; the Thomsons came often to see us, and we had other varieties enough to prevent us feeling lonely. . . .

Frankfort was a pleasant place to live in then, whatever it may be now. It had its romance—old houses within, and green glades without the walls; and yet it was well furnished with all things needful. I should be glad to return there and see if the cherries taste as sweet as ever, if the environs are as luxuriant as when we went out on an afternoon to see the Prince [Landgraf] of Homburg drive round his park, or the streets as gay as when there was a rush of lights at night.

CHAPTER II

CAMBRIDGE

On April 6, 1841, William Thomson, then in his seventeenth year, was formally entered at St. Peter's College, Cambridge, as a student of the University. The Admission Book entry is :—

> 1841, April 6th, Gulielmus Thomson, Doctoris Jacobi Thomson Filius, Scotus, ad mensam pensionarium admittitur.

He came into residence in October of the same year. St. Peter's, or Peterhouse, to give it its ancient and more familiar name, is not one of the great or wealthy Colleges, but it has always maintained an honourable tradition for scholarship of the best sort, and for an intellectual activity that would do credit to a larger and more richly endowed institution. In the 'forties it ranked about fifth or sixth in size. Hence the position of a pensioner of Peterhouse would in no sense be regarded as inferior to that of one resident in Trinity, King's, or St. John's. In the Tutor's Book it is recorded that he was recommended to the College by his father, who himself accompanied him to Cambridge to introduce him to personal friends—Challis, Gregory, Hopkins, and others.

Probably Professor James Thomson decided on the choice of Peterhouse for his son because of the fame of Hopkins, the mathematical coach, for whom he had a great admiration.

For some reason Peterhouse had from this period onwards a distinct following of Scottish students. Thomson's famous colleague Tait entered Peterhouse three years after he had left it; and two years later, it was to Peterhouse that Clerk Maxwell came, though he migrated after one year to the more highly endowed Trinity.

Of that period the present Master of Peterhouse has written the following notice in the *Cambridge Review*:—

But Cambridge had a claim of her own upon Lord Kelvin. She had possessed him during those incomparable years of life through which a man of genius passes, as through a golden gate into a region open only to a few—the region of great achievement.

When he came up to Peterhouse the Tutor of the College was Henry Wilkinson Cookson, who had taken his degree in 1832, and afterwards became Master. No man could have served his College, and I may add the University, more loyally and more effectively than Cookson, who knew it both *intus et in cute*; but there could not be much intellectual affinity between him and Thomson, as his private scientific tastes were mainly biological.

On the other hand, Thomson was, as an undergraduate, brought into immediate contact with Frederick Fuller, afterwards Professor of Mathematics at Aberdeen, who graduated only three years before himself, and subsequently succeeded Cookson as Tutor. He survives as one of the oldest members of a College which owes him a deep debt of gratitude; and it was a rare pleasure to find myself voting on the same side with him not very long ago. But in the early 'forties an emanation of mathematical glory was already proceeding from our ancient house, where William Hopkins, after graduating as far back as 1829, had already become one of the most successful private tutors known to the *ancien régime*, and where his distinguished name and unsullied memory are still justly revered. Tait and Steele, as again every one knows, headed the Mathematical Tripos in 1852, and both of them became Fellows in the following year.

The intimacy of Thomson and Tait, and the joint production of their great book, therefore, do not belong to their Cambridge years, though counting among the chief glories of Peterhouse. Routh's year, 1854, when Clerk Maxwell was second Wrangler, was another *annus mirabilis* for Peterhouse.

Canon Grenside, one of Thomson's contemporaries at Peterhouse, has narrated how he first met him at the wine-party given to freshmen by Mr. Cookson the tutor, shortly after the opening of the October term of 1841. "William Thomson, a slender, fair-haired youth, sat immediately opposite me," writes the Canon. "I noticed him particularly—especially his youthful appearance. Of course no words could be exchanged across the table in the august presence of the College Tutor. We soon became friends, and that friendship lasted to the end of his distinguished life, though meeting at rare intervals. He had not been settled in his rooms for more than three days. . . . Two days afterwards it was currently reported in the College that Thomson would be Senior Wrangler!"

Thomson had scarcely entered Peterhouse when his anonymous paper in defence of Fourier's

Expansions of Functions in Series appeared in the *Cambridge Mathematical Journal.* The secret soon leaked out; and it became evident that here was a student of unusual promise. In November 1841 he had a second article,[1] written in reality in the previous April, giving a new proof of the generality of Fourier's Solution of the Expansion in Series—a proof different from that already advanced by Poisson. This was followed in 1842 by two more papers in the *Cambridge Mathematical Journal,* still signed "P. Q. R.," of a much more advanced character.

Thomson's life at Cambridge differed little from that of the earnest and active undergraduates of his time, save perhaps in the intensity with which he threw himself into everything with which he let himself be occupied. He read, walked, boated, and even indulged in occasional dances and more occasional rides. The days during term time at Peterhouse were filled with varied activities. Thomson usually began his morning by a rapid walk or run, before breakfast, around the College Grove. Every day, almost without intermission, summer and winter, he used to take a dip in the waters of the Cam, sometimes making his way to Byron's Pool for a plunge. Lithe in figure, and wiry of constitution, he enjoyed other outdoor recreations, particularly rowing. Athletics had not at that date

[1] Copies of these two articles were sent in the New Year by James Thomson to Kelland, who replied: "I have to return you my best thanks for your kindness in sending me the papers of your son. I will only add that the early genius displayed in these and in all his papers promises to rank your son *soon* amongst the mathematicians of Europe."

swelled to the overweening proportions of later time, and occupied a more rational share in the life and outlook of the University man. How Thomson distinguished himself in play as well as in work we shall see.

Thomson's tutor for the first term was Cookson. In January 1842 he began to read with Fuller, but he worked for one term, and through the long vacation of 1842, without a tutor. After that he had William Hopkins as his private coach, "an excellent and sound mathematician and scientific man," as Thomson described him sixty years afterwards. In the Cambridge of those days, as since, the career of the student who was reading for the Mathematical Tripos depended greatly on the tutor or coach under whom he read. A tutor who could impart method and enthusiasm to the men working under him was sure to bring them forward. And Hopkins, who was also a very competent geologist, and who left his mark in more than one department of physics,[1] was assuredly capable of sympathising with the ardours of the youthful Thomson. He had, moreover, himself contributed to the investigation of a problem of particular interest to Thomson, the theory of the rigidity of the globe of the earth,—an exceptionally

[1] Hopkins had written in 1835 on Aerial Vibrations in Tubes. In the years 1839 to 1842 he had no fewer than three memoirs in the *Philosophical Transactions of the Royal Society*, on the Precession and Nutation of the Earth in Relation to the Fluidity of its Interior, and on the Thickness of its Crust. From 1843 to 1861 he wrote much on the theory of Glacier Motion, and from 1852 to 1860 on Terrestrial Temperatures. He was President of the British Association in 1855.

suitable guide, one would say, to direct the mathematical studies of the fervid youth.

Such letters as a young undergraduate writes to his family from the University, even if filled with the trivialities of the hour, throw much light not only on the life of the time, but on the development and character of the writer; and to this the letters of William Thomson are no exception. Happily a very large number of these have been preserved, as also those written to him by his father and his sisters, and his letters to his widowed aunt, Mrs. Gall, who at this period was housekeeper for Dr. James Thomson in the lively family circle at No. 2, the College, Glasgow.

On 21st October he tells his sister Elizabeth how he has been fortunate in getting comfortable rooms in College—a parlour, a bedroom, and a gyp's room. (He has to explain afterwards that the name gyp is derived from γύψ, a *vulture!*) Then he has had to make his own breakfast, succeeding very well, except that he forgets whether to put in the coffee after or before the water is boiling, so asks for the proper directions! Next he tells of the calls of the tradespeople, and of the hairdresser who asks him to contract for getting his "hair dressed at 2s. 6d. a term—very cheap"; which advantageous and tempting offer he declines, considering that hitherto his hairdressing has cost him only 2d. the half-year. He is surprised at the way the gyp[1] lays his

[1] This old famulus bore the name of Boning, and, to distinguish him from other college gyps of the same *gens*, was always known as "Gentleman Boning," because he always went about in a high hat, and wore gloves. He

table for breakfast and tea, and clears away the things afterwards.

To his sister Anna, on 23rd October, he writes telling of various events: of surplice-day at chapel; of his having gone to "take wine" with Cookson—a solemn occasion; of King's College Chapel, where he is struck with the roof as a problem in the equilibrium of structures; and he wants her to tell him how much tea he must use to make a cupful.

On 26th October he writes his father that he finds himself to have been partially anticipated by Liouville in one of his papers. He has been told by Cookson what books to read; and he has joined the Union. October 29th brings him a letter from his father narrating his return journey, and advising him as to personal economy. "You must keep up a gentlemanly appearance, and live like others—keeping, however, rather behind than in advance." He winds up by asking William for a solution of the problem to find the centre of gravity of a spherical triangle. William's reply gives an account of Mr. Cookson's first "lecture" (on Euclid), in which he laid down the University's ideas of education as opposed to modern "diffusion-of-useful-knowledge-society's ideas." He grieves that in his rooms he has fifteen yards of bookshelves

used in after-years to relate that when he was conducting father and son for the first time to Thomson's rooms he remarked, "Your son's very young, sir, to be coming to college"; to which the father replied, "He may be, but you'll find he's very well prepared." Mr. J. D. Hamilton Dickson, Fellow of Peterhouse, to whom Boning recounted this, has also told that when he was at college the hairdresser Bendall was still alive, though in old age. It was on his death that Shilleto wrote the poem "Ultimus Tonsorum," published in the *Cambridge Chronicle* of June 26, 1875.

and only half a yard of books; tells how his friend Grenside advises him not to join the boat-club because of the rowdy men in it; mentions the canvassing of votes for the President of the Union, and how he has promised to vote for Hardcastle *because* his opponent is a Johnian; touches on sundry mathematical problems, but has not yet found a solution to the one sent him. A day after, he writes again that he has received Chasles' book, where he finds another anticipation of his theorems, but by a different method.

November 15th finds him writing again to his father. Money matters are urgent; he wants to pay some bills at once so as to secure discounts. He has been to a second "wine" with Gregory. Then a message to his sister. "Anna says she was rather amused at my using the word *man* so much in my letters, but the reason is because I am so much amused myself at the great use made of it here. It is quite unprecedented to talk of going to see a friend, or a student, or a person, but the word used is universally *man*, and it certainly does sound rather strange to hear them calling me a man." "Letter-writing is nearly as fatiguing to me as mathematics," he adds. And, indeed, he was throughout his life a slow writer, laboriously penning a large script in which he loved to imagine each individual letter to be distinct. His friend Scratchley[1] is thinking of migrating to Queens', and he himself now raises the question whether, as the chances of

[1] Arthur Scratchley, graduated from *Queens' College* 1845.

a Fellowship at Peterhouse are limited, he had not better also think of migrating elsewhere. His letter, sputtered over with ink-specks, is written—as a postscript explains—with a quill pen, which he finds to be used at the examinations, and therefore he "must get into the habit of being able to write with them." A few days later he writes to his sister Elizabeth :—

I adventured myself to-day for the second time in a funny (or funey or funney), *i.e.* a boat for one or two people to row in. It is certainly rather a venture to go in them, as we can hardly stand upright in them for fear of upsetting them, they are so very light and narrow. I can manage it quite well, however ; and, besides, I would not care for an upset, except for my watch and the disgrace. In this College, and in all the others, there is a boat club which has one or more eight-oared racing boats which go out very frequently to practise the crews for the races. Our boat goes out every day, and will be at the head of the river in the next races, now that I [!] have come here, though it was not before. I have not joined the club, however, as rowing for the races is too hard work for getting on well with reading ; and, besides, the men connected with the club are generally rather an idle set.

His father is glad he did not join the boat club. William's next letter, of November 21, tells of his work, reading for both Cookson and Hopkins, and doing seventy lines of *Prometheus Vinctus* every other day. It gives him very little trouble. He has had the honour of a call one evening from Archibald Smith[1] and D. F. Gregory[2]—both Fellows !—

[1] Archibald Smith, of Jordanhill, near Glasgow, of *Trinity* ; Senior Wrangler, 1836 ; later a distinguished equity draughtsman of Lincoln's Inn ; author of the *Admiralty Manual on the Deviation of the Compass* ; died 1872.

[2] Douglas Farquharson Gregory, *Trin. Coll.* ; B.A., 1838 ; Fellow of *Trinity*, 1840 ; author of *Examples of the Processes of the Calculus* ; died February 1844.

who had discussed mathematics and even worked problems in his room. He narrates a festivity:—

On Tuesday night I went to Hopkins's party. . . . I went in at about eight o'clock, and was nearly among the first. A few wrangling-looking men soon began to drop in, and a great many freshmen, or raw materials for manufacture. Any to whom I spoke said they were going to read with Hopkins if, or as soon as, he would take them. There were no less than three of our freshmen present, besides myself, and one of our other men. Later in the evening some ladies, and older gentlemen, and among them Ansted of Jesus College, one of the proctors, came in. Mr. and Mrs. Hopkins and a young lady sang some glees, and Mr. Hopkins asked all of us whether we performed on any instrument; and when he heard that we did not, he said he was very glad to hear it. After music, conversation, and looking at a great many beautiful prints, we adjourned into another room for supper, which was in very splendid style.

On 6th December his father wrote to him: "Recollect my invaluable maxim never to quarrel with a man (but to waive the subject) about religion or politics," and added much good advice about wine-parties and avoidance of danger in skating. The reply of 12th December deserves summarizing: "I have gone to as few wine-parties as I possibly could, and at any to which I have gone there has not been the least approach to excess. . . . I have given no wine-parties, or indeed any parties yet, but I suppose I must return some of the invitations next term." "The separation of the freshmen of this College into the two classes of 'rowing men' (pronounced rouing, and meaning men who are fond of rows and 'rowing' parties) and 'reading men' has very soon become distinct. All my *friends* are among the latter class, and I am gradually dropping acquaintance with the former as much as possible. I find that even to know them is a very troublesome thing if we want to read, as they are always going about troubling people in their rooms." . . . Then he discusses the migration question: he has consulted Cookson how to beat Scratchley if Scratchley stays on at Peterhouse; the difficulty of choice of a college lies in finding one with lay Fellowships. He has now finished the "reading" of his first-year subjects, Euclid, etc. "My anti-short-sight glasses are getting on very well, and I certainly think I am very much less short-sighted than I should be if I did not use them." . . . "With regard to boating, you need not be in the least afraid. As I do not belong to the boat-club, I always row by myself in a funny (or, as it is called, *skulling*, for Alex. Crum's satisfaction), or at least go in a two-oared boat, with some friend with whom I should otherwise be walking. With regard to rowing in funnies, I think it a very useful thing, as it gives variety from mere walking, which alone is not the best exercise, and we never meet anybody except those with whom we go to row. Indeed, very few of the dissipated men row at all, except in the College boat, as they are always too much occupied, and the only objection I see to rowing without joining the club is the expense of going very often. I mean, however, when the fine weather comes, to make application to you with regard to how often I may take a boat." The letter closes with a discussion of some accounts, and of the costs of wines.

Christmas he spends at Gainscolne Rectory with Dr. Greenwood, the father of one of his fellow-collegians. Thence he writes to tell his father of Gregory's doings in finding the values of definite integrals "in a very curious way by the separation of symbols," and of a party at Challis's where he had met Cayley, "who is to be Senior Wrangler this year."

In an undated letter of this period William writes:—

> Hopkins has now given me two examinations, and he says, as the result, that he sees I know the principles very well, but that though I could probably read the subjects as well, or better, by myself, I may perhaps be the better of a tutor for a term or two before I read with him (which will be next October), to drill me in writing out a little. He says that if I stay up in the long vacation (which, he says, will be a great advantage) he will probably be able to direct my reading sufficiently so that I shall not require a tutor. . . . After the fourth-year men go away I am to get other rooms in the old court, which will be much better than these which I am in. . . . All the rooms in the old court are much cheaper in proportion to their excellence in the old court than in the new.

In the New Year of 1842 Dr. James Thomson writes to William, enclosing two bankers' drafts, and cross-questioning his son rather severely about his accounts of expenditure, the items of which do not tally with the total. He urges the importance of his acquiring "accurate business habits," and points the moral by recounting the financial straits of a colleague at Glasgow who had expended money recklessly on instruments, and was deeply in debt. William replies on January 15th explaining the items of the accounts. Bits of Cambridge news follow. Cookson and Hopkins have decided that Fuller is to be his tutor for the next term. "All the great mathematical men here are very much against the tutoring system. . . ." "You should get for the library a new French work on the Difl. Calc. by Moigno, which Gregory says is the best he has seen, and De Morgan's Difl. Calc. (in sixpenny Nos. by the Society for the Dif. of Useful Knowledge), which is very queer, but contains a great many good ideas." The criticism of De Morgan at this stage by the undergraduate, then in his eighteenth year, is curiously suggestive. A day or two later he writes again to his father asking him to send him his *Essay* (on the figure of the Earth; see p. 9), also his Fourier, Poisson's *Mécanique*, and Peacock's *Examples*, "and as many books of a lighter kind as you choose, as my library is so very scanty that I shall almost be obliged to buy books to fill the shelves." Then he tells how he has been measuring his strength in a preliminary way with the wranglers of the year. The Senate-house examinations being just over, he sat down to most of the papers to see how many questions he—a mere freshman—could do. "I found, on comparing with what some of the men had done who went in, that I got on tolerably well, especially in some of the problem papers, though of course I missed a great deal from not being very 'well up' with many of the subjects."

The safe arrival of the books was acknowledged in a letter of February 6th. He has got into new rooms. He has been rowing two or three times with Hemming.[1] He has got on well with Fuller, getting three papers a week from him to work; on low subjects so far, but next week to be on Analytic Geometry of three dimensions. On February 19th he sends home a long letter, with a surprise. Along with another man he has bought a boat for rowing, built of oak, as good as new, for seven pounds, the price new being twenty-four. The boat was decorated in blue and gold, and called the *Nautilus*.

The boat which we have got is made for only one person, and so we shall go down by ourselves on alternate days to row between two o'clock and four. I shall go down often along with Hemming who has a funny of his own. He is a very hard-reading and steady man, and will certainly be a very desirable acquaintance. He is very fond of rowing, but will not pull in the College boat on account of the kind of men of which the clubs consist usually. For his boat, which he takes by the year, he had to pay twelve pounds for this year, which is the first he has had her, and will have to pay six pounds a year afterwards, as long as he keeps her, so you see we have got a wonderful bargain. I have been going on reading steadily, about eight hours a day, and getting up perfectly regularly a little before six o'clock.

He adds that he thinks he may get a Gisborne scholarship, worth £30 a year. His father replied,

[1] George Wirgman Hemming, of *St. John's*; Senior Wrangler in 1844; Fellow; later Q.C. and Official Referee. Died in 1905.

expressing surprise at not having been consulted about the purchase of the boat, and saying roundly that he thinks his son has been taken in over the "wonderful bargain."

I think I told you to send me your accounts of expenditure from time to time. Any explanations, except those of importance, can stand over till I see you. Write them on slips of paper on one side, and you can cut them out as occasion may require. Use *all economy* consistent with respectability. Be most circumspect about your conduct and about what acquaintance you form. You are young: take care you be not led to what is wrong. A false step now, or the acquiring of an improper habit or propensity, might ruin you for life. Frequently look back on your conduct and thence learn wisdom for the future. . . . Have you been returning your parties? Tell me about anything of the kind. You must contract no debts except through Mr Cookson.

On February 25th William writes to explain further the purchase of which his father had disapproved. He tells how he has returned his invitations by giving two parties, both of which broke up about seven o'clock. His explanations must have had weight, for on March 3rd he writes again: "I was very glad that you do not object to the boat now, as I had been very uneasy since I received your first letter." "I am beginning to get very anxious to see all at home again, and am already looking forward with pleasure to the time when I shall be able to get away." "Our (Peterhouse) boat is at the head of the river."

On February 27th Elizabeth wrote to her brother that papa (he was always "papa" to his children) was reconciled to the purchase of the boat. But on March 12th William had to write that there were more accounts to pay. The College examination was now demanding all his time for preparation. "I have been thinking on writing a short paper on some points in electricity for the May number of the *Mathematical Journal*, but I do not know whether I shall have time till after the littlego. . . . I bathed to-day at Byron's Pool, for the first time, along with Hemming and Gisborne."[1]

Dr. James Thomson was, however, not quite satisfied. He wrote to Cookson to ask whether he approved of the way his son was conducting himself, and the reply was reassuring. Accordingly, on March 27th James Thomson sends his son, without further inquiries, £10, out of which he may pay for the boat; but he hints that Cookson doubts the propriety of the young undergraduate writing those advanced contributions to the *Cambridge Mathematical Journal*. Before this letter was received William had sent his father another batch of College accounts, which promptly evoked a call for further explanations as to unexplained items. On March 30th William admits in a rather crestfallen way his failure to account for the discrepancy of a few shillings, and explains the principal items of his College bills. On April 6th the father writes, hoping that his lecture on economy to his son has not been too severe, and tells of a visit of Archibald Smith, who does not agree that William should be discouraged

[1] Francis Gisborne, of *Peterhouse*; B.A. in 1845.

from writing in the *Journal*; also mentioning a dispute raging in the Senate of the Glasgow University, where he, James Thomson, was championing the abolition of religious tests against the party led by the Principal and Professor Fleming. On April 14th, acknowledging bank-notes from his father, William writes suggesting certain mathematical subjects for junior and senior classes at Glasgow. Archibald Smith's encouragement came, he says, just when he had taken down from its shelf his Fourier, and some notes made in Frankfort, which he now proposes to work up into an original paper. "The sculling is going on with great vigour, and is keeping me in excellent preservation. Every one now says that I am looking much better now than I did some time ago, and I find that I can read with much greater vigour than I could when I had no exercise but walking in the inexpressibly dull country round Cambridge."

On April 20th he writes about his original mathematical work for the *Journal*, for which he will have time in the summer in the house at Knock Castle (three miles from Largs), which his father has secured for the holidays; and referring to the College examinations adds, "of course, at present I have not much time for such 'dissipation.'" "Our classical lectures are on the 6th book of the Æneid (one of our littlego subjects) which will form the Latin part of our classical examination."

On May 6th he writes again that the College examinations are now nearing, and that he is reading hard in hopes of getting a Gisborne scholarship. He has now bought the other half-share of the boat. "I always go down along with Hemming and Stephen[1] (who is also one of Hopkins's men, and is 3rd of his year at St. John's). Budd, of Pembroke, another of Hopkins's men, and probably the only one of whom Hemming has anything to fear in the Senate-house, is thinking on joining our fleet in the long vacation." He asks his father to bring to Knock Fourier's and Poisson's *Theories of Heat*, as he will want to work at them. "You should endeavour to persuade Sandford to come to Cambridge instead of to Oxford. The well-taught, well-trained, and at the same time clever man is *the* man for Cambridge." He was now working very hard for the College examinations, working a mathematical paper each day, and spending the rest of the time on classics; rowing, however, from two to four with "the fleet." He distinguished himself sufficiently to earn the Gisborne scholarship. Cookson sent word to Dr. James Thomson: "Your son has passed an excellent examination, and has shown that he possesses talents which will enable him to obtain the highest honours in the University, if he goes on as he has begun. I thought it possible that there might be some slight deficiency as regards his qualifications for a Cambridge examination, but there appears to be little or none, and one may anticipate a very successful termination to his University career."

[1] James Wilberforce Stephen, of *John's*, Wrangler in 1844.

James Thomson wrote advising his son how to travel as cheaply as possible from Cambridge to Largs; and on June 30th the question of College expenses is again the subject of severe parental comment. The total cost of maintenance at College had been £230 : 7 : 8 since October 1841.

The summer of 1842 was spent by the united Thomson family very pleasantly at Knock; the event of the season being the engagement of Elizabeth Thomson to the Rev. David King.[1] William found time to complete for the *Journal* the two original memoirs which he had in hand.

The first of these memoirs of 1842, "On the Linear Motion of Heat," gave the solution in two different forms of the differential equation which expresses the linear motion of heat in an infinite solid, by which equation it is sought to find the temperature at some point at any distance, x, from a given zero-plane at any time t. This paper was a mathematical development of some intricacy on the lines of Fourier's work.

Again and again in later years Lord Kelvin would return to this paper as containing the germs of many of his subsequent ideas. In its concluding passage it contained a speculation as to the inference to be drawn if negative values are assigned to the time t; for obviously the theorems laid down hold good for negative values of t, as well as for positive

[1] Rev. David King, born 1806; minister of Greyfriars Secession (United Presbyterian) Church in Glasgow; LL.D. of Glasgow, 1840; one of the founders of the Evangelical Alliance, 1845; lived at Kilcreggan, 1855-60; minister of the Presbyterian Church, Bayswater, 1860-69, and of Morningside Church, Edinburgh, 1869-73; died in London, 1883.

values. In general it resulted that the temperature of any plane except the zero plane will be impossible for negative values of t; since the initial distribution of heat, assumed in the function, is in general not of such a form as to constitute any stage, except a first stage, in a possible system of varying temperatures. In other words, the state represented cannot be the result of any possible anterior distribution of temperature. Lord Kelvin used to declare that it was this mathematical deduction which convinced him that there must have been an origin to the natural order of the cosmos; that therefore natural causes could not be deduced backwards through an infinite time. There *must* have been a beginning.

A second part of the investigation on the linear motion of heat was published in 1843. It dealt with the solution of cases where the source was periodic in time; as, for example, the case of the propagation downwards into the earth of the periodic changes of temperature produced on the surface by the diurnal and annual variations of the heat received from the sun.

The second memoir, which is dated "Lamlash, August 1842," has for its title "On the Uniform Motion of Heat in Homogeneous Solid Bodies, and its connection with the Mathematical Theory of Electricity." It was subsequently reprinted (1872), as Article I. of Lord Kelvin's collected volume of papers on *Electrostatics and Magnetism.* In this memoir the leading idea is a certain analogy that had struck him when pondering the Faraday problem of curved lines of force. In the flow of heat through a solid conducting body, surfaces, called isothermal surfaces, may be drawn through all points that are at equal temperatures; and the stream-lines of the flow of heat as it passes from one isothermal to another will always intersect these surfaces normally. Again, if a conducting body be electrified, the charge of electricity at once distributes itself over the surface with such a distribution that the attraction on a point close to that surface, if oppositely electrified, will be perpendicular to the surface. The sole condition of equilibrium of electricity, distributed over the surface of a body, is that it shall fulfil this requirement. Consider a (closed) surface in an infinite solid to be somehow retained at a constant temperature from within, there being a steady flow of heat outwards across the surface. Next consider an electrically conducting body, bounded by a surface of identical shape, to be exercising forces on electrified points outside it. Then the electrical attraction at any point of surface, in the second case, will be proportional to the intensity of the flux of heat at a similarly-situated point in the first case; and the direction of the attraction will correspond to that of the flux. Farther, there follows this remarkable theorem, that if around a conducting or non-conducting electrified body of any shape, a surface be conceived to be described, such that the attraction on points situated on this surface may be everywhere perpendicular to it, and, if the electricity be removed from the original body and distributed in equilibrium over this surface, its intensity, at any point, will be equal to the attraction of the original body on that point, divided by 4 π, and its attraction on any point without it will be equal to the attraction of the original body on the same point. The possibility of thus replacing the actual system by an ideal distribution that should be equivalent to it so far as the production of forces was concerned, greatly facilitated the calculation of attractions in certain cases which previously were not amenable to mathematical treatment. The memoir went on to consider the special case of the uniform motion of heat in an ellipsoid. In the case of heat, where the isothermal surfaces are confocal ellipsoids, as Lamé had previously shown, they will meet the lines of flow orthogonally; so also will the lines of electric force in the corresponding electric case. The development of this conception, in mathematical form, was masterly, but the requisite integrations were stated quite simply; the theme presenting the appearance of a piece of physical insight mathematically stated, rather than that of an analytical investigation having a physical interpretation. After Thomson's paper had been some time in the hands of the editor of the Cambridge *Journal*, he discovered that he had been anticipated by M. Chasles, the eminent French geometrician, in two points, namely, in the ideas that led to the determination of the attraction of an ellipsoid, and in an enunciation of certain general theorems regarding attraction. He, therefore, when the paper appeared some months later, prefixed a reference to M. Chasles' memoirs, and to another similar memoir by M. Sturm. Still later, Thomson discovered that the same theorems had been also stated and proved by Gauss ; and, after all, he found that these theorems had been discovered and fully published more than ten years previously by Green, whose scarce work he never saw till 1845.[1]

Here was an undergraduate of eighteen handling difficult methods of integration readily, and with mastery, at an age when most mathematical students are being drilled assiduously in so-called geometrical conics and other dull and foolish devices for calculus-dodging. And not only was he handling with mastery the processes of the higher mathematics, but he was here attacking and solving problems, and laying down general and important theorems in physical science, to which three of the finest mathematicians in Europe had already independently been led. And yet his methods were not theirs. That of Chasles was geometrical rather than analytical, while Thomson had arrived at his by discussing Faraday's paradox of the curved lines of force at a moment when his mind was steeped in Fourier's treatment of the flow of heat.

October 1842 saw William Thomson back at Peterhouse to begin, under Hopkins, his higher mathematical training, the normal course of which should end in the Senate-house examinations in January 1845. He writes, on October 1st, that he

[1] See p. 113 for an account of this.

has sent to Gregory the paper he was writing at Knock; that Hemming and Stephen are both back, and sculling has begun again. By October 7 he knows that his paper is accepted for the November number of the *Journal.* He has now begun reading with Hopkins, who is giving them *viva-voce* examinations. "I can judge very little yet of any of the other men whom I meet with him, but I hope they are not extremely formidable." But there are more College bills to be met. He has, however, won a mathematics prize of £5, which he purposes to spend upon a Knight's *Illustrated Shakespeare.* His brother John writes to him, on October 6th, that Anna has been at Thornliebank;[1] also, that Margaret Crum and her sisters, Mary Gray and Jessie, came to Glasgow College yesterday for a call, and that Margaret was staying over the night. Then his father writes, asking why he did not buy, as his prize, Liouville's *Journal de Mathématiques,* instead of the Shakespeare. Next Elizabeth writes that Robert is ill with scarlet fever; and a fortnight later, when he is recovering, sends gossip about two young ladies, whom William will regret to hear are engaged to be married. On November 14 Anna writes: "We are all going on much as we did last winter. Our German studies resumed; Margaret Crum being in the class, as formerly, and John and papa have also joined us."

[1] The Rouken, Thornliebank, near Glasgow, the residence of Walter Crum, J.P., F.R.S., head of the famous calico-printing firm, and a great authority on all pertaining to cotton fibre. Walter Crum was a first cousin of Dr. James Thomson.

On December 7 James Thomson sends his son a piece of news. Dr. Meikleham, the aged Professor of Natural Philosophy, is seriously ill, and he is concerned as to the possibility of a vacancy. Who would be a suitable person to succeed him? Professor J. D. Forbes or Mr. Gregory?

William returned to Glasgow to be present at the wedding of his sister Elizabeth to Dr. King on December 15th. After that the winter seems to have gone uneventfully, though there are many letters sent to William from the family. His cousins the Crums, of Thornliebank, are often mentioned. An inquiry, of February 11, from Alexander Crum, "How is the cornopiston coming along?" reveals the fact that Thomson had fallen under the fascination of music, and had begun to practise playing the cornet. Of which more hereafter. A week later John writes that "Margaret Crum has been staying with us."

William had been fearful when he first went to read with Hopkins, that he might have a formidable rival in Fischer,[1] another of Hopkins's pupils. But as time went on William was reassured as to his own powers, and told his father so. On March 22nd James Thomson wrote his son a letter of worldly wisdom. "I am glad to hear that Fischer is not likely to be so formidable. Do not relax, however, as he or some of your persevering Johnian competitors may shoot ahead. I am also glad to find

[1] W. F. L. Fischer, of *Pembroke*; Fourth Wrangler, 1845; Fellow of *Clare*, 1847; afterward Professor of Mathematics at St. Andrews.

you have got acquainted with Walton.[1] Your having the favourable opinion of such people may serve you much hereafter. You never mention Aytoun[2] or Lushington,[3] and their friends are asking me from time to time whether you say anything of them. You should, by all means, cultivate Aytoun's good wishes, as you might thus, *as readily as in any other way*, secure the support of a friend of his in case of a certain event coming round. You should also pay attention to Lushington, walking with him, etc., if you can make it answer; and mention both frequently, if it were only to say they are well, or any other little matter."

This letter reveals, for the first time, the existence of a secret between father and son as to a certain event which might occur. In the precarious state of Dr. Meikleham the chair of Natural Philosophy at Glasgow might fall vacant; and Dr. James Thomson had now formed the ambition that his son might be qualified to succeed to it. As the months went on, and Dr. Meikleham rallied, and William continued to prove his remarkable original powers, not only in mathematics, but also in physical applications, this ambition became almost an obsession, as subsequently appears. Dr. Meikleham was an esteemed and trusted friend of the elder Thomson, and his son Edward Meikleham was an intimate comrade of the younger Thomson.

[1] Rev. William Walton, of *Trinity*; Eighth Wrangler and Third Senior Classic, 1836; Fellow of *Trinity Hall*, 1868; author of Walton's *Mechanical Problems*, and other works.

[2] Roger St. Clair Aytoun, of *Trinity*; Third Junior Optime, 1845.

[3] Franklin Lushington, of *Trinity*; B.A., 1846; afterwards Fellow.

Early in 1843 Thomson had begun to keep a diary of his doings; whether any earlier part was written is unknown. That which has been preserved extends over the Lent and Easter terms till October 1843. If it is ever published, it will be found to exhibit a striking picture of University life in the 'forties. A very few extracts bearing on Thomson's own career are here given.

EXTRACTS FROM CAMBRIDGE DIARY (1843)

February 13, 12 P.M.—Nothing remarkable to-day. Commenced rising at seven, after my last week's laziness, and mean to take shower bath to-morrow. Had a scull to-day with Hemming and Stephen. Though it was a glorious day, Stephen still grumbled very much about sculling. (Weighed 8 stones 10 lbs. in my jersey.) After hall walked with Barton [1] on business in town. Had half an hour's practice on the cornopean, before seven, when I commenced reading.

February 14, 11¾ P.M.—Had rather a long paper from Hopkins. After it, as it was a snowy day, practised the cornopean, partly along with Shedden [2] till hall time. After hall went to vote at the Union, and after that to Hemming's rooms, where I found Foggo.[3] Field came in afterwards, and we waited till chapel time. After I got to my rooms I practised a little on the corn[opean], and then read a little Paley, and looked over some of De Morgan's Diffl. Calc., on Geom. of 3 dimns. At 11 Barton came over with his knee cut and trousers shattered, having fallen in taking a corner on account of the frost. Fitzpatrick [4] came in, and interrupted any conversation we should have had.

[1] Richard Barton, of *Peterhouse*; B.A. in 1845.
[2] Thomas Shedden, of *Peterhouse*; B.A. in 1846; died 1906.
[3] David Foggo, of *John's*; B.A. in 1843.
[4] Richard William Fitzpatrick, of *Peterhouse*; B.A. in 1841.

March 5, 2 *h.* 25 *m.* A.M.—. . . Yesterday night I got foul of the orthogonal surface again, and sat till 12½ with my feet on the fender, but got no satisfaction. To-day after coming from Hopkins, I have got some new ideas, but not the ones I wanted. . . . After I had worked at Hopkins' problems till 11½, I commenced practising and summoned Tom.[1] About half-past 12, after we had been for about half an hour practising "We're a' Noddin'" and "Logie o' Buchan" in the lowest keys we could devise, and when I was in the act of playing "Adeste Fideles," at my reading stand, and Tom playing "Logie o' Buchan" at the chimney-piece, a gentle tap was heard at the door. "Come in," shouted Tom, and in walked Mr. Cookson. "Perhaps you are not aware, gentlemen, how much noise these horns make," etc. "We are very sorry," etc.

March 15, 1843.—This morning I got hold of my math. journal, and spent an hour at least in recollections. I had far the most associations connected with the winter in which I attended the Natural Phil., and the summer we were in Germany. I have been thinking that my mind was more active then than it has been ever since, and have been wishing most intensely that the 1st of May 1840 would return. I then commenced reading Fourier, and had the prospect of the tour in Germany before me. What a melancholy change has taken place with Dr. Nichol since then!

March 16, 11½.— . . I found Gregory reading "Piers Plowman," and spent a long time with him looking over it, and discussing old words. I asked him about where I could see anything on electricity, and we had then a long conversation in which Faraday and Daniell got abused.

March 20, 12¼.—. . . On Saturday night I got Shedden to mull some of the wine I had just received from Lynn, and got Greenwood over to help to consume it. We remained till 3 o'clock, and had a great deal of

[1] Thomas Shedden.

interesting conversation on metaphysics, dreams, ghosts, etc. . . . I was delighted to find that the passage which (the only one I ever read) disgusted me with Butler's *Analogy* had had exactly the same effect with him.

March 24, 11¾.—To-day I went to the Court before I had time to read at all. I remained for two hours or three hours listening to Kelly's speech about a will case. . . . If something else fail, I think I could reconcile myself to the Bar, though it would be a great shock to my feelings at present to have to make up my mind to cut Mathematics, which I am afraid I should have to do if I wished to get on at the Bar. . . . After hall I received a letter from papa (containing tin), advising me to see something of Lushington and Aytoun, and to mention them now and then in my letters. I accordingly set out and saw Aytoun, and asked him to wine to-morrow, and left a card for Lushington to the same effect.

March 26, *Sund.* 1½ A.M.—. . . My party went off seedily enough. Littlego and boats kept us barely in conversation. I read nothing after it except a chapter of Paley, but occupied myself with my cornopean.

March 31, 11 *h.* 10 *m.*—. . . This evening I have been working at Paley and Xenophon, keeping steadily before my mind the fear of being plucked. I have been corng. a good deal, to relieve my head from the seediness concomitant upon littlego subjects.

April 24, *Monday*, 10 *h.*—. . . On Sunday night, after I was left alone, I read *Evelina* till 2 h. 20 m., when I finished it (the first novel I have read for two or three years).

May 1.—. . . I went to Challis's first lecture to-day. He showed us prisms principally, and after lecture I saw the dark lines well.

Sunday, *May* 14, 1843.—. . . The boat racing has commenced in earnest. On Wednesday we had not much racing, but kept easily our place on acct. of the Johnians being bumped by Caius. Yesterday the odds were strongly in favour of Caius bumping us, but we astonished the University by keeping away. We had a glorious pull for it, and I shall remember for my whole life the work of seven minutes last night. My pleasure at keeping away was beyond anything I have ever felt. We shall have another hard pull to-morrow, as Caius means to bump us, and so I must have plenty of sleep.

October 23.—I have been reading *Faust* every evening after hall with Blackburn.

This last entry introduces us to Hugh Blackburn of Killearn, who later became Professor of Mathematics in Glasgow. Thomson had met him in his first term, and often he used to repair, after hall, to Blackburn's rooms in Trinity. It was here that they swung the famous Blackburn's pendulum. Some time in 1841-42 Blackburn's elder brother Colin (then in chambers in King's Bench Walk, Temple, London, afterwards Lord Blackburn), was asked by Archibald Smith to introduce him to Thomson. The introduction was effected at an informal dinner in Colin Blackburn's rooms, to which Thomson and Hugh Blackburn came up. Archibald Smith remained a firm friend of Thomson's for life, and influenced his bent towards the study of the phenomena of tides.

On March 24, William writes to his father that he has been pulling in the second Peterhouse boat, and that they want him to pull in the races next term. He will not, however, as he would be too sleepy in the evenings.

On April 9, James Thomson writes his son that Dr. Meikleham is much recovered, and, though he may be called away suddenly, he may survive for some time. He consults him as to books suitable for algebra teaching at Glasgow. Then he adds a few words of advice :—

Never forget to take every care in your power regarding your health, taking sufficient, but not violent exercise. In "your walk in life" also, you must take care not only to *do* what is right, but to take equal care always to *appear* to do so. A certain *censor morum et omnium aliarum rerum* [1] here has of late been talking a good deal about the vice of the English Universities, and would no doubt be ready to make a handle of any report or gossip he might pick up.

William replies on April 12, sending copies of papers and suggestions on algebra books. Adds that he won £6 last term in prizes. On April 20 he sends a solution of the centre of gravity of spherical triangles, and tells that he has been awarded the Clothworkers' Exhibition of £6 : 15s., and that he has been bathing before breakfast with Hemming.

The same day the father writes to his son on the turn which affairs are taking.

GLASGOW COLLEGE, *April* 20, 1843.

MY DEAR WILLIAM—Busy though I am, I cannot avoid writing to you on this the eve of our last penultimate Friday. On Monday forenoon Dr. William Thomson [2] called on me, the earliest time he could after the funeral of his daughter. He had been in Edinburgh, where your friend Gregory's brother-in-law, Alison, had met with him, and spoken to him about the N. P. Chair here for Gregory; and Dr. T. told me that he had that morning written to Forbes to hear whether he still looked to the

[1] This is a sly reference to Dr. William Fleming, Professor of Moral Philosophy, familiarly known to his students as "Moral Will."

[2] Professor of Materia Medica, see p. 21.

chair, as he told him that if he did not the electors ought to be aware, so that they might look out for other candidates, or something to that effect. I felt that in these circumstances I ought to mention to him my views regarding you. In doing this I asked him whether Dr. Nichol had ever conversed with him about the chair, and finding that he had not I told him about Dr. N's. views regarding you. He was naturally struck with the idea of your youth, etc.; but he received the proposition as favourably as could be expected. He asked about your *experimental* acquirements, particularly in Chemistry; and he mentioned Forbes as being *in this respect* one of the first men of the day, and as being of "European reputation." He seemed also to *wish* Gregory to be found to be a good experimentalist, as well as what he is acknowledged to be, a good mathematician, and he said that a mere mathematician would not be able to keep up the class. In the course of the evening I sent him a note, the first copy of the main part of which I enclose [see below]. I also wrote to Dr. N., requesting him, as the matter was thus opened, to call as soon as he could on Dr. T., and to state his opinion regarding you. This Dr. N. did not fail to do *the next day*, and he called on me after the interview. He told me Dr. T. received his communications very favourably, and said that, were it only to prevent objections, you ought to practise a good deal in performing experiments. I saw Dr. T. the next day (yesterday), when he spoke in a very friendly manner. . . .

Now I wish you to consider this subject seriously. Consider whether you can or should get any introduction to your professor of Chemistry, or whether you ought to be at the expense of some apparatus for experimenting in your rooms at your times of recreation. Dr. W. T. justly remarked, that while he had no doubt of your being able to lay before the electors here ample proofs of your being an accomplished analyst in mathematical and physical science, yet it would operate much against you, especially if Forbes were on the field, should you not be able to give evidence of your acquaintance with the manipulations, to a certain extent, of experimental philosophy.

Could you get a proper introduction to Cumming, you might tell him you wished to practise in some small degree in performing experiments (keeping, of course, your main object concealed from him and all others); and he, if you could get no means in his laboratory, would probably direct you regarding some simple apparatus and some suitable books; and a certificate from him or any such person on this subject might be of great consequence.

Turn the whole matter carefully in your mind, and write to me soon about it. Dr. W. T. would, I know, be glad to do a kindness to me or to any of my family when he could do so with propriety, and I feel it to be kind in him to offer such suggestions. At the same time, as I have told him, neither you nor I could think of carrying the matter, were it in our power, unless it were likely to be for the good of the establishment and of the public.

I shall shortly answer your last letter.—I am, your affectionate father, JAMES THOMSON.

The letter sent by James Thomson to Dr. William Thomson is as follows:—

COLLEGE, *April* 17, 1843.

MY DEAR SIR—I beg you to regard a part of our conversation to-day (about the part which I mean you cannot mistake) as strictly *confidential.* When you adverted to the subject, I felt it to be only candid, in the terms on which you and I are, to say what I did.

Having said so much, I only ask, that in a quiet and prudent way you will get, as occasion may offer, *from persons more disinterested than myself*, information regarding the *character*, the *qualifications*, and the promise of the person about whom I naturally feel a deep interest. On some convenient occasion I shall show you some private communications regarding him.

On April 24 William replies to his father :—

I feel very strongly what you say about the propriety of my endeavouring to get some practical experimental knowledge as soon as possible. I am, however, afraid that almost anything I could do in my rooms here could not be much more than trifling. I shall, however, look at some apparatus as soon as possible for the polarization of light, as there are many experiments in it which I might perhaps repeat with advantage, and without losing much time. I must, however, as long as I am here rely principally upon reading for getting experimental knowledge. I have of late, whenever I have had time to spare, been reading some of Lamé's *Cours de Physique*, which is an entirely experimental work. . . .

I must principally, however, depend upon getting some experimental practice in Glasgow. I should be delighted to have access to the laboratory, and I am sure I could improve myself very much. If there was any *immediate haste*, I might perhaps cut Hopkins for the long vacation and spend my time in Glasgow, but still I must not forget my principal object in being here. . . .

As soon as possible I shall speak to him [Gregory] about some papers for the May No. of the *Math. Journ.* As Mr. Cookson, however, has been "hoping I do not now lose any time with the *Math. Jour.*," I must endeavour not to attract his attention. All my papers as yet have been on physical subjects. I am sorry I cannot get some copies of my paper in the 14th No., as it contains demonstrations of some propositions, deduced *entirely from physical considerations*, which I could not prove analytically till after two or three years.

On May 5 William writes to his father, then expected shortly to visit Cambridge :—

At the beginning of this week we commenced reading with Hopkins for the term. . . . The first morning I went, I was agreeably surprised by his telling me that, if I improved a little in writing out my papers more explicitly, I should be sure of being Senior Wrangler. I had been beginning for a long time to think that he considered Fischer to [be] better than me, and so you may imagine that I was very much delighted with what he said. As he only said it to myself, however, I have not told anybody except you, and I think it should not be told to any one else.

On May 4 James Thomson sent his son two letters :—

GLASGOW COLLEGE, *May* 4, 1843.

MY DEAR WILLIAM—I send you the remaining half notes for £20, of which I have every confidence you will make the best use. Your bookseller's bill seems large. Purchase no books you can avoid. You can have the use of my books ; and as to the important object—the formation of a library of your own—you ought to postpone it for the present.

You will be glad to hear that I have succeeded in carrying the election of Sir Thomas Brisbane as Dean of Faculties. . . . We carried the election by only a single vote ; but we could even have spared that one. . . . Dr. Meikleham was brought out, and I am happy to say was so well as to be bandying jokes, and he seemed to be *more himself* than I have seen him for a long time.

In the present state of matters our party, *if we agree among ourselves*, and if we can carry the Rector and Dean with us, is *exactly equal* to the other. We have the advantage, however, that the Principal being chairman has no vote at an election except in case of an equality ; and in case of a *certain chair* becoming vacant, a vote would thus be lost to the other party. What you have to do, therefore, is to make character general and scientific so as to justify the Lord Rector, the Dean, and the other electors who usually act with me, in supporting you—a matter of difficulty on account of your youth.

I have told Dr. W. Thomson what you say about such experiments as you could perform in your rooms being only a kind of trifling. He says you are wrong, as the bringing of your *hand into practice*, at your time of life, is of great importance, were it only in the management of vials, and in other similar things of apparently an equally unimportant kind ; and Dr. Nichol is of the same opinion. In fact, your being able to get some certificates as to your having attended to such matters would help in neutralising, or at least meeting the objections sure to be brought forward by certain persons here, and what is of great consequence, would tend to secure the support of electors who would be friendly to you and me, but who might be afraid to support you on account of your youth. For your age your character here stands, I believe, excellently. You must strive to support it and to add to it. Take care to give a *certain gentleman here* (who, as to private affairs, is more nearly omniscient than any one I have known) no handle against you. Avoid boating parties of in any degree of a disorderly character, or anything of a similar nature ; as scarcely anything of the kind could take place, even at Cambridge, without his hearing of it.

I have more to write, but as I fear it will be too late for post, I must close, and am your affectionate father,

JAMES THOMSON.

The second letter related to the possibility of Professor J. D. Forbes becoming a candidate for the professorship. Forbes was a warm friend of the Thomsons, and his friendship was prized by William Thomson till his death in 1868.

In reference to this William writes :—

May 8.—I take the first opportunity of returning to you Prof. Forbes's letters, with which I have been *very much pleased.* As far as I can judge, I think it is pretty clear that he is very anxious for the situation, and, thinking himself sure of it, wishes to make his own terms before he accepts it. . . .

It was in this term that Thomson joined the crew of the Peterhouse boat in the College races.

We have seen that a year before he had taken to "sculling," and his earlier letters speak of men who had joined the group dubbed by their comrades "the Fleet." This was a coterie of five persons: Hemming of St. John's, who was Senior Wrangler in 1844, and who subsequently became a leading Chancery barrister; Stephen of St. John's, a Wrangler in 1844; Field[1] of St. John's; Gutch of Sidney Sussex, a Wrangler in 1844; and Thomson of Peterhouse. The "Fleet" was so styled because every day, fine or wet—unless very tempestuous,—at two o'clock, when other reading men usually went out for a walk, they five went out, as regularly as the clock, rowing in five boats.

Chatting of the "Fleet," Lord Kelvin gave me the following account of his achievements as a rowing man. "You know I won the Colquhoun silver sculls? Colquhoun was a Scotchman of Loch Lomond Side who gave a prize for Cambridge men for rowing. For the first years the race was rowed on the Thames, then afterwards was transferred to Cambridge. The first race at Cambridge was in November 1843. I never thought of going in for it.

"I had previously taken one term as oar in our Peterhouse boat, but had found it too exciting. All one's thoughts ran on the idea that whatever happened Peterhouse should not be bumped by Caius. In my first year Peterhouse was head of the river,

[1] Rev. Thomas Field, who took the Classical Tripos in 1844; for many years Fellow and Tutor of St. John's.

and we were all very proud of it. It was in the last term of my second year that I rowed for Peterhouse. Two or three of the men in the crew were leaving; we had to have two or three new hands. I had rowed by myself in my first year, so they wanted me, though a light-weight, in the eight-oar. There were three new men in the Lent term, of whom I was one. It was a foregone conclusion at the first race (there were six races, one every other day for a fortnight) that the first Trinity boat would bump Peterhouse, and so it happened: the first Trinity did bump. But what was feared was that Caius would bump us. Caius had gained ground in that first race, and it was desperately debated whether Caius would be able to bump. I pulled in number seven place in the first race, but, being found too light, was shifted to number one place for the other five races. I sculled quietly in the intervening days; but could do no reading. We had as cox an Irishman, Blake[1]; and Cobbold, the 'University steam engine,' was captain of the Peterhouse crew. He was afterwards an archdeacon and missionary in China. He and Blake arranged a plan: we were not to row ourselves out, but to row quite easily till Blake gave us a signal, and then we were to put out our strength. We rowed easy, just to keep a few inches ahead of the Caius boat, for about three-quarters of the course, till we got to 'the Willows'; then we laid out. We won

[1] William Gage Blake, who graduated at Peterhouse in 1844. He entered as pensioner April 27, 1840.

all the rest of the races, though the betting at first was ten to one that Caius would bump. At the last race, and for the previous two days, the weather was bad. All the colleges wanted to give up, except Caius, and Caius would not. So twenty-seven boats went out in a storm of wind and rain. Caius never got near the Peterhouse boat. During those three weeks of the races nothing occurring on the whole earth seemed of the slightest importance: we could talk and think of nothing else. It was three weeks clean cut out of my time for working at Cambridge; so I determined to do no more rowing.

"But six months afterwards I won the silver sculls. I had no intention, but simply rowed for exercise every day, as I found it better exercise in the time than walking. Three days before the sculls, Beresford of Peterhouse—he was uncle of Lord Charles Beresford, the admiral—met me and said: 'Now, old fellow, you go and rub your arms with opodeldoc, and go in for the sculls, and you'll win them!' I had been rowing in my own boat—a boat I had bought in my freshman's year, and which was then quite antiquated. I said I had not a boat fit to row in. Beresford said: 'Well, I have got Hemming's boat, and if I get bumped you shall have it.' It was so. I went down in the intervening days and tried it; shifted the thole-pins and got used to it—you know the boats in those days were rough things, without the modern appliances—and so felt my way about in the boat. As I had thus got used to it I bumped my men in each race.

The last race was a time race, going to the man who came in first at the winning-post, and I passed my rival about half-way up the course. The eight-oar race was at the end of my second year—the sculls half a year after." An eye-witness of the race, afterwards head-master of a large English school, has narrated that Thomson nearly fainted after reaching the winning-post, but soon rallied, amid the congratulations of his friends.

Dr. James Thomson visited Cambridge in May, on his way to London, whence, on 23rd, he wrote :—

> With my trip to Cambridge I have been much gratified. I am glad to say that what I saw and heard of you was very satisfactory. Your success in your studies, and in making the most valuable of all acquisitions —character—has afforded me great pleasure. . . . Whatever time or times you write, tell me about your races. You see, that though I consider it necessary you should give them up for the future, yet I feel an interest in them so far as you are concerned.

In June William went to stay at Horncastle with his friend J. B. Smith.[1] His father writes to him from Southannan, the house near Largs, which he has taken for the summer, but is called away to London in July by illness of the youngest boy Robert. Anna writes meantime to William that she is expecting Margaret Crum to stay a week with her at Southannan. "I am soon to lose her

[1] John Bainbridge Smith, of *St. John's*; Senior Optime in 1844 (son of John Bainbridge Smith, D.D., Head-Master of Horncastle Grammar School); later Professor of Mathematics at King's College, Nova Scotia, till 1855; died at Tunbridge Wells, 1904.

altogether now, as she goes to school to London in the end of this month, and she will be there two years." In August Professor James Thomson, still in London, writes to his son: "Dr. Meikleham has had another attack — a very bad one. He has weathered it, and is pretty well. In all human probability he will not survive another."

Thomson read with Hopkins in the long, but took a month working in the chemical laboratory at Glasgow, and another month at Southannan with the family, and returned to Cambridge on October 13th. His father wrote him:—

October 18, 1843.—"You were right about the propriety of my writing confidentially to Mr. Cookson. His reply is very kind, and I do not doubt his sincerity in the slightest degree. I am glad you have told him decidedly about your determination not to *pull* at races. If he have any innocent humours or peculiarities, you ought by all means to study them, and to gain not merely his approbation but his friendship." A week later he writes again that he has seen Dr. Meikleham, who appears, except for a slight increase of deafness, as he was last winter. "These matters are all favourable to you. Think, therefore, of every fair, honourable, and practicable way of preparing for the contest; and when you hear of the bad health of others, use every precaution in your power regarding your own. Try to become known to persons of name and influence—Liouville, for instance, etc., etc."

With November came the distractions of the

Queen's visit to Cambridge. Thomson writes home criticising the Latin speech of the Prince Consort. He has begun reading Optics with Hopkins. A merry party of men have been trying mesmerism in Greenwood's rooms. "Hemming operated upon Gisborne, and Greenwood on a freshman, a friend of his, and both were perfectly successful in driving the patient into the most exquisite hysterics."

At the middle of November he sends Anna a letter telling how Hopkins had charged him full fees for the long vacation, and how their elder brother James was now visiting him. His own news relates to the Colquhoun sculls. "I am practising now every day for a great sculling race which takes place on Tuesday. As there are fourteen *candidates* the decision will not take place till after two or three days' pulling. The winner will get a cup of about fifteen guineas' value, as well as the honour of holding a pair of silver sculls in his hands for a year. I do not, however, aspire to such an honour, and I shall be very well satisfied if I come in second or third.

"Blackburn and I went on very regularly with *Faust* till James came, but since that time we have been rather interrupted. I have very seldom time now to take out my cornopean, but after the sculling is over, I mean to miss going down the river one day at least in the week, and to have some practice on the cornopean."

Anna's reply lets him into the secret that she has

had a proposal, from a Belfast friend (Mr. William Bottomley); but she has a message to deliver.

"Papa bids me enclose a few lines . . . telling you that he is a good deal annoyed to hear of your joining in the boat races you mention in your letter to me; on different accounts he would much rather you would not enter into any such things in future; it has a little of the same spirit as gambling, he thinks, and also you will be very likely to get so much excited by the racing as to injure your health instead of improving it, which you know was the cause for which papa allowed you to get the boat." To William's answer Anna's further reply was: "I got your letter to-day containing all your reasons for having joined the boat races, which has one good effect at least—that of convincing us all that you are a most excellent logician. . . . However, though I am not *quite sure* yet that I entirely approve of the thing, my sisterly vanity is too great for me not most heartily to wish you success in all that you undertake, even boat racing, and I shall certainly be delighted to learn that you have carried the prize. I suppose by this time the contest is over, and from what you say, I should think it very likely that you have won. . . . Your silver cup will be quite a handsome ornament to your room if you get it." Her next letter is full of the motion for petitioning Parliament for the abolition of tests. The motion was carried in the Glasgow Senatus by 11 votes to 7. "What would your Cambridge friends say to all the movement in the Scottish

Universities? I am sure they must think it very radical, and if they heard of papa being such a leader, you would be expelled from the college for being a Chartist, so I suppose the less you say about it the better, only do not let yourself become imbued with Toryism." On November 20th she sends him congratulations on having won all his races and carried off the silver sculls. On December 5th Elizabeth writes to William announcing definitely Anna's engagement; and on the 9th he sends her congratulations, and mentions that he will leave for Glasgow on the 15th. He stayed on into January, returning to Cambridge *via* London in order to see his brother James, then training as an engineer in the Isle of Dogs, visiting also his cousin Margaret Crum at her school; of which visits he writes to Anna, telling her also that Hugh Blackburn has given him a copy of Goethe's works. Anna is glad to hear of his safe return to Cambridge. "The delightful company of *Faust* and *Punch* would have made even a far longer journey pleasant. . . . I am glad you saw Margaret Crum. Does she not look quite well that you say you hope she is not working too hard? Papa has been reading Moigno with great energy." "Are you going to become a great student of Goethe now that you have got his works?" February runs on, and Anna's wedding is fixed. Their brother James writes to tell him how sorry he is that William will miss the marriage. Gregory is dead; and advice is wanted from William about two Glasgow students,

Porter and Traill, who think of coming up to Cambridge. In reply William writes: "I think it is no possible advantage to a man to have the pleasure of novelty done away with before he comes up in October along with the others of his year. As far as regards reading, I think three years of Cambridge drilling is quite enough for anybody." Accounts of Anna's wedding are sent him by Aunt Gall, by John, and by James, who is just returning to London, and says that Mrs. Crum has given him a parcel to take to Margaret, whom he will see on Friday. William writes to his father saying he wants more money, and adds: "If you have any occasion to write to Cookson, thank him for all his kindness and attention." He is now revising first and second year subjects in readiness for the Senate-house. Hemming wants him to take a trip to Ireland to see Killarney; but papa puts a veto on the suggestion on the score of expense.

In April William ran down to spend a week-end with his brother James at Millwall, and sends his father a history of his doings. To-morrow being Good Friday, he will ride with W. Ker on an omnibus to Richmond. He has been with Anna and her husband to hear *William Tell* at Covent Garden, and to Exeter Hall to hear the *Messiah*. He had been with Archibald Smith to call on Colonel Sabine (afterwards Sir Edward Sabine, P.R.S.), and is invited to dine with them. Archibald Smith is getting on as a barrister, and this fact revives the old idea that William may also turn to the Chancery bar. "When I said that I might not be able to get anything else to do, he (A. Smith) answered that if Dr. Meikleham would live a little longer I might be appointed his successor. We then had a good deal of conversation about the Natural Philosophy class, and on scientific subjects in general, with which Smith seems still to be very much engrossed, even though almost all his time is occupied with business." Aunt Gall writes that papa is particularly pleased that William has made acquaintance with Colonel Sabine. Then William writes home that he wants O'Brien's *Calculus*, and Woodhouse's *Isoperimetrical Problems*, and De Morgan, and sends papa some problems about tangent circles.

On April 20 Dr. James Thomson sends William some more money, and writes :—

> Keep the matter in mind, therefore, and think on every way in which you might be able to get efficient testimonials. . . . Do not relax in your preparation for your degree. I am always afraid some unknown or little heard of opponent may arise. Recollect, too, that you might be thrown back by illness, and that you ought therefore to be *in advance* with your preparation. Above all, however, take care of your health.

To this William replies on 22nd: "I am very sorry to hear about Dr. Meikleham's precarious state. I have now got so near to the end of my Cambridge course that even on my own account I should be very sorry not to get completing it. For the project we have, it is certainly very much to be wished that he should live till after the commencement of next session." William has been attending Challis's lectures on Practical Astronomy, and will follow these by an experimental course on Natural Philosophy.

This is the appropriate place to mention the part which Thomson took in the foundation of the Cambridge Musical Society, which originated in an informal private amateur concert at Peterhouse. Shortly afterwards the undergraduates of Peterhouse held a supper, winding up with certain hilarious proceedings on the College roof, which nearly wrecked the enterprise. On the next occasion, when the concert was to be given in the Red Lion Hotel for want of an adequate room in College, the Master refused permission unless the Peterhouse Society would call itself the University Musical Society; and, as the *Cambridge Chronicle* records, the concert was given to a large audience, consisting almost entirely of gownsmen, on December 8th, 1843. The band of eleven instrumentalists performed Haydn's *First Symphony*, the overture to *Masaniello*, the overture to *Semiramide*, the *Royal Irish* Quadrilles, and the *Elisabethen* Waltzes of Strauss —Thomson playing the French horn. But the first official concert given by the Cambridge University Musical Society, as such, was on May 2nd, 1844, with the following programme:—Haydn's *Surprise Symphony*, the overture to Mozart's *Nozze di Figaro*, the overture to Auber's *Les Diamants de la Couronne*, two glees, a violin solo of De Beriot, the *Aurora* Waltzes, the *Troubadour* Quadrilles, and in addition a comic song, *The Nice Young Man*, sung by Dykes, with an encore, *Berlin Wool*. The Society prospered, and received at last official recognition from the University authorities.

In 1888, for the occasion of the celebration at Cambridge of Dr. Joachim's jubilee in March 1889, Lord Kelvin wrote to Professor (now Sir Charles) Villiers Stanford, the following account of the rise of the C.U.M.S.

Sir WILLIAM THOMSON to Professor C. VILLIERS STANFORD

December 20, 1888.

DEAR MR. STANFORD—I am sorry that I have been prevented by incessant occupations from sooner answering your letter of the 26th November. I should like exceedingly to be present if possible on the 14th of next March at your celebration of the fiftieth anniversary of Joachim's entry to public life, but my University duties keep me very closely bound here at that time and I fear I may not be able.

The origin of the University Musical Society in 1843 is to me an always happy recollection. The real founder of the society was G. E. Smith, who entered Peterhouse as a freshman along with me in 1841. He was greatly devoted to music, but he died quite young—before taking his degree, I believe. William Blow[1] was our strongest man, being a really first-rate violin player. He died a few years ago. Alfred Pollock (whose two or three published songs you probably know) came up a few years later, and helped the Society greatly as hautboy player and secretary. He was, I believe, a really first-rate hautboy player, and for many years after he went to London to become a solicitor used to play frequently by particular

[1] Rev. William Blow, of *Peterhouse*; B.A., 1845; Rector of Layler Breton, Kelvedon, Essex; died 1886.

request in the Crystal Palace and other bands. He was son of Sir D. Pollock, Chief Justice of Bombay. C. M. Ingleby helped us as tenor singer, in which he was really good. He too was a solicitor, and was the author of several books on philosophical subjects. He died three years ago. John Dykes is, I believe, still alive, a clergyman, I am not exactly sure where. You know his name well, of course, as author of some beautiful hymns, and, I daresay, other compositions. He was, I think, our first conductor after we left Peterhouse and became the University Musical Society.

Coombe[1] and Cridge,[2] both sizars of Peterhouse, played the tenor and violoncello, and were very good musicians. Both became clergymen, but I do not know whether they are still living. A clergy list would tell, and I am sure they would both be greatly pleased to be asked to join your celebration. We use to have delightful quartette evenings in the rooms of Sutton[3] of Jesus (a brother of the present Sir Richard Sutton, I believe). He was very mediæval in his tastes, approving of nothing more modern than Gregorian music. But we had delightful Mozart, Handel, Beethoven, Weber, and Mendelssohn often in his rooms. He had an organ in his rooms which had belonged, I believe, to Handel. That is all I can recollect of these happy times so far as the C.U.M.S. is concerned. The practisings and the concerts of the C.U.M.S. were a great pleasure, I believe, to all who took part in them.

Wishing you a happy Christmas and New Year, I remain, yours very truly, WILLIAM THOMSON.

Several reminiscences of this time and of the musical set which Thomson joined are to be found in the memoirs of the late Doctor Dykes,[4] who was

[1] Rev. Chas. Geo. Coombe, of *Peterhouse*; B.A. in 1846, and Fellow. Died about 1899.

[2] Rev. Edward Cridge, of *Peterhouse*; B.A. in 1848; now Bishop of the Reformed Episcopal Church in Victoria, British Columbia.

[3] Later Sir John Sutton, Baronet.

[4] *Life and Letters of John Bacchus Dykes, M.A., Mus. Doc.*, edited by the Rev. J. T. Fowler, London, 1897.

a contemporary, having entered St. Catherine's College in October 1843.

In 1844 Dykes wrote home :—

I had a deputation almost as soon as I got up, to ask me to take the presidency of the University Musical Society, now vacant by the departure of Blow. I asked for time to think about it, and then sent and declined it, in consequence of the reading, etc. (Was not this good of me?) I am very glad, however, that I did, for we have prevailed on that splendid fellow Thomson, of Peterhouse, to take it. It will be no end of a feather in our caps to have such a man as our representative in the University.

In April 1844 Dykes wrote to his sister Lucy :—

I am bringing home [1] a friend who plays the cornet in our band and rejoices in the very *un*common name of Thomson. He is a great friend of mine, and a very nice fellow indeed, and what is more, a most gentlemanly man.

The biographer of Dykes adds :—"The friend came, and delighted his hosts with his simple, good-natured kindliness. They have a lively recollection of how he made a flying-machine with umbrella whalebones, persuaded an egg to stand alone, and performed wonderful experiments, besides joining in the family concerts."

The following letter, addressed by Thomson to the sister of Dykes, preserves some interesting reminiscences.

Lord KELVIN to Mrs. CHEAPE

THE UNIVERSITY,
GLASGOW, *February* 23, 1896.

DEAR MRS. CHEAPE—The first president of the Peterhouse Musical Society was G. E. Smith, who remained

[1] *I.e.*, to Wakefield, where his father, Wm. H. Dykes, was Bank Manager.

president until his death in 1844, when it had become the C.U.M.S. He came up to Peterhouse as an undergraduate at the same time with myself, October 1841.

Blow, an undergraduate of a year later, succeeded him, and retired in 1845, when I became president.

Blow was a splendid violin player, and he continued as first violin in the C.U.M.S. till 1846. He became a clergyman, and lived unto about 1874.

When I left Cambridge in 1846 to enter on my professorship here, your brother John succeeded me as president.

I came up from Glasgow in the May term, and continued my part (as second horn) in the orchestra, till (as far as I can recollect) your brother retired in 1847, on leaving Cambridge. He was succeeded by E. W. Whinfield, of Trinity, an excellent violoncello player of whom your brother thought very highly.

A Jubilee Commemoration of the first fifty years of the C.U.M.S. was held at Cambridge in 1893, and I find a menu-card of a dinner given on the occasion at King's College Hall on the 12th June, with some pencilled notes on the back of it for a reply to the toast "Prosperity to the C.U.M.S."

"Founded twenty-two years after the birth of *Der Freischütz*.

"Seventeen years after the birth of *Oberon* and death of Weber.

"Sixteen years after the death of Beethoven, Mendelssohn still alive.

"These were our gods.

"1843: G. E. Smith, cornet, founder and first president.

"1844-45: Blow, violin and president.

"1845-46: Thomson, second horn and president. Macdonnel, of Magdalen, first horn and secretary.

"1846-47: Incomparable John Dykes, musician and president.

"1847: Whinfield, 'cello and president."

I well remember my first visit to your father's hospitable house, in Wakefield, with your brother, in the Easter vacation 1844. I can never forget the kindness I received from all your family, including the extreme good-nature of your father in giving me some instructions in the French horn, and allowing me to play on it in his study when he was out at the Bank, and the, if possible, more extreme good-nature of the rest of the family in tolerating the noises that came from that room during many hours of each day of my visit. I also remember well being taken to a friend's house in Wakefield, and being delighted with selections from *Don Giovanni*, instrumental and vocal, in which your father played the horn part.

I am glad to hear that there is to be a published memoir, and a collection of your brother's hymn tunes, of which the general public knows just enough to have some appreciation of their extreme beauty.

Will you give my kind remembrances to your brother and sisters, and accept my warm thanks for your kind letter and congratulations.—Believe me, yours very truly,

KELVIN.

Another reminiscence has been given by the late Rev. J. A. L. Airey, Rector of St. Helen's, Bishopsgate, in 1844 an undergraduate of Pembroke, and second wrangler in 1846.

One evening in the May term I was invited to attend an instrumental performance of a Musical Society then in its infancy under the direction of a Mr. Sippel, a musical-instrument seller, and teacher of music in the King's Parade. This subsequently became the "Cambridge Musical Society." Then, for the first time, I heard the violin and cornet-a-piston played to perfection, the former by a Mr. Blow, an undergraduate of Peterhouse, and the latter by a Mr. Smith, who produced such a soft and mellow tone from his instrument that I made up my mind from that time forth that the cornet was to be *my* instrument. Forthwith I borrowed one from Mr. Sippel, and took a course of lessons from him, until I was prepared to take my place as "second cornet" in the Musical Society at the commencement of my second year, and played regularly in the concerts for two seasons. John Bacchus Dykes, of St. Catherine's College—afterwards *the* Dr. Dykes of sacred music reputation—was the conductor, and Thomson, of Peterhouse, my yoke-fellow, at the same desk for horns. He is now Lord Kelvin. Both he and "Johnnie Dykes," and some few other musical friends, remained my intimate companions for the rest of my University life in all things pertaining to musical affairs and pleasant evening gatherings.

Here I should like to record a little incident when Thomson and I, a short time afterwards, who were fellow-pupils in Mathematics under the then distinguished mathematical "coach," Mr. Hopkins, were reading the *Theory of Sound*. In our cornet practice the fundamental note, as ordinarily accounted and played, was an octave higher than the theoretical fundamental note of an open tube. How was this? Thomson's inquiring mind could not rest until he had reconciled the apparent anomaly; and one evening, as I was sitting quietly in my room, enjoying my after-dinner pipe, he rushed in wildly, crying, "Airey, Airey, I have found it; I have found it! Where is your horn?" Whereupon, after sundry efforts, he succeeded in producing the *real* fundamental note of the open tube, and was satisfied. In fact, this note is so low, and practically impossible of production, that it is not quoted at all as *the* fundamental note, but the octave above takes its name and place in ordinary parlance. This was an early indication of the strong powers of investigation and research in the student, who afterwards became the distinguished scientist and electrician.

The printed rules of the C.U.M.S. for the Lent Term 1846 gives a list, showing that the Society had then grown to number 426 adherents, the Committee being W. Thinn, *Pet. Coll.*, President; C. G. Coombe, *Pet. Coll.*, Secretary; B. Cridge, *Pet.*; J. B. Dykes, *Cath.*; J. D. Hodgson, *Pet.*

Amongst the "full members" were Airey, J. A. L., *Pemb.*; Blow, W., *Pet.*; Frost, Rev. P., *John's*; Pollock, A. A., *Pet.*; Shedden, T., *Pet.*; Walmisley, Prof., *Trin.*; Whinfield, E. W., *Trin.* Amongst the "half-members" (non-performers) were Bashforth, F. D., *John's*; Blackburn, H., *Trin.*; Hemming, G. W., *John's*; Hopkins, W., *Pet.*; Scratchley, T., *Queens'*; Stephen, J. W., *John's*; Venables, H., *Jesus*; Waddington, W. H., *Trin.* The last-named was afterwards French Minister and Ambassador at the Court of St. James's.

The prominence of Peterhouse will be noted. Rule 7 runs: "That Peterhouse men be not subject to the ballot."

Lord Kelvin retained throughout his life his love for music. Though he ceased from 1847 to take any part in the concerts of the C.U.M.S., he frequently came up to listen to the performance in May. While he set down Weber, Beethoven, and Mendelssohn as "our gods," he almost equally admired Mozart, Haydn, and Schubert. Beethoven's *Waldstein* sonata was a special favourite with him.

The spring of 1844 had been a very busy time for Thomson. He was reading hard with Hopkins at the appointed routine of mathematical tasks; but a mind like his could not dwell within routine. The entry of March 15th, 1843 (see p. 50) in his diary shows that he had then, and ever since Frankfort days, kept a private mathematical note-book in which he set down the ideas that came to him. Though the first volume of this journal cannot now be found, two volumes exist, one extending from December 1843 to June 1845, the other from July 1845 to July 1856. In this journal, on January 11th, 1844, he sketches out his mathematical paper, "Note on some Points in the Theory of Heat" (see p. 185), an investigation which he finished in two days. It was inserted in the *Cambridge Mathematical Journal* for February 1844. On May 1st we find the entry that he has been looking at Bertrand's paper, but "I have not yet had time to read his Memoir, as I have been quite engaged with examinations and coaching on Astronomy, etc." On May 4th he writes to his father that Hopkins is intending to go to the seaside—to Cromer—in the long vacation with a party of reading men, his pupils, to go on there with their preparation for the coming ordeal of the Senate-house examination. William, who goes to read with Hopkins at 7 A.M. every other day, proposes to join this party. "I always bathe, before I go, with Blackburn; and on intermediate days I bathe before going to chapel, so that I am up every morning at six. The boat races commence on Thursday, but I have nothing to do with them now, as I have no time to spare from reading." From his brother-in-law, Wm. Bottomley, he gets a letter telling how the writer has heard report of a certain dinner to the musician Hullah, at Cambridge, where, "on going upstairs to coffee, the conversation became highly scientific: one gentleman was demonstrating a geometric proposition, using a muffin and several triangular pieces of toast for a diagram." So do music and mathematics mingle. The reading party at Cromer has now been fixed for the end of May. It is to end about the middle of August. He will have a month in Scotland, returning to Cambridge in September, after which the course will be clear to the Senate-house.

On May 13th the father writes to the son :—

In going to Cromer, or elsewhere, you must of course do as others do with regard to expense for lodgings, etc. You may have it in your power, however, consistently with propriety, directly or indirectly to check extravagance in such things, and you ought to do so.

And now another matter of importance looms upon the scene. William writes of it to his father :—

ST. PETER'S COLLEGE,
Sunday evening, *June* 2, 1844.

MY DEAR FATHER—Our examination has been over for a week now, as I suppose you know, but I have not written to you about it, as there was nothing connected with it particularly exciting to me. I have been taking an idle week since, and on Tuesday I start for Cromer along with Blackburn and Fischer. After we have been eight weeks with Hopkins, we shall leave, and so Blackburn and I shall be able to get to Scotland about the first of August. We shall have to leave again about the middle or end of September.

Ellis[1] has been here for some time, having come up to set some of the Trinity examination papers. I have seen a good deal of him last week. We have been talking of a plan which I proposed for enlarging the

[1] Robert Leslie Ellis, of *Trinity*; Senior Wrangler, 1840; Fellow of *Trinity*; editor of the *Cambridge Mathematical Journal*. Died at Cambridge, May 1859.

Mathematical Journal, so as to make it something of the nature of Liouville's if possible. Of course the two great difficulties will be to get contributors of memoirs, and money enough to defray expenses, but as mathematical study is considerably on the increase, here at least, we are in hopes that in a few years' time we may have succeeded in doing something for the object. If the plan be carried out at all, the great object of course would be to make the journal as general as possible in this country, and to get it made known on the Continent. I have been speaking to Cayley since, and he quite enters into the plan. One great assistance he thinks would be, that there is at present no journal of the kind in this country, and that the want is very much felt by mathematical men. Of course, I cannot spend any time on such plans till after my degree, and Ellis proposes that we should let the Journal go on, on its present plan, till this time next year, when the fourth volume will be finished, and then we may think on enlarging it.

My address will be Cromer, Norfolk, which I suppose will be sufficient, as it is a small place. To prevent mistakes, however, it might be as well to put on the address of the first two or three letters, "of St. Peter's College, Cambridge," after my name.—Your affectionate son, WILLIAM THOMSON.

He spent the first week at Cromer, before serious reading began, partly in reading Airy's and O'Brien's mathematical tracts, and partly with Blackburn in experimenting on spinning bodies, humming-tops, and pebbles gathered on the beach, and hoops, working at problems of their rotation. Thomson and Fischer had lodgings in a house which, because of its proximity to the edge of the crumbling cliff, was called "Jeopardy College." Charles Buxton was also of the party. Thomson writes again :—

CROMER, NORFOLK, *June* 13, 1844.

MY DEAR FATHER—I have again to write to you on the same pleasant business that I had to write to you about so lately, which is to say that my money is again all gone. I spent nearly all the money you sent me in paying my Cambridge bills before I left, and the journey here, which was rather expensive, and my first week's expenses here, have exhausted my stock, so that I have only half-a-crown left.

Blackburn, who has been on several reading parties such as the present, states, as results of his experience, that the expenses are from 12 to 15 shillings a week for each person, besides lodging. Three of us—Fischer, Blackburn, and I—have lodgings together, and for the three of us we pay two and a half guineas a week, so that each will have to pay 17s. 6d. a week. Thus, before I go away I shall have to pay £7 for lodgings, as I shall have been here about eight (or possibly nine) weeks. This, besides my weekly expenses (which we have, as yet, been paying at the end of the week), will be all I shall require, in addition to my travelling expenses to Scotland.

I hope this will reach you before you leave Glasgow, though, from what John says in his letter of June 5th, I am rather afraid it may be too late.

This is a very pleasant place for England, and especially for Norfolk, which is rather remarkable for its dulness. The whole country is elevated on the top of high, steep cliffs of sand, and it is encroached upon every high tide by the sea. The house we are in cannot survive another winter, I think, unless great care is taken to protect the cliff below it with a wall. This was attempted to be done some time ago, but the first rough weather at spring tide carried a great part of it away. We have been bathing regularly since we came here, and walk about a great deal, so that we are making the best use of the sea air we can. We go to Hopkins every second day to have an examination paper, and on the intermediate days I go to him at seven in the morning to have my papers looked over. This will be our system all the rest of the time we read with Hopkins till we take our degree, as we have now gone over all our subjects. Since we came here, we have been commencing our second year subjects (Diff. and Int. Calc., Diffl. Equations, etc.).

Shedden arrived here on Tuesday.—Your affectionate son, WILLIAM THOMSON.

A letter to Walton, of date June 25th, shows that, amongst other recreations, he is helping him to revise the proof-sheets of Gregory's posthumous book which Walton is completing, and he writes giving Walton some novel demonstrations.

On July 4th Aunt Gall writes to tell William that the family has gone down to Gourock for summer quarters, and adds, good soul, "I hear Margaret Crum is home and quite well, and looking well." His fair cousin's health seems to have been not without interest.

CROMER, *July* 10, 1844.

MY DEAR FATHER—I heard from Aunt that you have got pleasantly and comfortably settled at Gourock, and I hope soon to be able to join you there.

The very earliest that I can leave this is the 7th of August, and even with remaining so long I shall not be able to get my whole time with Hopkins, as he does not return to Cambridge till the beginning of October. However, he says that if I come up about the middle of September and read by myself, I shall have plenty of reading before January. He recommended me, however, to take care and not give up preparation until the very time of the examination, as even though I may have gone over all the subjects carefully, the only plan to be sure of "doing myself justice" is to keep up preparation till the very last. We are still going on with examination papers every second day, and have now got through

Diff. and Int. Calc., and Statics and Dynamics, and to-morrow we commence the first year subjects, beginning with Algebra and Trigonometry. I have been investigating, as far as I have been able in the time I have been able to spare, the theory of spinning tops and rolling hoops, which is very curious and difficult. Blackburn and I have been making a great many experiments on the subject, and have collected quite a cabinet of ellipsoids of various proportions, which we find on the beach, besides having got a teetotum, humming top, and peery. Some of the results we have obtained are very curious. . . .

My best way from this will be by an Edinburgh steamer, if the weather be sufficiently calm to enable us (Blackburn and me) to get on board from this. If the sea be rough this will be impossible, and we shall have to find out some other way of going.—Your affectionate son, WILLIAM THOMSON.

One of the incidents of the Cromer time was, as Prof. Blackburn relates, their dining, of a Sunday, at Northrepps Hall with Sir Thomas Fowell Buxton, and there after dinner each guest had to spout a bit of poetry, and each was rewarded by the genial old baronet with a crown piece! Blackburn recited Perdita on the gardener's art. On a second Sunday Thomson obliged the company with Longfellow's "Village Blacksmith." Sir Fowell delighted them by letting them ride his horses for recreation.

When the reading party broke up, Thomson and Blackburn took a boat and rowed out to sea, and intercepted the G.N.S. steamer *Trident*, and so went to Leith—a course, says Blackburn, "obviously suggested by Thomson." From Glasgow Alexander Crum accompanied him to Gourock. Thomson pays visits to the Crums at Largs. He also goes to

Belfast to see his sister, Mrs. Bottomley. Papa, ever on the alert, sends him word of the suspicious circumstance that Prof. Forbes has been at the British Association at York. Blow writes to him to ask how the cornet gets on. "You will, I suspect, give a solo on it at one of the concerts, and I would accompany you—what say you?"

The mathematical note-book affords further glimpses :—

Aug. 14, GOUROCK.—Left Cromer Aug. 8, having finished all the first and second year subjects with Hopkins, and Hydrostatics, Optics, and $\frac{1}{3}$ of Astronomy (*i.e.* one paper out of three) of the third year subjects. I had not much time for additional reading. This morning the following theorem has occurred to me. I hope it has not been proved by Gauss, and if not, it is highly improbable that any one else has. . . .

GOUROCK, *Sept.* 10, 1844.—I have long been entertaining a project of writing a series of essays on the mathl. theory of electricity, commencing with the fundamental principles, and giving all the applications of the general theorems relative to attraction, which are of use in giving a comprehensive view of the subject. Though my plan is not quite settled yet, and I shall have no time to think upon it till after I take my degree, I commence writing down some of my ideas in a disjointed manner. If my projects relative to the *Cambridge Math. Journal*, which I mentioned to Ellis last June, at Cambridge, succeed, I shall probably be able to get the essays printed in successive Nos., and possibly afterwards get them reprinted and published in a separate form, and thus fulfil my old thoughts of writing a treatise on the Math. Theory of Electricity.

To the foregoing Thomson four months later added the following notes :—

Jan. 27, 1845.—I have just met with Green's memoir, which I first saw referred to in Murphy's first Memoir, on Definite Integrals with physical applications, which renders a separate treatise on electricity less necessary. (Paris, Feb., Green's memoir creates a great sensation here. Chasles and Sturm find their own results and demonstrations in it.)

In the note-book, under date September 10, we find five pages of a draft outline of a treatise :—

"On the Mathematical Theory of Statical Electricity" (historical sketch and general explanations).

"The object of the mathematical theory of statical electricity is to determine," etc.

Sept. 25.—Arrived in Cambridge, Monday, Sept. 23, having left Gourock on the preceding Monday for Belfast. . . . I shall probably be hardly able to open this book till I take my degree, as all my time must now be spent in preparation.

Nevertheless we find entries: Sept. 27 (parabolic orbit), 4 pp.; Oct. 14 (orbital velocity), 1 p.; Oct. 16 (principal axes of curvatures), 5 pp.; Oct. 25 (*ibid.*, and application to *n* dimensions), 4½ pp.; Oct. 26 (variable elements in planetary theory), 3 pp.; Oct. 27 (problem of the hoop), 5¾ pp.; Nov. 6 (*ibid.*), and gives a theorem afterwards set by him, May 1, 1845, in Peterhouse examination!; Nov. 15 (lines of curvature), 1 p.; Nov. 23 (axis of tangent cone to ellipsoid), 3 pp.; Nov. 26 (reduction of general equation of second degree, afterwards put in *Math. Journal*, Feb. 1845), 2 pp.

No sooner has Thomson returned to Cambridge than his father sends him the latest intelligence :—

GOUROCK, *Sep.* 22, 1844.

MY DEAR WILLIAM—When in Glasgow on Friday at the High School examinations I saw Dr. William Thomson for a short time. He seemed much gratified when I told him that you would probably take his advice about going to Paris. He said he had conversed with Dr. Nichol on the subject, who highly approved of the idea. He (Dr. T.) lately met with a young man who had been studying in Paris, and from him he took down for your express use the accompanying notes regarding the Parisian lectures on the back of a note that had been addressed to his brother. (Preserve the memorandum.) He still speaks emphatically about the necessity of your giving very great attention to the experimental part as soon as you can; as he says no one will have any doubt as to your mathematical attainments, but that some may even think them to be such as to make you neglect the popular parts of Natural Philosophy; and he wishes you to be able to obviate such an objection, should it be started by any of the electors. He is also equally anxious about your following out his hint about writing some lectures or discourses of a popular kind, and he remarked smiling (and, I suppose, alluding to Lushington), that a Cambridge education did not always give the power of easy expression or of commanding the attention of an audience. He added also that for such reasons it would be very desirable both for your own improvement and for strengthening the hands of your friends, that you should give great attention to such matters. I told him that till January you could not attend to such matters, but that I was sure you would do so then. I am sure he is extremely friendly, and that he will feel a pride in contributing to bring you in; but at the same time he would wish you as far as possible to justify your friends in the eyes of the public in making the choice.

I saw the Principal also. He told me of his trip to Arran. He says Dr. M. is greatly fallen off, but still he rises at one o'clock, and takes an airing before dinner and one after it; and he says his power of conversation is quite gone, and that he answers only in monosyllables. I said I supposed the class would be taught by David Thomson as usual, and he answered in rather a hesitating way that he supposed it would.

In the steamboat on my return I was introduced by Mr. Patrick Mitchell to Mr. Bouverie, member for the Kilmarnock burgh, and son to Lord Radnor, the great Whig peer. He seems a very nice, agreeable, intelligent young man. I am sure he must be a Cambridge man (inquire and tell me); but so short was the time after the introduction till he and Mr. M. left the boat at Bowling, that I did not find time to ask him amid the rest of the conversation which flowed very rapidly. Mr. M., on "Tom Shedden's" authority, immediately told him that you were to be senior wrangler, as there was no one in Cambridge who could "touch" you. Mr. B. then asked me had you a private tutor? "Yes." "Is it Hopkins?" "Yes." "Oh, then, your son must be a crack man. Do you really expect him to be *Senior* Wrangler?" "Mr. Hopkins writes that he thinks he will." "Then I shall watch the papers in January with great interest to see him gazetted as such." He spoke of the hard case of Sylvester having been refused his degree, because as a Jew he could not subscribe to the Thirty-Nine Articles. On this Mr. M. mentioned our anti-test movement in Scotland, and Mr. B. hoped we would go on with it, and said he would give all the support he could.—I am, your most affectionate father, JAMES THOMSON.

On October 7, William writes back wanting money. Hopkins, he says, has returned, and is giving them trial examinations in physical astronomy and optics; and then we are to "commence for the *n*th time revising our subject." The request so soon for money surprises the father. In sending his son the halves of bank-notes for £100 he raises the question of expenditures.

The following are the amounts of your three years' expenditure :—

Up to Sept. 26, 1842	. .	£240 : 11 : 7
From that date till Nov. 1, 1843	.	238 : 15 : 0
" " Oct. 12, 1844	.	295 : 0 : 0
		£774 : 6 : 7

. . . How is this to be accounted for? Have you lost money or been defrauded of it, or have you lived on a more expensive scale? Do consider the matter and state fully and clearly what you take to be the reason. With regard to what you now receive and what you may get in future you *must* exercise the *strictest* economy that shall be consistent with decency and comfort, not expending even a penny unnecessarily.

He then adds that Dr. Nichol strongly recommends William's going to Paris.

William replied on October 14 that he is quite surprised at the account of his expenditures. He has latterly been living less expensively, and cannot account for the increase. He has certainly neither lost money nor been defrauded. He agrees that it would be most useful to him to go to Paris, the expense being the only objection.

At the end of October Mrs. Bottomley sends news to William that Dr. M. has again had a slight attack of illness, but that in the Glasgow session now opening David Thomson will take the Natural Philosophy class again for the winter. William writes to Mrs. King on November 6 :—

I have very little time now for practising the cornopean, and make no progress at present; but knowing Blow very well, I still manage to hear a good deal of fine music. Blow is now conductor of the Musical Society, which is going on with great vigour. The performers are getting too proud for quadrilles and waltzes, with which of course the Society commenced, and they are now to give us only symphonies and overtures.

On November 11 William hears from his younger brother John that papa has seen Dr. M., and thinks he cannot long survive. Aunt Gall next sends word that papa has disclosed privately to the new Lord Rector (Marquis of Breadalbane), "his intention regarding *you* and the N. P. chair."

The date for the Senate-house examination was fixed for January 1, and the impression seems to have been pretty general that William will emerge from the ordeal as Senior Wrangler. Thomson's brothers and sisters all expect it, and his Cambridge friends do not conceal their opinion. On December 11 he writes his own views of the situation to Aunt Gall :—

The subjects for examination are pressing in on all sides, and seem very formidable at present; but the greater the pressure is now, the greater will be the relief this day four weeks when all will be over. I do not feel at all confident about the result, but I am keeping myself as cool as possible, and I think I shall not be very much excited about it when the time comes. One thing, at least, I am sure of is that if I am lower than people expect, I shall not distress myself about it; and if any of you lose any money on me I shall consider it your own fault for giving odds.

Next week he sends word to John that he and Hugh Blackburn "made an experiment on animal heat by bathing last Saturday," — in the middle of the frost. Whereupon papa writes :—

GLASGOW COLLEGE, *Dec.* 18, 1844.

MY DEAR WILLIAM—Having no proof sheets and no very urgent business on hand to-night, I sit down to drop you a note. . . .

David Thomson is a candidate for the Nat. Phil. chair in Aberdeen, now vacant by the death of my old friend Dr. Knight. I think he has a pretty good chance of success.

So your examinations commence this day fortnight. You cannot but feel anxious till they are over, and so will your friends. In the meantime think not of them, but of the preparation for them. By so doing I have no fear that you will be cool and retain your self-possession. Above all things be careful of your health. I much doubt the propriety of bathing out of doors on such a day as last Saturday was here. Experiments on yourself regarding animal heat are very questionable. I suppose you will not know how matters stand regarding your examination till the result is finally announced. Can you contrive to drop us a note each evening telling us about your health, how you get on, etc., or to get some one else to do so? Let the note be as brief as you choose, and let it be merely *legible*. I am told the best men often do worse than the inferior ones, during the first and second days, from having forgotten the earlier and more elementary subjects. Are you made up on those subjects? . . . Till your examinations drop a short note now and then. The *first* part of the new edition of Euclid is out, and is in the hands of the students; the second part is half done.—Ever most affectionately yours, JAMES THOMSON.

A silence of expectancy falls upon the family circle, and until the testing time is over there are only a few briefest notes to keep the anxious father informed. His tutor Cookson wrote after the second day:—

ST. PETER'S COLLEGE, *Jany.* 3, 1845.

MY DEAR SIR—I write you a few lines to say that your son continues in good health and spirits during the examination, and that he does not complain of being tired.

He has shown me what he has done in each of the papers up to last night; but I am unable to draw any great inference from the information.

The papers in which he may be expected to do so well in are those of this afternoon, and the three remaining days of the examination.

He does not appear to have been much pleased with Mr. Blackall's problem paper yesterday afternoon.—With best wishes I remain, my dear sir, yours very faithfully,
H. W. COOKSON.

Dr. James Thomson.

Hopkins also wrote encouragingly. Then the father begins again to elaborate plans :—

GLASGOW COLLEGE, *Jan.* 8, 1845.

MY DEAR WILLIAM—I cordially congratulate you on the close of your examinations, which you must feel as a great relief. I am perfectly satisfied that you have done your part well, and whether you get the place we all wish you or not, you cannot be blamed either by yourself or your friends.

From Mr. Cookson's and Mr. Hopkins's letters we have great hopes. Still there is no surety as yet, and I do not think you are more anxious on this subject than we are here. Relieve our suspense, therefore, as soon as you can. I thought the announcement would have been made much sooner than Friday, the 17th inst.—the time Mr. Hopkins and you mention. You must have got on most satisfactorily on Monday. Tell me when you write whether you will get credit and advantage for having finished the papers in less than the prescribed time. In *justice* you ought.

I still think it best for you to go to Paris; and as soon as your place at the examination is announced, I shall set about procuring you introductions. You, too, must try to get some at Cambridge to the leading science men in Paris. I shall apply to Sir David Brewster, Sir Th. Brisbane, Dr. Chalmers (who is a member of the Institute), Mr. Crum, and *perhaps* Professor Forbes; also to Mr. Ogilby of London, Dr. Patterson's brother-in-law, etc.; and Dr. W. Thomson is laying plans for getting introductions from Lord Brougham and Lord Jeffray. Colonel Sabine also might do something. Whewell, Peacock, Babbage, Herschel, Airy, could serve you much, if you could "get at them," which I think you might *after* your exams. Take down a list of all the persons you can think of, and consider and purge it at your leisure.

I write to Longmans' people to send copies of the Algebra to your care for Walton and Fischer. Deliver them without delay with my compliments.

Robert is going on well, but, as his sick-nurse expresses it, he is "a perfect anatomy."

At Hamerton Academy my Algebra has superseded De Morgan's, and my Euclid Chambers' edition of Playfair.—I am, most affectly. yours,

JAMES THOMSON.

Thomson's private mathematical note-book bears the following entry:—

Jan. 11*th*, 1845.—The Senate-house examination is now over, and has been very satisfactory to me. The result will be known on January 17th. One of Goodwin's problems suggested some consideration about the equilibrium of particles acted on by forces varying inversely as the square of the distance.

And so at once he sits down to write an original paper, a "Demonstration of a Fundamental Proposition in the Mechanical Theory of Electricity." It appeared in the *Cambridge Mathematical Journal* in

February 1845. The eight pages of notes conclude thus :—"If I succeed in putting it in French I shall send it to Liouville," and forthwith there follow fourteen manuscript pages of French.

While waiting for the fateful 17th of January, William sends modest but hopeful letters to his brother James and to his father.

ST. PETER'S COLLEGE, *Jan.* 9, 1845.

MY DEAR JAMES—I received papa's letter yesterday, in wh. he told me about Robert's serious illness, and about your pulse. I had been afraid about Robert, from what I heard previously, and am glad to hear he is recovering. I hope to hear that you are getting on well too. I am sure you must be glad to be at home for some time after so long being away.

Tell papa that the Smith's Prize examination commences on Monday week. It will last for four or five days.

Nothing can be heard about the result of the Senate-house examination till Friday week, when the list will appear. Of course I shall let you hear at the earliest possible time. I feel quite easy myself about it, not from being confident, as that depends on others as well as on myself, but because I have succeeded quite as well, or in fact much better than I expected, and I have therefore a very slight amount, in comparison to what is usual to me, and I suppose everybody, of regrets hanging about, after the examination. It is certainly rather annoying to remember blunders, and to think how many things you might have done and did not; but as I know I could not possibly do better again if I were to go in again to a similar examination, I feel tolerably comfortable.

I have been enjoying the time to the utmost since the examination, wh. is a great weight removed. I have read *The Chimes* and half a novel, and to-day I had a ride with Shedden and Buxton, wh. is the first time I have been on a horse in Cambridge.

I shall be ready to leave Cambridge on Saturday fortnight, and to go to Paris if papa thinks proper. I shall get my pecuniary matters settled as soon as possible, and write when I have done so.—Your affectionate brother,

WILLIAM THOMSON.

ST. PETER'S COLLEGE, *Jan.* 14, 1845.

MY DEAR FATHER—I received the newspaper yesterday containing Rutherford's speech. I am very glad to see that he took up the subject of tests, and the manner in which he treated it. I hope when the matter is so well advocated that it will soon be settled, as it must be ultimately.

I called on Cookson to-day, and had a long conversation. He is going to get me some introductions in Paris if he can, and he recommends me to ask Professor Challis for some as soon as the Smith's Prize examination is over, which will be about Thursday week. On Monday week I shall leave this if I can get all settled, and shall lose as little time as possible in going to Paris. I have applied for rooms here at the middle or end of May, and Mr. Cookson is to get me some pupils if he can, from the first of June till the first of August. Of course there will be nothing binding about either rooms or pupils, as I can easily change my mind if any good reason should occur before June. I think Blackburn will possibly accompany me to Paris, as I have been trying to persuade him to do, but he has not made up his mind. I wish you could hear of a professorship for him at a Scotch University. I think if there were any more professorships in Natural Philosophy he would answer exceedingly well. I think he is to be second or third on Friday morning, but of course it is impossible to say yet. The Johnians are talking confidently of their hero, but I feel quite certain that nothing has fallen from the examiners yet, and that reports merely arise from what each man says he has done in the Senate House, himself. I have not been making myself in the least anxious on the subject, but have commenced reading again, with more satisfaction than I have been able to do for three years, as I have now no more examination cares.

I shall not be able to leave this without some of the sinews of war, as my money is all spent, and I have still about £10 to pay here. I thought I should have had enough to clear me, but I found unexpectedly that I had to pay £8 : 2 : 6 of fees to the University Registrar and Junior Proctor, before taking my degree, and so my funds fell unexpectedly. If you choose I can leave my debts to be paid by the Smith's Prize, wh. though not in the hand, yet as there are two in the bush, I think that at the worst I shall get one of them. However, as I know in Hemming's case, the money was not paid till after two or three months, I should have to leave my debts till I come here in May, wh., however, I could easily arrange to do.

I received to-day the copies of your Algebra for Fischer and Walton, wh. I am just going to deliver. I may say in advance that they are both very much obliged to you, and I am sure that they will make very good use of their presents.—Your affectionate son,

WILLIAM THOMSON.

P.S.—I was very glad to hear that Anna has got well over her *examination.* Whatever may be the result of mine, I hope you will write to Cookson thanking him for the kind way in wh. he has always done everything he could to help me.

To his Cambridge training Thomson owed much, as he himself was always ready to acknowledge. It compelled his impetuous and somewhat erratic genius, which was apt to rush into untried methods, to put itself, in measure at least, under restraint. It schooled him to test generalities that otherwise he might have taken for granted; to seek for exact proofs rather than remain contented with partial solutions. But if it did this for him, the ponderous Cambridge system of mathematical training of those days must have been truly galling to any student of real genius. To secure a leading place in the Senate-house the candidate must have at his finger-tips the solutions of known problems, polishing up his methods of writing out the solutions, so as to be able to get through an immense quantity of book-work in a short time. Drilled for months by his coach or his tutor in this species of gymnastic, as if he were being trained for a race, the candidate indeed acquired facility in dealing with particular classes of problems in the particular mode then in vogue. But such a training was not adapted either to cultivate originality or to advance mathematical science. It set up a false ideal of attainment, and not infrequently caused the coveted position of Senior Wrangler to be awarded, not to the best mathematician of his year, but to the one who had been best groomed for the race. As we know the respect for it fell, and the Senior Wranglership is now consigned to the past. At this time the other great mathematical coach at Cambridge was Hymers of St. John's. He had mastered the art of training men to work the ordinary mathematical problems with precision and celerity, and could tell to a nicety the kind of work that would "pay" in the Senate-house examination. Hopkins, who was a far abler man, of much wider outlook, doubtless did his best to train young Thomson in the way then considered orthodox, and to fill him out with the book-work that could be reproduced in the examination-room by effort of memory, and rapidly written down. But Thomson, to whom much of the book-work must have been child's play, and who had worlds of his own to conquer, could never—happily for science—have been kept to this soulless routine. He had, as we have seen, already broken ground in certain higher branches of mathematics, the convergency of Fourier's series, the application of it to determine the age of an assigned thermal distribution, and the analogy between the conduction of heat and the attractions between electrified bodies. He had also either published, or prepared for publication, papers on Curvilinear Isothermal Co-ordinates, on the Intersections of a Triple System of mutually Orthogonal Surfaces, on the Reduction of the General Equation of Surfaces of the Second Order, and on the Lines of Curvature of these surfaces. Only two months before his final examination he had published a paper on the gravitational effect at the surface of a steadily revolving ellipsoidal mass of homogeneous fluid—a geo-metric problem of a class which continued for many years to have a species of fascination for his mind.

It is a sufficient commentary on the then Cambridge system, to observe that devotion to scientific activity of the highest order such as this acted as a hindrance rather than a help to University honours. Hopkins might well declare that he had in Thomson a candidate whose mathematical abilities would outshine those of any man in England; but Hymers was reported to boast that he had a candidate whom he would guarantee to beat any man in Europe. And so it came about that at the Senate-house examination of January 1845, to the amazement and chagrin of Thomson's friends, the names were announced, as Senior Wrangler, Parkinson[1] of St. John's; as Second Wrangler, Thomson of Peterhouse. The disappointment was, however, mitigated by the circumstance that in the immediately following award of the Smith's Prizes, made by the examiners under conditions where mathematical power counted for more, and mere acquisition and memory were of less moment, Thomson far outdistanced his competitor.

The examiners in the Tripos of 1845 were Samuel Blackall of St. John's, Harvey Goodwin (afterwards Bishop of Carlisle) of Caius, Robert Leslie Ellis of Trinity, and John Sykes of Pembroke. The papers set that year differed in no special feature from the Senate-house papers of the time. Ellis and Goodwin, at least, were men of mark in Mathematics, having been respectively Senior and Second Wrangler in 1840. Goodwin in the biography of Ellis, which he wrote for the volume of Ellis's collected papers, gives the following reminiscence:—"It was in this year that Professor W. Thomson took his degree; great expectations had been excited concerning him, and I remember

[1] Rev. Stephen Parkinson, later D.D., F.R.S., and Tutor of St. John's College for many years. Author of Parkinson's *Treatise on Optics*. Died 1889.

Ellis remarking to me, with a smile, 'You and I are just about fit to mend his pens.'" It was assuredly, therefore, not from any want either of mathematical capacity or of appreciation of a candidate's merit that the examiners awarded the Senior place to another than the man to whom one of themselves could pay this tribute. There are several accounts current to explain Thomson's failure to secure the highest position. One of these ascribes Parkinson's success to the extraordinary speed[1] with which he threw off the answers to the bookwork, thus gaining more time for the problems. Another legendary story is that amongst the questions set were certain theorems taken from Thomson's own published papers, and that these Parkinson was able by pure memory to prove, whilst Thomson tried in vain to reconstruct his own arguments.[2]

Thomson took his defeat quietly. To his father he wrote :—

[1] In the Rev. F. Arnold's *Oxford and Cambridge, their Colleges, Memories, and Associations* (pub. 1873), p. 380, is the following narration : "The 'pace of Parkinson' has at Cambridge almost passed into a proverb. It was said that the successful candidate had practised writing out against time for six months together, merely to gain pace. Mr. Ellis, who examined that year, said it exercised quite a snake-like fascination on him to stand and see this young Johnian throw off sheet after sheet. He could scarcely believe that the man *could* have covered so much paper with ink in the time (to say nothing of the accuracy of the performance), even though he had seen it written out under his own eyes. There was a tremendous scene in the Senate House when the disappointed favourite took his degree."

[2] Sir Joseph Larmor has given the following comment on this story :— "Another form of the tale is that in the Smith's Prize examination two of the candidates answered a question in such striking and identical terms that investigation was made ; when it turned out that the answers were taken from Thomson's path-breaking paper of four years previously, which had appeared under his customary signature of P.Q.R. As a fact, Earnshaw did set a question asking for a development of the general analogy between the theory of attractions and the conduction of heat."

ST. PETER'S COLLEGE,
January 17, 1845.

MY DEAR FATHER—You see I was right in cautioning you not to be too sanguine about my place. If neither you nor any others of my friends will be less disappointed with the result than I am, all will be right. The only thing I feel in the least degree about it is that it may make it more difficult to succeed in getting the professorship in Glasgow.

In a mathematical point of view the disappointment is absolutely nothing when I see Gregory fifth, Green fourth, with Brummel above them, and Sylvester second with Griffin above him, and when I see a man like Pierson above such men as Fischer and Blackburn.

I hope you will not think I have misspent my time here. I feel quite satisfied that I have spent as much time on reading and preparation as I could, consistently with higher views in science. The Johnians give themselves up to one object, and it is fair that they should have their reward.

The Smith's Prize examination commences on Monday, but I have always had much less hopes from it than from the Senate-house. However, I hope I shall get one of them, and the prize is the same for each.

I have just got two papers ready for Ellis, with which I have been composing myself in the week of suspense, wh. I hope will please him better than the papers I sent him in the Senate-house.—Your affectionate son,

WILLIAM THOMSON.

P.S.—As I cannot get a printed list in time for the post as I had intended I transcribe a short extract.

Parkinson, *Joh.*; Thomson, *Pet.*; Pierson, *Joh.*; Fischer, *Pemb.*; Blackburn, *Trin.*

ST. PETER'S COLLEGE,
January 18, 1845.

MY DEAR FATHER—I hope by this time you have recovered from the shock of what I am afraid you must have considered very unpleasant intelligence. I have felt perfectly tranquil and resigned ever since I came out of the Senate-house, though, of course, there was a slight temporary intermission in the feeling yesterday. I have been with Ellis a good deal yesterday and to-day, though, of course, since he accepted the office of examiner till the result was known we had to cut one another. I have (and had previously) been speaking to him about my chance of getting the professorship of Natural Philosophy, and to-day he said, quite of his own accord, that he "has no doubt but that I should be a much better professor than if I had spent my time the way in which it would have been necessary to beat Parkinson," which, you see, confirms what I said to you yesterday. Excepting for public opinion, I could not possibly wish to be better trained for the Senate-house examination than I was, as during the whole examination I was never once in doubt about any principle, or even fact, in any of the subjects that were set, though at the same time I can see easily how I might have gained marks in many cases by being more ready in writing things out. As I have the readiness in remembering things and applying them when I wish, which is, of course, very desirable, I do not see in the abstract why I should wish to be able to write them down in examination form with a little more rapidity or in a little shorter space; though, however, it would have been much pleasanter to be first than second.

Ellis tells me, and does not hesitate to tell others of his friends, that even without previous opinion he could see by my papers that I am better than Parkinson, but that I fell short of him in quantity. I believe, though I am not quite sure, that in the last day's papers I beat him. I shall, however, find out if possible.

I am afraid you must have thought the beginning of my yesterday's letter rather strange. The fact was that I wrote it with the idea that I should receive the printed list, wh. I now enclose, in time to send along with it, wh., however, did not come soon enough: *item*, I was a little confused as I suppose you saw.

I do not know whether I have spoken to you yet about the British Association wh. is to meet here in June. I have asked Cookson whether you could have rooms in College, if you come up, and he says you could have them without the least difficulty. I hope you will come and visit me then. I shall be here at any rate, as it will be prudent for me to be about Cambridge till July, when the Fellows are elected.

Ellis also to-day asked me whether you would not come up then.

Has David Thomson succeeded in getting the situation he was trying for? If he succeeds I suppose another assistant would be wanted in his place, for wh. I might perhaps become a candidate. Would it be possible in such a case to get an arrangement made with Dr. Meikleham to get a new professor appointed? Of course, if I were to succeed, Dr. Meikleham could keep his house, provided no other professor would use his right of moving into it. As far as regards money I think an arrangement might easily be made, as for the present a very little would be enough for me.

I have forgotten, among other things, to tell that I have got through the dreadful and imposing ceremony of having my degree conferred, and that I am pronounced to be[1] a young man suited, as well in manners as in doctrine, for answering the question, to do wh. I shall therefore be admitted sometime or other when I am in Paris, at wh. instant I shall be a total bachelor, wh. I now am partially.

Having exhausted egotistical subjects and finished the second sheet of paper I must stop.—Your affectionate son, WILLIAM THOMSON.

Another letter, which Thomson sent on January 20th to his sister Elizabeth (Mrs. King) shows that while he felt the disappointment more for his father's

[1] [The formula of presentation to the Vice-Chancellor of a candidate for the degree of B.A. runs: Presento vobis hunc juvenem quem scio tam moribus quam doctrina idoneum esse ad assequendum titulum baccalaurei in artibus designati.]

sake than for his own, he yet feared the possible effect on their cherished ambition. He wrote :—

I have this moment received your letter with congratulations on my approaching triumph which is now passed, though it was not perfectly satisfactory, yet it was quite well enough, as I never relied much on examinations nor built many hopes on them. Of course, the only thing I care in the least about is the effect on my future prospects, and I think my place will do well enough in that respect, at least as far as regards a fellowship here; and, I hope, for the professorship in Glasgow also.

The principal thing that I care about in the result of the examination is the disappointment which I am afraid Papa must feel, as I am afraid he had rather raised his hopes about it, though I tried to keep him from expecting too much before the examination, as I knew the uncertainty. Will you write to me and tell me what he has said? I suppose he will write, but I should like to hear from you what he thinks.

Elizabeth replied :—

I was very sorry in reading your letter, which arrived on Sunday, when I came to the part where you say you are afraid papa will think you have misspent your time at Cambridge. He does no such thing: he is very proud of his son, and not in the slightest degree less pleased with him since the small humiliation he has met with.

His father himself replied :—

January 19, 1845.

MY DEAR WILLIAM—The place you have got at the examination is an excellent one, and you and all of us ought to be well satisfied with it. In point of *name* the next higher place would have been desirable; but coupling with your place all other distinctions that you can claim, we can and will make out a good case for you.

No doubt you will do what you can at the examination for Smith's Prizes. I fear, however, the *cramming* which succeeded before will do so again; but try what you can do.

Dr. W. Thomson and I have had one of our private consultations. He is most friendly and kind. He says it is no matter with sensible people about the Senior Wranglership, but that you must endeavour to do more in future than you otherwise might. He is to write to Prof. M'Vey Napier of Edinburgh, editor of the *Edinburgh Review*, to get Lord Brougham to give you introductions to Paris, which he says Lord B. is very capable of giving with good effect. Failing Prof. Napier, he will apply to Lord Jeffray; and I am sure his head is at work to serve you in other quarters. He is, in fact, to write to his brother, Dr. Allen Thomson, one of the Edinburgh professors, to endeavour to get you an introduction to some good medical student in Paris, who might assist you regarding lodgings, etc. etc.

I will send you money in due time to free you in one way or other of Cambridge, and I will send you the means of getting a letter of credit in Paris from Coutts, or some other bankers in London.

You will be glad to hear that John has got a *unanimous* appointment by the Faculty to the Brisbane bursary for four years at £50 a year.

Communicate with me as frequently as you can, particularly about your examinations for Smith's Prizes—not, however, so as to fatigue you or to interfere with your preparations. . . . Your very affectionate father,

JAMES THOMSON.

[*P.S.*] When will Smith's Prizes be decided?

Whewell wrote to Forbes: "Thomson of Glasgow is much the greatest mathematical genius: the Senior Wrangler was better drilled." This verdict being shown to Dr. James Thomson, he copied it out and sent it with great joy to his son.

Cookson wrote, too, expressing regret at the

result of the examination, and adding that his opinion of William's talents was not affected by it. Hopkins also sent a sympathetic letter to the father to express his very great disappointment, and to assure him that, nevertheless, the circumstance had not affected in the slightest degree the high opinion in which he held the talents and acquirements of the son.

Then the father writes again to the son:—

GLASGOW COLLEGE, *Jan.* 22, 1845.

MY DEAR WILLIAM—Griffin sells the Glasgow Newton in two large 8vos (the same as your prize) at 14 shillings in boards, and he will send it free of expense to London. If Fischer agrees to take it let me know, and I shall pay it here, and he can pay you in Cambridge. It could, probably, be sent for no or small expense to Cambridge, perhaps in a parcel to some Cambridge bookseller. If it is to be sent, state what is to be done with it when it reaches London.

Hopkins's letter has done you *great* good here. The man that wrote it has a *heart* and a *head*. Mr. Buchanan read it to-day in our dining-room, *apparently* and I think *really*, with great satisfaction, and at one part of it exclaimed, "That is the kind of man we should have!" I do not see any meaning to put on this except what at once occurred to myself, and what will readily occur to you.

I gave this letter to Mr. Ramsay, and he, without asking me whether he might, sent it to Mr. Maconochie. Dr. Nichol, who has been dining with Mr. Lushington, has just called, and he and Lushington have been greatly delighted with Hopkins's letter, which I had sent to L. before dinner.

From Dr. Chalmers I have got introductions for you to Biot and another. Sir D. Brewster writes to say that he *will send after you*, what he is too busy just now with the Royal Commission at St. Andrews to write at present, introductions to Biot, Arago, and another whose name I cannot read. Mr. Crum will introduce you to Dumas. From Forbes I have not yet heard.

I am sending money to Fisher and Son, part of which is to carry you to Paris after paying for your passport, etc. etc. Of this, however, more again. You must be brushing up your French.—I am, your affectionate father,

JAMES THOMSON.

Happily the shadow of disappointment quickly passed away, for at the Smith's Prize examination the tables were turned, and the son could write :—

ST. PETER'S COLLEGE, *Jan.* 24, 1845.

MY DEAR FATHER—The result of the Smith Prize examination has just been announced to me this moment, and I am successful in getting the first. Parkinson has the second. I have no more time now, but I enclose you the remaining papers.—Your affectionate son,

WILLIAM THOMSON.

[*P.S.*] I received your letter this instant, which was no small addition to my happiness.

Ellis wrote to Thomson most cordially :—

MY DEAR THOMSON—Let me congratulate you on the result of the Smith's Prize examination. The papers were, I should think from the result, of a higher character than they have usually been. As the Smith has *ex professo* especial reference to natural philosophy, it will necessarily tell upon the minds of people in Glasgow, more one would be apt to believe, than the degree. I look upon the professorship as the point in which university honours may be of some service: I hope they will do all that you can wish in this matter.

Further, the kindly tutor and the sympathetic coach sent their congratulations in glowing terms.

Cookson wrote hastily on January 24th, "I have just seen your son, who is overjoyed at his success. The intelligence will, I am sure, be very gratifying to you." The next day he wrote again :—

ST. PETER'S COLLEGE, *Jan.* 25, 1845.

MY DEAR SIR—I wrote a few lines to you yesterday to communicate the joyful intelligence of your son being first Smith's prizeman. Since then I have heard something of the sort of examination which your son passed, and I think that you will be glad to know it.

In the first place, the decision was *unanimous*. The examiners were Dr. Whewell, Dr. Peacock, Prof. Challis, and Mr. Earnshaw, and your son beat all his competitors very decidedly in *all* their papers. In two of them the "marks" were in the proportion of *three* to *two*. In the other two my informant (one of the four examiners) told me that your son was decidedly the first, though he did not know the proportion of marks. It is certain, therefore, that in this examination your son has proved himself decidedly superior to the Senior Wrangler. It was the unanimous opinion of the examiners that he was so.

It was also stated by the examiners—though perhaps this is not a matter to be made too public—that the candidates were the best they had ever examined.

My informant speaks of your son as a mathematician of very first-rate powers, and expressed himself much pleased at the manner in which he had made light of his most difficult questions.

There can now be no longer any doubt that your son has reinstated himself in the opinion of the University, and he may continue his career in the conviction that he will never be looked upon as a beaten candidate.

I hope that the present is only the commencement of a *glorious* career.

Allow me at this time to say how much pleasure I have had in considering him as one of my pupils. The absence of all conceit and self-will from his disposition has made our intercourse agreeable, and rendered it additionally grateful to listen to the constant praise bestowed upon his talents. In this respect I cannot speak in too high praise of him, and I feel assured that the same good qualities would always earn for him the same good opinion.

Accept my best thanks for the kind expressions contained in your last letter.

I feel that I have but very imperfectly conveyed to you the information I received respecting the examination. The terms made use of towards your son were of the most flattering kind, and showed the extremely high opinion which was entertained of him, much higher, I believe, than you would infer from my account.

The college to-day presented him with twenty guineas as a prize[1] for books, a small and insufficient acknowledgment of the honour done to us, and a testimonial of the esteem we have for him.—Believe me, my dear sir, yours very faithfully, H. W. COOKSON.

Dr. Thomson.

Hopkins also wrote :—

CAMBRIDGE, *Jan.* 29, '45.

MY DEAR SIR—It is only want of time that has prevented my writing to you sooner on the gratifying result of the examination for the Smith's Prize. It has made us all quite happy again. The examination, as you are probably aware, is altogether of a higher character than that of the Senate-house, being, in fact, intended to furnish a higher test of the merits of the first men. The high philosophical character of your son's mind and acquirements found here much more room for development than in the Senate-house, and the consequence was that he beat his opponent with ease—he was the *facile princeps*. None of the four examiners had the smallest hesitation in placing him decidedly first. The

[1] The College books record this gift to William Thomson "in consideration of his great mind and of his exemplary conduct."

result, I assure you, has given great satisfaction to a great number of persons here, as having restored your son to that pre-eminence to which they believe him to be entitled. He has had to contend with a most formidable opponent, with whom he has now fairly divided the highest honours of the University, having himself obtained unquestionably the highest, though not that which, out of the University, confers the most popular reputation.

Your son himself, I may venture to say, is perfectly satisfied, and set out in high spirits last Monday morning for Paris with his friend Blackburn.—Yours very truly,

W. HOPKINS.

And so Thomson prepares to bid for a time adieu to Cambridge, and writes to his father as to his immediate movements :—

ST. PETER'S COLLEGE, *Jan.* 25, 1845.

MY DEAR FATHER—To-morrow morning at six o'clock I start along with Blackburn for London, where we mean to stop as short as possible on our way to Paris. I received the second halves of the notes yesterday, and have got matters all right.

On account of the seasonable cheque I received yesterday I have been able to pay everything here. If you choose, I need not buy books with the money yet, as I can do that afterwards. . . .

I called on Prof. Challis yesterday to bid farewell. He told me of his own accord that the examiners had been unanimous.

Yesterday I got some separated copies of various memoirs from Hopkins, and among them a most valuable one by Green, with wh. I am greatly delighted. He is to have an introduction ready for me to Élie de Beaumont, one of the French geologists, who, however, has an extensive acquaintance among all kinds of scientific men.

I shall attend carefully to Prof. Forbes's commission and to his general directions.

I shall not be able to write again till I reach Paris probably, and so farewell.—Your affectionate son,

WILLIAM THOMSON.

P.S.—If you have not paid Hopkins for the long vacation, £40 will be due, that being the payment for the long vacation and last term.

This last sentence renders appropriate some further reference to finance. Thomson had gone to Cambridge in 1841 at his father's expense, and without any scholarship to maintain him. In 1842 he obtained the Gisborne Scholarship, worth about £30 per annum, and this he retained during the remainder of his undergraduate course; he appears in the University calendars as Gisborne Scholar for 1843 and 1844. No scholars were admitted at Peterhouse in the years of Thomson's residence, except the Gisborne Scholars; but certain exhibitions, called scholarships, were from time to time awarded at the Tutor's discretion. In 1842 his College awarded him a mathematical prize of £5 and another of £1; in 1843 again a mathematical prize of £5 and the Hale Scholarship, amounting to £9:7:1; in 1844, likewise, a mathematical prize of £5 and £6:10:8 for the Hale Foundation. In 1845 he received numerous small scholarships, the list being as follows: Gisborne Scholar (half-year), £15; Hale Scholar, £6:10:8; Perne Scholar, £1:12:5; Parke Scholar, £4:17:9; Blythe Scholar, £1:12:5; Whitgift Scholar, £1:5:6; Woodward Scholar, £3:17:7. Also, under a College order of Jan. 25, 1845, he received the special prize of £21 on taking his Tripos. These contributions, however, as we have seen, fell far short of the expense of maintenance of an undergraduate who took an active part in sports and festivities of the time. At the College audit of 1845 it is also recorded that he paid £3:6:8 *pro argento*, as a Foundation Fellow—his contribution towards the service of plate at the Fellows' table.

Once—in the summer of 1906—when Lord Kelvin was in a chatty mood, I asked him point blank how it occurred that he was not Senior Wrangler. His blue eyes lighted up as he proceeded to explain that Parkinson had won principally on the exercises of the first two days, which were devoted to text-book work rather than to problems requiring analytical investigation. And then he added, almost ruefully, "I might have made up on the last two days, but for my bad generalship. One paper was really a paper that I ought to have walked through, but I did very badly by my bad generalship, and must have got hardly any marks. I spent nearly all my time on one particular problem that interested me, about a spinning-top being let fall on to a rigid plane—a very simple problem if I had tackled it in the right way. But I got involved and lost time on it, and wrote something that was not good, and there was no time left for the other questions. I could have walked over the paper. A very good man Parkinson—I didn't know him personally at the time—

who had devoted himself to learning how to answer well in examinations, while I had had, during previous months, my head in some other subjects not much examined upon—theory of heat, flow of heat between isothermal surfaces, dependence of flow on previous state, and all the things I was learning from Fourier." Then he went on to explain how he had had his head in these problems before coming to Cambridge, and told me he wished he could find his note-book of the Senior Greek lectures of his last year at Glasgow, when he was supposed to be listening to Lushington on the Hippolytus of Euripides, for the notes would show that he was all the while working at his ideas on the uniform motion of heat, and on the Boscovichian idea of force acting independently of intervening matter. Then he drifted back to his own early writings on Fourier, and pulling from the shelf a copy of his Mathematical and Physical Papers, vol. i., pointed to page 15, where he gave the mathematical inference, as the result of assigning negative values to the time t, that there must have been a creation. "It was," he continued, "this argument from Fourier that made me think there must have been a beginning. All mathematical continuity points to the necessity of a beginning—this is why I stick to atoms . . . and they must have been small—smallness is a necessity of the complexity. They may have all been created as they were, complexity and all, as they are now. But we know they have a past. Trace back the past, and one comes to a beginning—to a time zero, beyond which the values are impossible. It's all in Fourier."

All through his life Thomson continued to cherish his youthful enthusiasm for the men who had inspired him. Fourier and Green in the domain of mathematical physics, Faraday in that of experimental science, were the *Di majores* of his veneration.

CHAPTER III

POST-GRADUATE STUDIES AT PARIS AND PETERHOUSE

THOMSON'S degree-course at Cambridge having terminated, he prepared to widen his outlook by spending some months in Paris. Down to this date he had never seen Green's *Essay*,[1] his only knowledge of it being derived from a curt reference in a memoir in the Cambridge *Transactions* by Murphy, on "Definite Integrals with Physical Applications." In vain had Thomson searched the College Library, and inquired at the bookshops for Green's thin quarto. Pacing "Wranglers' Walk" with Hopkins on January 25, the evening before his departure, he mentioned the matter to

[1] Green's remarkable work, *An Essay on the Application of Mathematical Analysis to the Theories of Electricity and Magnetism*, a quarto tract of 72 pages, was printed at Nottingham in 1828 by private subscription. Probably fewer than one hundred copies were printed, and the work is now exceedingly rare. In 1850 Thomson sent Green's *Essay* to be reprinted in full in *Crelle's Journal* (vols. xxxix., xliv., and xlvii.), prefixing to it a short introduction, which has never been included in any of his collections of papers. In this introduction he mentions a number of independent investigations made by subsequent authors on the same subject, two by Chasles, one by Gauss, one by Sturm, one by Despeyrous, and three of his own. A brief biography of Green follows, and a list of his contributions to mathematical physics.

The printer of *Crelle's Journal*, Reimer of Berlin, reprinted the complete *Essay*, with Thomson's introduction, in the year 1854.

Green's Mathematical Works, including the *Essay*, were collected by Ferrers, and printed for Caius College in 1871.

Green's *Essay*, translated into German, was republished in Berlin in 1895 in the series of *Ostwalds Klassiker der exakten Wissenschaften*.

his tutor, who, to Thomson's joy and surprise, told him he had at least one copy at his rooms, whither they repaired. It turned out that Hopkins possessed three copies, two of which he gave to Thomson, one for himself, the other for Liouville.

Thomson took Green's *Essay* to France with him, assimilating its contents with marvellous and rapid insight, and working out its bearing on the problems that occupied his brain. Before he had been a month in Paris he sent to Liouville's *Journal* (vol. x. p. 137, 1845) a paper, entitled "Démonstration d'un théorème d'analyse," which two years later he expanded in the *Cambridge and Dublin Mathematical Journal* (vol. ii. p. 109, March 1847), under the title: "On certain Definite Integrals suggested by Problems in the Theory of Electricity." The importance of this investigation will be considered a little later.

Thomson was accompanied by his friend Hugh Blackburn to Paris, which they reached on January 30. They took rooms in the Rue Monsieur le Prince, No. 31, near the Odéon, on the fourth floor. In the heart of the old Quartier latin, before it was cut up by the great Boulevards, their lodging was conveniently close to the Sorbonne and to the Collège de France. His note-book bears the following entry:—

RUE M. LE PRINCE 31, PARIS, *Feb.* 4, 1845.

I am settled here with Blackburn, and we have been getting to work. We have only heard one lecture yet (Dumas, on Carbonic Oxide, etc.), as there are holidays for a few days (Mardi gras being to-day). I called on Liouville on Friday, Jan. 31, and he received me very freely. . . . I have been studying Green's memoirs on Attraction, and am just beginning to see through them. I saw his memoir on Electricity for the first time on Saturday (two days before I left Cambridge), and the same day I got two copies in a present from Hopkins (one of which I have given to Liouville), along with several other memoirs published in the Cambridge Transactions."

The same day his father wrote :—

GLASGOW COLL., *Feb.* 4, 1845.

MY DEAR WM.—I am glad you have reached Paris in safety. You lost the sweets of all our congratulations about Smith's Prize, as the letters have all been returned in consequence of your having left Cambridge. In addition to them, we have received, at different times, various other letters of congratulation addressed to you at Cambridge. Those we have opened lest they should contain anything requiring immediate attention. These are from J. B. Smith (Horncastle), Howard Parry, D. Foggo, Jr., Wm. Walton, Henry M. Fletcher (Oxford), Gisborne (Berkeley Square, London), Wm. Nixon (I think it is), Northrepps, who writes at the desire of Sir Fowell Buxton, who is himself in progress of recovery from a severe illness, and from John Mitchell (St. Vincent St., Glasgow). Your success has given great pleasure to your friends both here and elsewhere. Cookson and Hopkins have written in great spirits, and most kindly. Their letters will very likely be of much use hereafter. They speak in very strong terms about the completeness of your victory, and the high opinion expressed by the examiners regarding your style of answering. . . .

Sir D. Brewster has promised to give an introduction to Arago, Biot, and Babinet. To the last two you have introductions already. Shall I send you the one to Arago, or will you get at him otherwise? Kelland has sent one for you to Cauchy, but as you have one already I do not send it. He mentions to Cauchy the death of his wife, with whom Cauchy and family must have been acquainted, as he expresses his trust that *she and they will meet in heaven.* Say nothing of this, but be prepared to speak of her, and you can say that though I just knew her slightly, I thought highly of her. She died last autumn. Dr. Wm. Thomson says you should spend your 20 guineas in purchasing books in Paris, where you will be able to get much for your money, and he says you should spend much of it in purchasing books on *La Physique Expérimentale.* To the lectures on this he says you ought to pay *the greatest attention*, and is glad Blackburn is with you, and he says you and he will be able to repeat, as it were, the lectures to one another in the evening ; and, above all, he says you should be writing discourses or lectures in the plainest and most attractive terms in your power, and improving your elocution by constant, free, and open practice. I think it likely he will write you on the subject. He says people may think you too *deep* to have *popular* talent. Do all in your power to obviate this impression. Use all economy consistent with comfort and respectability.—I remain, your aff. father, JAMES THOMSON.

P.S.—You will go to the meetings of the Institute. I may be able to send you further introductions.

PARIS, RUE M. LE PRINCE, NO. 31.
Feb. 10, 1845.

MY DEAR FATHER.—I received your letter on Saturday with additional introductions. I have not, as yet, had time to present any of the introductions, as we have had enough to do to get settled here, in addition to attending lectures and private reading. To-day Blackburn and I are going to present an introduction to Élie de Beaumont which we received from Hopkins, and by which we are to hear anything we may about lectures, etc., if we wish it, probably get introductions to lecturers whom we may attend. If he can give us an introduction to Pouillet we shall get it. There has been only one of his lectures yet, as there were holidays last week (Mardi gras, etc.). We attended it and found it exceedingly interesting. The subject was common electricity, which he illustrated beautifully with his apparatus, which was exceedingly good and on a very extensive scale. As we know the subject very well we did not get much new information, but he is going on to electro-magnetism, which will be exceedingly interesting. (*February* 11).—We have also been attending Dumas' lectures, which are also very interesting, and exceedingly well illustrated by experiments. All the things which are required are prepared with great care beforehand, so that he has always a great many experiments to show. There is another lecturer on "Physique," Regnault, at the Collège de France. He is a very young man, but nevertheless I hope to profit by his lectures. We have only heard one of them, which was on thermometry. We have also attended a lecture of Sturm on mechanics, but as he seems merely going over the mathematical parts with which we are quite familiar, and as we have very little time to spare, we shall not go on with them. To-day we commenced attending a course of lectures of Libri's on elliptic functions, a subject of which neither of us knows much. We shall continue attending him as long as we find it profitable. I called on Liouville, shortly after I arrived here, with a paper which Cayley had given me to carry to him.

We very soon became acquainted, and he began directly to work with pen and paper at various subjects of conversation, and when I went away he invited me to return again "pour causer de toutes ces choses." I called again, bringing with me another paper which I had received from Cayley. I found Chasles there, to whom Liouville presented me, and we had a long conversation on mathematical subjects. I had lent Liouville a memoir of Green's, of which I had received two copies from Hopkins before I left Cambridge, in which I found that he (Green) in 1828 had given almost all the general theorems in attraction which have since occupied Chasles, Gauss, etc., and I also lent him my copy of the *Camb. Math. Jour.*, as far as it had appeared, which he wished to look over. Liouville showed them to Chasles, who seemed very much struck with both. Liouville asked me to come and dine with him on Thursday along with Chasles.

My French is exceedingly bad in conversation, but I hope it will improve by practice. I am writing at present a short paper for Liouville, which will also serve for a French exercise. I shall continue to do so when anything appropriate occurs. I have also several subjects on hand which I am working at whenever I have time, of which, however, there is very little to spare. . . .

After I received your letter I bought Pouillet's *Traité de Physique*, of which I had not a copy before, and besides which has been republished a few months ago, being now the 4th edition. I have besides renewed my subs. (30 f.) for the *Journal de Mathématiques*, and as I have only £7 left I must take care not to spend much money on books or anything else at present, as it would not do to be without. The journey here cost, from Cambridge, between £5 and £6, so that I had just £11 when I got settled here. I am afraid I shall have to wait some time for the Smith's Prize, as it is often a long time in coming when the University is poor. Sometimes I believe it is not received. Have you seen Gregory's and Walton's book yet? It is published now.—Your aff. son, WILLIAM THOMSON.

Liouville's friendship meant much to Thomson. To him Thomson opened his heart on many of the deep problems of Geometry and Natural Philosophy with which he was occupied. The great philosophers of the French school of Mathematics were gone; Laplace, Legendre, Poisson, Fresnel were no longer living. Arago he never met, nor Fizeau, though the latter lived on nearly half a century afterward. But Biot[1] he met, and Cauchy, who tried hard to convert him to Roman Catholicism, and Chasles, and Sturm; also Foucault, of whom he spoke in terms of admiration.

One night about three weeks from their arrival, when Thomson was sitting with Blackburn over a wood fire in their chilly lodging, eager steps were heard without, and with a hasty rap upon the door a panting visitor rushed in. It was Sturm, in a state of high excitement. "Vous avez le Mémoire de Green," he exclaimed, "M. Liouville me l'a dit!" The *Essay* was produced, and Sturm eagerly scanned its contents, turning over page after page. "Ah! voilà mon affaire," he cried, jumping from his seat as he caught sight of the formula in which Green had anticipated his theorem of the equivalent distribution. Prior to this, the work of Green had been as completely unknown to the French mathematicians as it had been at Cambridge. Liouville, in particular, seems to have been impressed with the enthusiasm of Thomson, and with the importance of the ideas which were fructifying in his mind as the result of his devotion to Fourier and to Green. To Liouville he confided his ideas about electric images, of which more presently; and to him he gave for publication the previously-mentioned paper (Démonstration d'un théorème d'analyse), carefully dating it "Paris, le 21 février, 1845." This habit of dating his researches he observed with

[1] It was the aged Biot who, taking him literally by the hand, introduced him to Regnault.

scrupulous precision, extending even to the jottings in his note-books—a habit which in his later life became almost an obsession.

PARIS, RUE M. LE PRINCE 31,
Feb. 23, 1845.

MY DEAR FATHER—I delayed answering the last letter I received from home (James's) till I should be able to tell you about the result of my introductions. I have now presented all which are to scientific people. The others I reserve for some time yet, as at present I have no time for visiting or going out to parties. However, I shall take care to present them before I leave, and they will, I am sure, be of a great deal of use to me. I was much obliged to Mr. Crum for his letter to Dumas, and to Dr. R. D. Thomson for one to Pelouze (whom I had met at dinner at Liouville's), both of which I presented without delay. I called on Dumas at his house at the Jardin des Plantes, and left the letter with a card, as I did not find him in. I called again next day, but was still unfortunate; but on the day after I saw him at his lecture, and he asked me to call on him next day (yesterday) at eleven o'clock, to have some conversation. When I went, he asked me about what lectures I wished to attend, and gave me useful information. He wishes to be remembered to Mr. Crum the first time I write to Scotland. He asked me to go to-night, or any Sunday night, to his "beau-père," M. Brogniart, when he has a soirée, at which he (Dumas) and Pouillet (who is a relation), and other scientific people are generally present.

I called on Cauchy on Friday (I had previously called, but was told by the porter that he is only visible on Friday between 3½ and five) and presented Prof. Forbes's introduction. When he read the letter he commenced asking me questions about how much mathematics I had read, the first being whether I knew the Diffl. Calc. When he found that I knew enough to be probably able to understand him, he commenced telling me what he was working at, and various things which he had done, some of which were very interesting. He gave me two copies of the paper, which I enclose, one of which he wished me to send you, in which he gives an account of a memoir he had presented to the Academy containing a method as complete as that of Sturm for separating the roots of equations. From the explanations he gave me, and from what is contained in the paper, I think his method must be complete, though it is much more complicated than Sturm's. He says that he has since applied it to the imaginary as well as real roots, but I have not seen any of the memoirs of which he spoke. I am to go to his soirées if I choose, given by his "belmère" on Tuesday evenings.

I had a letter accompanying two memoirs from De Morgan to Sturm, which Mr. Ogilby gave me. I did not find him in, but left a card with the letter and memoirs. As he had heard from Liouville that I had a copy of Green's memoir, he called the same day here when I was out, and left a card, "C. Sturm, Membre de l'Institut," and said he would call again next morning. However, about ten o'clock the same evening Blackburn and I were astonished by his entrance. He soon began talking on mathematical subjects, and did not lose much time in asking about Green's memoir, which he looked over with great avidity. When I pointed out to him one thing which he had himself about a year ago in Liouville's Journal, he exclaimed, "Ah! mon Dieu, oui." I have also seen M. de Blainville, to whom Mr. Ogilby gave me an introduction, and who was very kind in trying to put me in the best way of improving my physical state, as I told him that was my principal object in coming here. I have also seen Biot, who is *the* person, as even M. de Blainville says, for that purpose. He is to introduce me to Regnault (Prof. along with Biot at the Coll. of Fr.) at the Institute to-morrow, who he says is the best physicien here. If it be possible and advisable for the short time I am here, I am perhaps to work in his laboratory. The other people to whom I had introductions (except Elie de

Beaumont, to whom Blackburn and I had an introd. from Hopkins) were all out, but I shall call again when I have time.

I finished my paper yesterday and took it to Liouville, but as he was not well I could not see him. It is very short, but as I have had very little time to give to it each day, I have had it on hands for a considerable time. We are attending Dumas' and Pouillet's lectures regularly, and are going also to attend those of Pelouze and Regnault at the Coll. de France. We are both much interested with them, and talk over them a good deal. I take a great many notes on Pouillet especially. I find I know almost all he says (which is not the case with Dumas, however), but I mark down the particular experiments he makes, and how they succeed, and what seem to be more appreciated by the audience, which is very numerous, and "popular."

I have just room enough on this sheet, and time enough before going to the soiree, to tell you about money matters. . . . I have, besides, spent 16 fr. on Pouillet's *Physique*, 30 fr. for Liouville for this year, and 20 fr. for the *Comptes rendus* for this year, which I receive every week (two quarto volumes at the end of the year), and which is necessary to enable me to know what is going on in the Institute.—Your affectionate son,

WILLIAM THOMSON.

To his brother Robert, on March 5th, he gave a further account of his doings.

PARIS, RUE M. LE PRINCE 31.

MY DEAR ROBERT

. . . . . .

On Monday Biot introduced me to Regnault (the professor of Natural Philosophy at the Collège de France), and told me to go to M. Regnault at the end of his lecture any day, and that he would show me his cabinet de physique (*i.e.* apparatus room). I went yesterday, and after hearing some explanations, which he gave to a number of the students who waited at the end, on the subject of lecture, he sent his assistant to show me all the apparatus. I was greatly interested in it, and saw a great many pretty things, of which they have a great abundance here, as the Government gives them a great deal of money for apparatus for popular experiments and historical illustrations in the lectures. A German (or rather Swiss, I believe), who had been speaking to Regnault along with others, accompanied us, and we had a good deal of conversation. He had been for some time working along with Regnault, assisting him, in the way I should like to do, if possible, and so I got as much information as I could from him on the subject. There are no pupils working with him, as in a chemical laboratory, and he has a person assisting him at present (besides the regular *préparateur*, who has the same office as *Dan*), and so I am afraid will not want any more at present, and besides, would probably want a person of experience. . . .

I have no more time to write at present, as I have to go to Dumas's lecture almost immediately. Tell papa that if I do not get work in Physique, I am thinking on entering Pelouze's laboratory for the time I am here. I have seen it, and am sure it would be very useful. Pelouze has, besides, an essaying laboratory at the mint, as he holds the office of "directeur des essais," and yesterday he showed me one of the regular processes.—Your affectionate brother, WILLIAM THOMSON.

Thomson was thus brought into contact with Victor Regnault, who had a short time previously begun his marvellous series of researches on the principal laws and numerical data which enter into calculations upon steam-engines,—researches which still remain classic amongst the works of experimental science. Regnault was a master of the art of minute and accurate experiment.

The mathematical note-book shows the following entries:—

March 7, 1845.—I have been reading Jacobi's *Nova Fundamenta* and Abel's 1st memoir on *Elliptic Functions*, but have been rather idle on the whole. The following curious case of orthogonal isothermal curves occurred to me a few evenings ago:—[Here follows a draft of the "Note on Rings and Brushes seen by Polarized Convergent Light in certain Crystals"—Thomson's almost solitary contribution to formal optics—which was published in the enlarged *Cambridge and Dublin Mathematical Journal*, vol. i., in 1846.]

March 8.—I have been thinking on spinning and rolling again, and speaking to Blackburn on the subject. [Seven pages of the Mathematics of Bodies in Rotation here follow.]

Then he writes to tell of his admission to Regnault's laboratory.

PARIS, 31 RUE M. LE PRINCE,
March 16, 1845.

MY DEAR FATHER—I received aunt's letter a few days ago, containing your message about Jerrard's work. I have not been able to see Liouville yet to tell him, as he has been unwell, but I communicated it to Sturm yesterday, who received it with great avidity.

I have been looking on at Regnault's work yesterday and the day before, and I offered my services yesterday. He seemed to be quite willing to let me come as often as I choose, and I suppose I may now and then have a job in the way of holding a tube for him when he is sealing it, or working an air-pump, as I had the privilege of doing yesterday.

The lectures are almost all stopped now in consequence of the Easter vacation, and will not commence again for a fortnight. I have taken the opportunity of the intermission to commence taking lessons for a short time in the cornopean. The teacher I have got is, I think, as far as I can hear, one of the best in Paris, and I think I shall be much the better of the lessons. Besides, I have the benefit of a tolerable lesson in French each time I go to him, as he speaks no English, and we have necessarily a good deal of conversation. The terms are very nearly the same (a very little less) as Mr. Hilpert's, in Glasgow.

Blackburn is going away in about three weeks with a large party of his family, who are coming here on a short visit in a fortnight. I shall be here for a month after he leaves, at beginning of which time I shall present my introductions to the *haute société*. Blackburn had an introduction to some very desirable people in the fashionable world, but he has not presented it yet, and will not do so at all, I think, though I was commissioned by one of his brothers to make him present it.

I see Cauchy very often, and when I call on him he always has a great deal to say, and tells me what he is working at, and all the fine things he is discovering. He has either one or two memoirs for the meeting of the Institute every week.

The reason I have not followed the plan you recommended of making a journal is that I have already two regular journals going on containing everything of the kind you mention. One of these is my note-book, which contains a register of all the lectures, etc., that I attend; and another, my mathematical journal, in which I put down everything I am working at, and anything else that may be interesting. In my letters to you I have space enough to put down everything of any interest in the way of news, and I daresay a great deal more. For the last week there has been nothing particular to tell about. There have been rather fewer lectures, but otherwise we have been just as usual, and read in the evenings, as before, with the exception of two evenings, one of which we went to the comic opera to hear "Fra Diavolo," and another to the Italian, to hear "Il Barbiere," either of which, if we had missed hearing, we should probably not have had an opportunity of hearing again, and I think nothing else we could have heard would be so good of its kind.

. . . .

At that date there was no provision made in the University of Paris—or, indeed, in any other university—for the systematic teaching of physics in a properly organized physical laboratory. So Thomson must take his part in the research work which Regnault had in hand. In the investigation then proceeding upon the density of gases,[1] two large glass globes, one filled with gas, the other empty, were weighed against one another. Thomson's chief work in the laboratory was to work the air-pump when Regnault gave the command to make the vacuum. At another stage he was given charge of the operation of stirring the water in a calorimeter in which determinations of latent heat were being made.

At that time Regnault had, as assistants or pupils, several men destined to make their names known in physics: Izarn, Bertin, Grassi, Bertrand, Lissajous, and Silbermann.

The mathematical diary again tells of the day's doings:—

March 15.—I am occupied the whole day in Regnault's physical laboratory at the Collège de France. At spare times I have been reading Poisson's memoirs on Electricity, which I find among the Memoirs of the Institute in Regnault's cabinet. I have applied my

[1] See Regnault's paper, "Détermination du poids du litre d'air et de la densité du mercure," *Ann. chim. phys.*, série 3, tom. xv. p. 348; and *Rélation des expériences*, tom. i. p. 154.

ideas on induction in spheres and the principle of successive influence, and get a very simple solution, in the form Poisson gives it, for two spheres. I think I can work it out for *i* spheres. . . . The *image* of an exterior point, in a conducting sphere, is a p. in the interior, with opposite electr.

March 27, 1845.— . . . Yesterday I called again on Liouville at 12 (when MM. Regnault and Izarn were at breakfast).

He stayed four hours discussing Faraday's objections to the theory of electrical action at a distance; and Liouville engaged him to write a memoir for the Institute answering them.

James Thomson was delighted that his son was engaged on experimental work, and wrote to encourage him :—

I think that were it only to hold a tube or work an air-pump, you should by all means go on in Regnault's cabinet. You will see what instruments he has, and you should take lists of them as far as you can. Besides, *certificates* from him, Pouillet, Dumas, etc., with reference to *practical* matters might serve you much.

Neither the diary nor the letters mention the visits of Thomson and Blackburn with Edward Armitage to his brother's studio in Paris, or their little Bohemian dinners with Armitage's artist friends in the unconventional surroundings of the Quartier latin.

Blackburn was about to leave Paris; and Thomson had agreed to return in May to assist Fuller in Peterhouse examinations. He wanted meantime to fulfil his promise to Liouville, and wrote home for books.

PARIS, RUE MONSIEUR LE PRINCE 31.
Sunday, March 30, 1845.

MY DEAR FATHER.— . . . I should like to buy Becquerel's Electricity if I can afford it.

It has been a very successful plan for me going to Regnault's laboratory. I always get plenty to do, and Regnault speaks a great deal to me about what he is doing, and has of late employed me in working, along with him, some of the formulas necessary for the reduction of the experiments. Besides, I have access to all the books he has there, including all the Memoirs of the Institute, the *Annales de chimie et de physique*, and a great many elementary physical works, of which advantage I have made extensive use. I sometimes go to the laboratory as early as 8 in the morning, and seldom get away before 5, and sometimes not till 6. I generally breakfast before I go, and so have the time free when Regnault and M. Izarn, his assistant, go to breakfast. On Thursday I made use of that time by going to see Liouville (at 12 A.M., for the French breakfast very late), whom I found recovered, as he said, from his illness. He did not let me away till 4½, as he had a great many things to speak about, besides, working a good deal, and reading the *Nova Comm. Petr.* through. He asked me to write a short paper for the Institute, explaining the phenomena of ordinary electricity observed by Faraday, and supposed to be objections fatal to the mathematical theory. I told Liouville what I had always thought on the subject of those objections (*i.e.* that they are simple verifications), and as he takes a great interest in the subject, he asked me to write a paper on it, and said he would get it translated if I choose to write it in English. I should be obliged to you if you could get John, or any careful person who has time, to make a list of all the papers or notes, with *dates* and ref^{ce} by volume (not *very* numerous) of Faraday on the subject of common (statical) electricity, including those of Snow Harris, which have been published in the *Philosophical Transactions* or in the *Phil. Mag.* I have seen

a collection of Faraday's in his *Experimental Researches on Electricity*, but he has published something (at least some notes) since. What I am most anxious about is to hear whether there is anything on the subject within the last six months, and whether there is anything except in the *Phil. Trans.* or *Phil. Mag.* Arago, it seems, has recently heard of Faraday's objections, and the uncertainty thus thrown on the theory prevented, as Liouville told me, its being made the subject for the mathematical prize of the Institute this year, and instead of it Abelian functions have been proposed. However, as Poisson before he died wished Liouville to do anything he could for it, I think it will very likely be proposed again. Liouville said he would lend me Chasles' *Précis historique de la géométrie* if I would go back for it again. I went yesterday, and he worked through a memoir of Jacobi's with me.—Your affectionate son,
WILLIAM THOMSON.

His sister, Anna Bottomley, has read his letters from Paris, and on April 2 writes in a bantering tone :

I expect to find you quite a Frenchman when you return, with Frenchified manners — lacing tightly — speaking broken English—and wearing a nice becoming moustache and dark blue trousers. . . . You do not tell me what sort of young lady Mdlle. Dumas is ! Beware of the captivating Parisienne. I shall long very much to see you when you return to England, and you must arrange to pay us a visit as soon as possible. I have had a letter from Margt. Crum this morning. She is going to Germany for the summer with her father, and will spend good part of her time with Liebig.

Then his father puts in a word of sage advice :—

. . . Do not spend a sixpence unnecessarily. Dr. W. T. is much pleased to hear that you have got fairly into Regnault's cabinet, and hopes you will be able to get a good testimonial from him and others, also one from Dumas and others regarding your knowledge of

Physique, and showing that you are not merely an expert x-plus-y man. He still says you should try to get practice in the mere *manipulations*, so as to acquire expertness in the mechanical operations.

Again the diary shows how dynamical notions are shaping his ideas about electricity :—

April 8, 1845.—To-day, in the laboratory (of Physique at the Coll. de France, M. Regnault, Prof.), I got the idea, which gives the mechanical effect necessary to produce any given amount of free electricity, on a conducting or non-conducting body. If m is any electrical element, v the potential of the whole system upon it, the mechanical effect necessary to produce the distribution is Σmv If the body be conducting this exp^{n} becomes vM. This enables us to find the attraction or repulsion of two influencing spheres without double integrals. Also the theorem of Gauss that Σmv is a minim. when v is const., shows how the double int^{l} which occurs when we wish to express the action directly, may be transformed into the diff. co. of a simple int^{l} taken with reference to the dist^{ce} between the two spheres. . . . This has confirmed my resolution to commence experimental researches, if ever I make any, with an investigation of the absolute force of statical electricity. As yet each experimenter has only compared intensities by the dev^{ns} of their electrometer. They must be measured by pounds on the square inch, or by "atmospheres." Also the standard must be the greatest intensity which can be retained by air of given density.

Then he replies to his father's suggestion about testimonials :—

PARIS, 31 RUE M. LE PRINCE,
April 14, 1845.

. . . As matters have turned out, having got my degree before the situation is vacant, I shall be able to get testimonials from the four examiners for the Smith's

Prize, as well as from Cookson and Fuller, at Cambridge, and here probably from Liouville and Regnault, and I think that it would now do more harm than good to ask for any from people who do not know me, or who are not in the way of science. I suppose if there is not a vacancy before October, David Thomson will be appointed again, and if so I shall continue at Cambridge taking a few pupils to support myself, and working as much as possible. I have been planning some experimental investigation in electricity, which, however, I could not commence till I have accurate apparatus at my command, as it is precise measurements I wish to make.

James would probably be interested to hear of the things Regnault is doing, as he is working at present, or rather about to continue some investigations on the latent heat of steam, and its pressure at high temperatures, for which he has been employed by Government. For this he has a tower erected by which he can measure, with mercury, a pressure of 36 atmospheres, and he is just getting his arrangements completed to commence the experiments. He has already made them on steam up to 6 atmospheres, which James will see described in the *Annales de chimie et de physique*. Besides this, Regnault is making "for his own amusement" researches on hygrométrie, part of which I think has been published. I am now regularly employed, and work regularly helping him, the other assistant being employed in graduating tubes for the experiments on steam. . . .—Your affectionate son, WILLIAM THOMSON.

PARIS, RUE M. LE PRINCE 31,
April 26, 1845.

MY DEAR FATHER—I got my paper on Electricity finished yesterday. I wrote it myself in French, and got M. Izarn at the laboratory to look it over. I left it at Liouville's house, but I have not seen him, and so do not know what he will do with it.

The regular work at the laboratory will be broken in upon next week by the preparations for the experiments on high pressure steam, and I shall have a little more time free than usual, which I shall require to prepare examination papers. . . .

. . . I am going this evening to try and see Dumas at M. Brogniart's. I have called several times in the forenoon at his own house, but have not found him in. I still intend to start on Saturday or Monday, so as to arrive in Cambridge on Wednesday, May 7th.—Your affectionate son, WILLIAM THOMSON.

How fertile Thomson's brain had been while he was in Paris his diary shows. His rapid assimilation of Green's neglected work, and the memoirs actually wrought into shape while under the inspiration of the hour, attest not only his own mental agility, but the greatness of the forgotten mathematical genius.

But before Thomson left Paris he had another forgotten hero to rehabilitate. Doubtless inspired thereto by the work proceeding in Regnault's laboratory, he had been reading Clapeyron's paper[1] in the *Journal de l'École Polytechnique*, vol. xiv., 1834, entitled "Mémoire sur la puissance motrice du feu," in which Clapeyron, a pupil trained in the great school of French mathematics, expounded the now famous doctrine familiar to every engineer under the name of Carnot's Cycle. Sadi Carnot,[2] a young French engineer, had in 1824 published[3] a short

[1] A translation appeared in the first volume of Taylor's *Scientific Memoirs*, 1837, p. 347.

[2] Sadi Carnot (born 1796, died 1832) was the son of the more famous Lazare Carnot, the "organizer of victory," one of Napoleon's generals. Another son, Hippolyte Carnot, subsequently edited his brother's tract. Sadi Carnot was uncle of the late President of the French Republic.

[3] This essay is exceedingly rare, very few copies having been printed. It might have remained quite unknown but for Clapeyron's paper and Thomson's

tract entitled *Réflexions sur la puissance motrice du feu et sur les machines propres à développer cette puissance.* The perusal of this *mémoire* incited Thomson to refer to the original tract of Carnot. In vain did he inquire for it in the Library of the Collège de France. No one could tell him even where a copy might be seen. But a copy he must have, even if he searched all Paris to find it. Years afterwards he narrated[1] how in 1845, when in Paris, he sought in vain for this work. He even wandered around amongst the old book-stalls on the quays of the Seine. Of Lazare Carnot and his works on fortification, and of Hippolyte Carnot's tracts on political and social affairs they knew, but not even of the name of Sadi Carnot. This is Thomson's own account of the matter :—

> I went to every book-shop I could think of, asking for the *Puissance motrice du feu*, by Carnot. "Caino? Je ne connais pas cet auteur." With much difficulty I managed to explain that it was "r" not "i" I meant. "Ah! Ca-rrr-not! Oui, voici son ouvrage," producing a volume on some social question by Hippolyte Carnot; but the *Puissance motrice du feu* was quite unknown.

Not until the end of the year 1848 did he see the book, a copy of it being sent him by Professor Lewis Gordon. But of this more hereafter.

Thomson returned to Cambridge after a residence

subsequent advocacy. It was reprinted in 1872 in the *Annales de l'École Normale*, tom. i. pp. 393-457; the editor stating in a footnote that it had hitherto remained "presque inédit." It was republished in 1878, with a biographical notice. It was translated into English by the late Professor Thurston, in 1890, and is published by Macmillan and Co.

[1] The *Fortnightly Review*, March 1892, in an article "On the Dissipation of Energy."

of about four and a half months in Paris. He plunged at once into work as assistant-examiner in his College, and began to take pupils to coach in mathematics. His father writes him about the projects for removing the Glasgow College to a new site, and about the struggle for the abolition of tests and the Bill then in Committee. There is also talk of Dr. James Thomson being invited to take the post of Principal of the new Queen's College at Belfast; which his son does not approve. When he was eventually offered merely the Vice-Principalship he promptly declined it.

In June came the British Association meeting at Cambridge. Amongst those present was Faraday, whose acquaintance Thomson made, and Joule (then an unknown young man), whom he did not meet, however. Thomson read a paper to which reference is made later (see p. 144). Scarcely was it over when Thomson had great news to send home.

ST. PETER'S COLLEGE,
Jun. 28, 1845.

MY DEAR FATHER—I have been elected Foundation Fellow this morning, very much contrary to my expectations this day week when I became candidate. You must know very well how glad I am of it, and so I need not say any more. I hope you have had a pleasant journey. —Your affectionate son, WILLIAM THOMSON.

His father replied congratulating him on the Fellowship, adding an autobiographical note: "At your age I was teaching eight hours a day."

The Fellowship was worth about £200 per annum, with rooms in College. He held it till September

1852, when he vacated it by his marriage; but on Oct. 29, 1872, he was re-elected a Fellow of his College, under Statute 22 of the College Statutes of 1860, which enabled the College to elect eminent men to the unusual privilege of a life-Fellowship notwithstanding marriage. For the last dozen years of his life he returned to the College £100 of the annual dividend of his Fellowship.

On July 9, 1845, his brother-in-law, Dr. David King, then newly returned from Geneva, wrote to Mrs. King:—

I went express this morning from London to see William, and have just returned to the metropolis. . . . Suffice it now to express the joy with which I learned from William that he has already got a Fellowship and a number of pupils! He has all the appearance of being destined to fill a very exalted place in the estimation and, I may add, in the affection of his age and country.

Thomson was apparently concerned about his father's health, for on June 28 he writes to his aunt, Mrs. Gall:—

I think if I was to succeed in getting the Professorship of Nat. Phil. that we could easily manage together to get the business of both classes done, even when he may not wish to work so hard as he has done hitherto.

Thomson adds that he is much occupied with his pupils and with a quantity of correspondence about getting the *Mathematical Journal* going on an enlarged scale. Ellis, in fact, decided to resign the editorship to Thomson, who now assumed full control of the publication to which hitherto he had contributed over the signatures of "P.Q.R." or

"N.N." Letters from Boole, Sylvester, Sir William Rowan Hamilton, and others, still preserved, show his activity in engaging new contributors to the venture, which henceforth, and until merged in the *Quarterly Journal of Mathematics*, was known as the *Cambridge and Dublin Mathematical Journal.* He himself began at once to prepare fresh matter for its pages. He began a new volume of the private mathematical note-books, and its contents show great activity from July 4 to the end of August. He fills seven pages with integrals of the attractions between electrified planes and spheres. On July 12 he suddenly conceives the idea that the permeability of soft iron in magnetism, and the specific inductive capacity of sulphur and other media in electrostatics, play the same part in the respective mathematical theories as conductivity does in the theory of heat. On August 18 occurs the following entry:—

> I have been a good deal occupied for some time with my pupils, and with making up my paper on the element[y.] laws of electricity for the *Journal*, and have not had much time to work. I have, however, got an immense addition of light on the subject of dist[n] of elect[y] [on] spheres and planes since the beginning of last week.

This led to the development of the principle of electric and magnetic "images," and the analogy of them with the optical problem of the kaleidoscope. His training in physical thought was fructifying fast. If he had left Cambridge an accomplished mathematician he returned as an enthusiastic physicist; and the communications which he made during the next few years dealt with the applications of mathematical analysis to Physics, and especially to Electricity.

To make clearer what was the substance of Thomson's contribution to the abstract science of Electricity during the transition period of his career, it is necessary to make some digression as to the state of things in the immediately preceding period.

The attractions between electrified bodies had been examined at various times in order to discover the laws which govern them; and the distribution of an electric charge standing in equilibrium on a conductor of any form had been investigated both experimentally and theoretically. The complications arising in any experimental disposition from the interfering effects of the induced charges produced by influence in neighbouring conductors, had, however, been such as to render the processes of measurement, for the comparison of fact with theory, very crude. It had long been considered probable that the force of attraction (or repulsion, according to the signs of the charges) between two elementary quantities of electricity would conform to a law similar to that which Newton demonstrated to be the fundamental law of gravitation, namely, that it should vary inversely as the square of the distance between them. And, indeed, as this is the fundamental law which is necessarily true for actions that radiate *from points*, it should hold as good for electricity as for light or sound, provided the electric charges should be effectively concentrated so as to act from definite centres, as, for example, in the case of two charges upon insulated spheres that should be small as compared with the distance between them. Cavendish had (in 1771) indeed demonstrated to a high degree of approximation that the electric force between two such elements of electric charge varied inversely as the square of the distance between them, but his main researches remained unpublished till much later. In France Coulomb (1786-87) had with minute care experimented, and applied calculation to show how exactly this law of force was verified by the experiments. As to the nature of electricity itself, Du Fay had proposed a two-fluid hypothesis, and Franklin a one-fluid hypothesis, both of which were compatible with the facts of experiment. The progress made in the application of theory down to the epoch under consideration was recorded by Whewell in his "Report to the British Association[1] on the Present Condition of the Mathematical Theories of Electricity, Magnetism, and Heat." His brief summary is as follows:—

> Electricity, after being brought under distinct conceptions by Franklin and his contemporaries, was formed into a mathematical science by Æpinus; the theory of Æpinus was reformed by Coulomb; the calculations which Coulomb could not execute, Poisson in our own time has performed: such are the main steps in the history of electricity as a mathematical science.

[1] Report of B.A. meeting at Dublin in 1835, p. 1.

Amongst the matters considered by Coulomb was the distribution of electric charges upon two spheres in contact. He obtained approximate solutions to these problems by indirect geometrical considerations of restricted generality, which he applied with ingenuity, comparing the result of his calculations with experimental observations, substituting a two-fluid theory for the single-fluid theory of Æpinus. Poisson's advance consisted in finding more general solutions by direct analytical methods, in particular, by the use of spherical harmonics, or, as they were formerly called, " Laplace's coefficients." Non-mathematical readers may gather a faint idea of the root notion of this method by the following rough illustration from the attraction exercised by a shell of matter upon a unit of matter placed at some point outside it. Let the shell be considered of variable thickness, distributed over a sphere according to any given law, but symmetrically with respect to some axis of the sphere; while the position of the point is given in terms of its distance from the centre, and its angular distance from the stated axis. Then the attraction exercised by the shell upon the unit of matter at the given point can always be expressed, by certain complicated integrations, in terms depending only upon the position of the point and on the law of the thickness of the stratum. This attraction may then be resolved into a series of terms or particular solutions; each such term containing two factors, one dependent only on the position, the other on the thickness. The power of the method for dealing with problems of attraction and of distribution depend on this resolution into terms, and on the mathematical properties of the factors. It can be extended to ellipsoids and other geometrical figures. Such was the analytical method by which Poisson was able to deduce in the form of a functional equation, to be solved by means of definite integrals, the distribution of electricity over two spheres; and it sufficed also to explain why, as an electrified sphere is brought near to an unelectrified one, a spark should pass between them. Poisson's theories confirmed Coulomb's experiments. But new experiments had been made by Snow Harris[1] with his "unit jar," and by Faraday,[2] who had discovered that in all the cases of electric attraction, and of influence between bodies at a distance from one another, the action is dependent on the specific nature of the medium or "dielectric." Poisson's theory had stopped short without accounting for this, and indeed left unsolved many of the problems of the distribution of electricity in equilibrium. "It might be worth while," said Whewell in 1835, "for some new Poisson to examine the rest of Mr. Harris's results." The "new Poisson" was not long in appearing.

For precisely at this stage Thomson, fresh from his own discovery of the similarity between the surfaces of electrical equilibrium and the isothermal surfaces in the flow of heat, stepped into the field.

[1] Sir Wm. Snow Harris, "On some Elementary Laws of Electricity," *Phil. Trans.*, 1834.

[2] *Experimental Researches*, eleventh series, arts. 1161 to 1306.

He absorbed with instant appreciation the fundamental theorems of Green, some of which, as we have seen, he had discovered for himself. He eagerly seized upon Green's physico-geometrical notion of a potential function. He was now to demonstrate, with many interesting applications to actual problems, that, by the light of the potential theory, the elementary laws established by Coulomb were not inconsistent with the discovery of Faraday that the electric forces between bodies were dependent on the nature of the intervening medium.

In the paper which Thomson sent to Liouville's *Journal* on his arrival in Paris, he was considering the problem of the distribution of electricity on an infinitely extended plane subject to the influence of an electric charge situated at a neighbouring point. This he perceived to be mathematically connected with the kindred problem respecting the flow of heat with reference to a body bounded in one direction by an infinite plane. The demonstration given in the first part (sent to Liouville in February 1845) was effected by a method suggested by Green; the analysis in the second part (published only in 1847) led to formulæ that would occur by following the method of Fourier; the third part (sent to Liouville in October 1845) was an evaluation by a direct process of reduction that had been suggested to Thomson by the beautiful geometrical conception of electric images.

It was at this period also that Thomson set himself to prepare a series of articles on the Mathematical

Theory of Electricity in Equilibrium, intended, as he himself stated, to be printed in the *Cambridge and Dublin Mathematical Journal.* The earliest of these to appear, though second in logical order, was communicated to Liouville's *Journal*[1] under the title, "Note sur les lois élémentaires de l'électricité statique." This article begins by a reference to Coulomb's experimental work on the distribution of electricity, and on electrostatic attractions and repulsions, and to the theorems in Green's *Essay* as completing the elements of Coulomb's Theory. He then refers to Snow Harris's experiments and to Faraday's work on the modifying effect of the dielectric medium. Faraday, who had rejected the notion of action at a distance, and regarded all electrical forces as being propagated by the action of contiguous particles, had, ten years before, demonstrated that the electric capacity of a conductor was augmented by surrounding it with sulphur or resin in place of air. In one of his most elegant experiments he constructed two equal spherical condensers, each consisting of a metal globe surrounded by a larger concentric envelope of metal. If in both instruments the space between the inner and outer globes was full of air, their capacities were equal; and if either of them was charged and then made to share its charge with the other, it was found that the charge divided itself equally between them, each taking half of the original charge. But this was no longer so if one of them

[1] Vol. x. p. 209.

was filled with a solid or liquid dielectric in place of air. Moreover, his observation that the inductive action between electrified bodies takes place usually in curved lines was supposed to conflict with the theory of Coulomb. Thomson now argued that there was no real conflict, but that the theory of electric distribution might be regarded equally truly from two different standpoints. Starting from the laws that Coulomb had deduced, and applying the analytical methods of Laplace and Poisson, it could be shown—as he himself had shown in 1841—that every problem of electric distribution is identical with a corresponding problem relative to the flow of heat in solids, which is unquestionably a case of the action of contiguous particles. If one supposes a solid dielectric medium to be replaced by one, itself of the same properties as the air, but containing uniformly distributed a vast number of small conducting spherules,—a hypothesis similar to that made by Poisson in his theory of the magnetism of soft iron,—the influence of the dielectric would be perfectly explained. When he came to rewrite this part of his paper, in November, for the *Cambridge Journal*, he amplified this part, pointing out that every portion, however small, of a dielectric subjected to electrical influence possesses *polarity*, and therefore the laws which Poisson had deduced for magnetic polarity might be generalized, without resorting to any mechanical hypothesis, to include a mathematical theory of electrical influence. He accepts without comment Faraday's view that gases cannot be so polarized, though more recent research has now shown that such an effect exists. When Faraday, in the twenty-sixth series of his Experimental Researches in 1850, returned to this subject, he had grasped the conception of a medium being possessed of a "conducting power" for lines of force, whether magnetic or electric. But this notion was identical with the analogy which had guided Thomson's thoughts since 1841.

To the British Association, which met in Cambridge in 1845 from June 21 to 28, Thomson contributed a paper[1] indicating a solution to the problem of calculating the mutual attraction between two electrified spherical conductors. From the published abstract it is clear that this paper went largely over the same ground as that communicated to Liouville's *Journal*, but it included, *inter alia*, the remark that Snow Harris's observations of the striking distance of an electric spark between two charged conductors afforded, within certain rough limits of accuracy, an absolute standard of "electrical intensity." At the end of the paper he returned to the point, calling attention to the establishment of a standard of intensity as an object for experimental research, and suggesting a standard case. He also observes that if the laws of Coulomb are assumed, then "by very simple analysis first given by Green, we arrive at the laws of Faraday as theorems."

[1] *Brit. Assoc. Report*, 1845, Proceedings of Sections, p. 11, "On the Elementary Laws of Statical Electricity."

Thomson was present in person, on June 23, to read this paper, and, according to the *Cambridge Chronicle* of June 28, at its conclusion Professor J. D. Forbes complimented the author.

The paper also includes the following remarkable paragraph which has not been reprinted in any of Lord Kelvin's collected volumes :—

> There are, besides, some remarkable questions relative to the physical state of dielectrics, which present themselves as objects for experimental inquiry. Thus it may be conceived that a dielectric in motion might present properties analogous to those discovered by Arago in magnetism, and exhibited in his experiment of the revolving disc. As however a very distinct element, that of electrical currents, enters in the latter case in a way which could probably have no analogy in the former, it could hardly be expected that any remarkable agreement of the phenomena presented by the bodies in motion should be found to exist.
>
> Another question, which can only be decided by experiment, is whether a transparent dielectric in a highly *polarized* state affects light transmitted in the same manner as a uniaxal crystal.[1]
>
> All analogy would certainly lead us at least to look for such an action in a plate of glass in which the particles are kept in a constrained state by means of opposite electrical charges in the two faces, especially when we consider that the constraint may be elevated to such an extent as to make the substance be on the point of cracking.

The first of these paragraphs shows that Thomson foresaw some such effects as those investigated thirty-five years later by Röntgen, and the possibility of dielectric hysteresis. The second is a prevision of the "Kerr effect" discovered in 1876, and fruitlessly sought by Faraday. After the reading of Thomson's paper Faraday had some

[1] "Since this paper was read the author has found that Mr. Faraday had previously proposed and examined experimentally the question here suggested, arriving only at negative results. See *Experimental Researches*, § 955."

talk with him, and a month later sent him a memoir of Avogadro. Thomson replied in the following letter :—

ST. PETER'S COLLEGE, *August 6th*, 1845.

DEAR SIR—I beg to thank you for your kindness in sending me the Italian memoir which you mentioned to me when I saw you here. I have to apologise for not acknowledging it before, but I did not wish to write till I could say something on the contents, as you were so good as to ask for my opinion. I shall return it to you almost without delay.

The memoir is entirely occupied with the determination of the distribution of electricity on two equal spheres, in contact, which had been examined experimentally by Coulomb, and calculated mathematically by Poisson according to the general theory of Coulomb. The hypothesis which M. Avogadro makes is that the intensity of electricity at any point of the surface of an electrified conductor is proportional to the portion of sky which can be seen from the point projected orthographically upon the tangent plane at the point, or, as he stated it, to the sum of all the portions of a large spherical surface, described round the two spheres, each multiplied by the cosine of the obliquity of the direction in which it is seen. Thus if the two spheres were black, and were exposed to a sky uniformly bright in every direction, above and below (as might be produced by laying them on a perfect mirror, placed horizontally), the intensity of electricity at any point of either would be proportional to the quantity of light which would be received by a small, white piece of paper laid upon the surface at the point. You will readily from this conceive whether the hypothesis is even analogous in any respect to the actual physical conditions of the problem. The numerical determinations differ very widely indeed from the measurement of Coulomb, but the differences are attributed by M. Avogadro to his having neglected the curvature of the lines of inductive action. For an experimental investigation of the curved line of induction, he refers to your eleventh series. The only numbers which he gives are the ratios of the intensities at 30° and 60° from the point of contact, to the intensity at 90° (which latter he, of course, on account of his hypothesis, finds to be the same for all of the unopposed hemispheres). For these ratios he finds ·6 and ·95. Coulomb's measurements give ·21 and ·80; Poisson's calculations ·17 and 74. The measurements are, as Coulomb himself considers, very uncertain, and may differ considerably from the truth on account of the excessively delicate and precarious nature of his most difficult experiments.

I am at present engaged in preparing a paper, of which I read an abstract at the late meeting of the British Association, for the first number of the *Cambridge and Dublin Mathematical Journal* (a continuation of the *Cambridge Mathematical Journal*), of which a principal object is to show that in all ultimate results relative to distribution, and to attraction or repulsion, it agrees identically with a complete theory based on your views. If my ideas are correct, the mathematical definition and condition for determining the curved lines of induction in every possible combination of electrified bodies are very readily expressible. The distribution of force (or in Coulomb's language, of electrical intensity) on a conductor of any form may be determined by purely geometrical considerations, *after* the form of the curved lines has been found. Thus, let A be an electrified conductor, placed in the interior of a chamber, and let S be the interior surface of the walls, which we may consider as conducting. The lines of induction will of course be curves, leaving the surface of A at right angles, and terminating at S, cutting it also at right angles. By means of these lines let any portion *a* of A be projected on S, giving a corresponding portion *s*. The quantity of electricity produced by induction on *s* will be exactly equal in amount, but of the opposite kind, to that on *a*, according to your theory (or according to Coulomb's, as follows from a general theorem on attraction). If now we suppose S to be a very large sphere, having A at its centre (and it may be shown that the distribution on A will be very nearly the same as if this were the case, provided every portion of S be very far from A, whatever the form of S), the distribution of the induced electricity on A will be very nearly uniform. Hence the problem of the determination of the distribution on A is reduced to the purely geometrical problem of the determination of the ratio of the *s* to *a*. The great mathematical difficulty is the determination of the form of the lines, when curved, as they will always be, except when A is a sphere. In some cases, as, for instance, when A is an ellipsoid, then the curved lines are found with great ease. In other cases, such as that of the mutually influencing spheres, the problem admits of an exceedingly simple solution if attacked from another direction.

It was from the connection with the mathematical theory of Heat (*Mathematical Journal*, vol. iii. p. 75) that I was first able to perceive the relation which the lines of inductive action have to the mathematical theory.

I have long wished to know whether any experiments have been made relative to the action of electrified bodies on the dielectrics themselves, in attracting them or repelling them, but I have never seen any described. Any attraction which may have been perceived to be exercised upon a non-conductor, such as sulphur, has always been ascribed to a slight degree of conducting power. A mathematical theory based on the analogy of dielectrics to soft iron would indicate attraction, quite independently of any induced charge (such, for instance, as would be found by breaking a dielectric and examining the parts separately). Another important question is whether the air in the neighbourhood of an electrified body, if acted upon by a force of attraction or repulsion, shows any signs of such forces by a change of density, which, however, appears to me highly improbable. A third question which, I think, has never been investigated is relative to the action of a transparent dielectric on polarized light.

Thus it is known that a very well defined action, analogous to that of a transparent crystal, is produced upon polarized light when transmitted through glass in any ordinary state of violent constraint. If the constraint, which may be elevated to be on the point of breaking the glass, be produced by electricity, it seems probable that a similar action might be observed.—I remain, with great respect, yours faithfully,

WILLIAM THOMSON.

Faraday replied :—

R. INSTITUTION, 8 *August* 1845.

DEAR SIR—I hasten to acknowledge and thank you for your letter. I reply thus speedily only in reference to your inquiries in the latter part of it.

I have made many experiments on the probable attraction of dielectrics. I did not expect any, nor did I find any, and yet I think that some particular effect (perhaps not attraction or repulsion) ought to come out when the dielectric is not all of the same inductive capacity, but consists of parts having different inductive capacity.

I have also worked much on the state of the dielectric as regards polarized light, and you will find my negative results at paragraphs 951-955 of my *Experimental Researches.* I purpose resuming this subject hereafter. I also worked hard upon crystalline dielectrics to discover some molecular conditions in them (see par. 1688, etc. etc.), but could get no results except negative. Still I firmly believe that the dielectric is in a peculiar state whilst induction is taking place across it.—I am, my dear sir, yours very truly,

M. FARADAY.

Wm. Thomson, Esq.

Later in the month Thomson wrote again, proposing to call to learn more of the experiments; but Faraday was leaving town.

In the final form which the paper on the Theory

of Electricity in Equilibrium took in the *Cambridge and Dublin Journal*, in November 1845, there were introduced additional paragraphs. One of these contained a formula giving a solution for the case of two equal spheres, one only of which is insulated, with numerical results calculated for four different distances. The formula was given without proof, and the method by which Thomson had arrived at it was not published[1] till 1849; but it is of extreme interest, inasmuch as it is probably the first case of the deduction of physical formulæ from energy principles. These formulæ were analytically deduced from the following fundamental proposition:[2]—The "*mechanical value*" *of the distribution of electricity on a group of insulated conductors may be easily shown to be equal to half the sum of the products obtained by multiplying the quantity of electricity on each conductor into the potential within it.* But before this demonstration was sent to Liouville, Thomson had already found another by a different method—the method of electric images.

In the second paper[3] of the series on the Theory

[1] Sent to Liouville in July 1849. Republished in enlarged form in the *Philosophical Magazine*, v. pp. 287-297, April 1853, and reprinted in art. vi. p. 92, in *Electrostatics and Magnetism*.

[2] In 1853 Thomson added the following footnote to the paper:—"This proposition occurred to me in thinking over the demonstration which Gauss gave of the theorem *that a given quantity of matter may be distributed in one, and only one, way over a given surface, so as to produce a given potential at every point of the surface*, and considering the mechanical signification of the function on the rendering of which a minimum that demonstration is founded. It was published, I believe, by Helmholtz in 1847, in his treatise, *Ueber die Erhaltung der Kraft*, by the translation of which, in the last number of the *New Scientific Memoirs* [Taylor's], a great benefit has been conferred on the British scientific public."

[3] *Cambridge and Dublin Mathematical Journal*, vol. iii. p. 131, March 1848, and *Electrostatics and Magnetism*, art. iv. p. 42.

of Electricity, which did not appear till three years later, there was a systematic statement of the principles on which the mathematical theory is founded. This was intended to be introductory, being published for the avowed reason that there existed "no published work in which the principles are stated in a sufficiently concise and correct form, independently of any hypothesis, to be altogether satisfactory in the present state of science." In this document great stress is laid on clear definitions. The development of the theory to embrace the internal electrical polarization of solid or liquid dielectrics was postponed; a footnote referring the reader to the earlier article of 1845 in Liouville's *Journal* (vol. x. p. 209), and adding that a similar view had been subsequently taken by Mossotti.

The third article of the series was an exposition of geometrical notions regarding solid angles and potentials, and of theorems about electrical influence.

The doctrine of electric images, several times mentioned, seems to have been in Thomson's mind before he left Cambridge; but it developed itself with surprising rapidity during his sojourn in Paris, and shortly after his return to Cambridge he communicated it to Liouville in three letters, the substance of which was printed[1] in the *Journal de mathématiques* in October 1845 and June 1846. This method, which even now is not adequately appreciated, is so elegant, and so fruitful in its application, that it merits some attention from the

[1] Reprinted in *Electrostatics and Magnetism*, art. xiv. pp. 144-154.

scientific reader. It furnishes a most cogent illustration of the value of that which has been aptly termed the cross-fertilization of the sciences, the value of introducing into one branch of science the concepts that have arisen in another. To explain the concept of electric images, it may be well to begin with a concrete and familiar case of optical images. Every one knows that if a candle is placed in front of an ordinary plane mirror, the image of it, seen by looking into the mirror, is, as it were, another candle of equal size and brightness[1] with the first, situated apparently exactly as far behind the mirror as the real candle is in front. In fact, if the mirror were to be removed from its frame, and a second candle set in the position where the image previously appeared, the amount of illumination received at any given point in front of the mirror from the two candles would be identical with the amount received from the one candle and its image. The limitations of the frame would prevent the image, or the substituted candle, from being seen from every direction. If the mirror be conceived to be indefinitely extended, and its frame to be infinitely wide and high, these limitations would be removed. If, instead of a plane mirror, a spherical mirror or silver ball were substituted, the image of a candle placed in front will appear to be a smaller candle situated now within the spherical surface,

[1] Of equal brightness, that is, on the assumption that the mirror is a perfect reflector. As a matter of fact there is a small loss due to imperfection. Even the most perfectly polished silver mirror reflects only about 95 or 96 per cent of the light that falls on it. We shall, however, assume perfect reflectivity in the argument.

and at a distance behind the silvered face not equal to the distance of the candle in front, but nearer to the face. The apparent size and distance of this image can be calculated by easy geometrical rules; and again it will be true that if the mirror be removed, and there be substituted an equivalent smaller candle in the place where the image appeared, the illumination at any point in front will remain unaltered by such substitution.

Now, consider a small insulated conducting body, such as a small metal sphere, charged with positive electricity. If placed in the middle of a large empty room it will project its influence into the space around it in radial straight lines, acting as though the electric charge upon it were concentrated at its centre. Now, suppose a very large sheet of copper, or some other good conductor of electricity, to be placed at the side of the charged sphere at a short distance from it. It can be shown that, for all points of the space in front of this sheet, the sheet acts as an electric mirror; the effects in this region being the same as if, in the absence of the sheet, there had been a second charged sphere placed exactly as far behind the sheet as the actual sphere is in front of it. But there is this curious peculiarity, that in order to procure an equivalent effect this "image sphere" would have to be charged with an equal quantity of *negative* electricity. The "image" of a positive charge is a negative one, and *vice versa*. If we suppose the sheet removed and the "image" replaced by a small actual sphere, negatively electrified to an equal degree, the space between and around the spheres will be traversed by a system of curved lines of electric force emanating from one and terminating at the other. If we remove the second sphere and replace the sheet, the lines of force in front, emanating from the positive sphere, will still curve round exactly as before[1] until they reach the sheet. The sheet will, in fact, receive (by induction) a negative charge; and this charge will be distributed over its surface according to a certain law, the density of the charge being strongest at the point of the surface nearest to the sphere, and weaker over the surface outwards from this centre.[2] This case illustrates not merely the conception of electric images, but also the much more refined conception which Thomson had arrived at in 1842 (and Green before him) of the replacing of an electrified body or point by an electric charge distributed over a surface in such a way as to be in equilibrium, as explained above on p. 141. If, to take another case, the positive electric point charge is placed in front of a large uninsulated metal sphere, to act as a spherical mirror, there will again be an electric "image" formed within the sphere, and this "image" will be negative; but its distance behind the reflective surface will not be equal to the distance of the electric "object" in

[1] In all this the charges on the balls are supposed to be distributed uniformly over them, the sheet to be uninsulated; and the distance from ball to sheet so great that no spark can jump across, and in fact much greater than the radius of either ball.

[2] The density at any point of the surface varies, in fact, inversely as the cube of its distance from the influencing electric charge.

front; and though simple rules[1] can be given for finding the position and magnitude of the electric "image," these rules differ from those for the finding of the conjugate images in optics. An electrical image is, then, an electrified point, or system of points, on one side of a surface, which would produce on the other side of that surface the same electrical action which the actual electrification of that surface really does produce.[2] The method of images can be extended to many applications. For instance, if an electrified point is situated between two parallel conducting sheets, not only will it generate a primary "image" in each sheet, but each primary image will generate a secondary image of itself in the other sheet, and these will generate tertiary images, and so on in an infinite row, exactly as a luminous point does between two mutually facing mirrors. The resulting electric distributions can be calculated by the superposition of the distributions equivalent to the successive members of the series. Again, if an electrified point is placed anywhere between two indefinitely extended planes that meet one another like the two leaves of a sheet of paper, there will be a number of images symmetrically situated if the angle between the planes is 90° or 60° or any integral fraction of 360°. This, as Thomson pointed out, is the analogue of the optical toy called the *kaleidoscope*. Further, if two electrified

[1] In fact, the radius of the reflecting sphere is a geometric mean between the respective distances of the point-object and point-image from the centre of the sphere.

[2] Clerk Maxwell, *Electricity and Magnetism*, art. 157.

spheres that are not small are brought near to one another, reacting on one another's charges so that the distribution of the charges over the spheres is distorted out of uniformity, the distribution of the charges and the resulting force of attraction or repulsion can be calculated by considering the whole effect as a succession of images that are superposed on one another.

In communicating the principle of images to Liouville, by letter, on October 8th, 1845,[1] he showed that the production of images in spheres is closely connected with the principle of geometrical inversion, applied (if not actually invented by him) to the calculation of the reciprocal positions of object and image, and he announced certain special applications to the distribution of electricity over a circular disc or a segment of a sphere intercepted by a plane. In June and September of 1846 he followed up his communication by two other letters announcing further extensions, which his preoccupation with new duties in Glasgow prevented him at the time from elaborating.

In October 1845, the earliest opportunity of giving him office after being elected to the Fellowship, Thomson was appointed, on the proposition of Cookson, to be College lecturer in Mathematics at Peterhouse—a post which he held to the end of the Easter term 1846. In November his paper on the

[1] Liouville's *Journal*, x. p. 364, 1845; reprinted in *Electrostatics and Magnetism*, art. xiv. p. 144. The subsequent letters on the same subject, in Liouville, vol. xii., are reprinted in continuation on pp. 146-154, followed by Liouville's appreciative commentary congratulating "ce jeune géomètre," and assuring him "de l'estime que j'ai pour son talent."

"Laws of Statical Electricity" (p. 143 above) appeared in the *Cambridge and Dublin Mathematical Journal*, as also another article, a "Note on Induced Magnetism in a Plate." This is an application of the principle of images to the case where a magnet is held in front of a plate of soft iron, the resulting magnetic distribution being shown to be equivalent to a series of pairs of image-magnets of intensities decreasing in a geometrical progression. He remarks that (if the magnetic inductive capacity of the iron could be regarded as indefinitely great) in the extreme case the effect of the plate is to destroy all action behind it, so that it will act as a (magnetic) screen; adding, "This we know to be the case when an infinite conducting screen of any form is placed before an electric body." To this question of screening he often returned in after-years; for example, in a Royal Society paper of 1891, reprinted at p. 681 of his *Baltimore Lectures*. It will also be remembered that he employed an iron shell to shield his marine galvanometer from disturbance by external influences.

Thomson remained at Cambridge for the long vacation of 1845, coaching four or five pupils. In August he wrote to his father, at Knock, that he was anxious to get home to hear what he had to say about the important matters in consideration at present. "As far as I am concerned, I think I can spend my time profitably and probably pleasantly here as long as there is any chance of the professorship of Natural Philosophy being in the present state." He tells how he is to be employed to lecture in the October term, besides taking pupils to earn money, and when he comes down to Knock for September he will bring Fuller with him.

In October he is back at Cambridge, and there learns that David Thomson has been appointed to Aberdeen. His father writes, saying that one of the Glasgow professors had asked whether William would take his place as *locum tenens*, and that he had demurred, saying that William could not with propriety give up his Cambridge engagements for any temporary post. A substitute was found for the session. On November 1st William writes of himself:—

> I have as many pupils as I wish this term, and I have managed to arrange so as to have my private pupils only three times a week, leaving the other days clear. I have College lectures every morning (saints' days, of which this is one, being happily excepted) at eight o'clock in the morning. I have charge of the third year men who are reading for honours. There are only three of them, one of whom is reading high subjects, and the other two are only at conic sections. I have, therefore, two classes, and I have one of the second year men in the conic section class. I find such lecturing pleasanter than taking private pupils, as I can class the men together more readily. Otherwise it is very similar.
>
> Last night I played for the first time in the orchestra of the University Musical Society, taking the cornopean parts.

The winter of 1845-46 seems to have been uneventful. An autobiographical note in the mathematical diary, dated July 1st, 1846, tells all that is known. It begins by reverting to the work of the previous summer when he was working on electric images.

During the remainder of the term I was at Cambridge I drew pictures of successive images in spheres, etc., etc. When I was at home (at Knock) I was principally engaged (when I did any work) in the consideration of the dist^{n} of elect^{y} on circular plates subject to the influence of points in their planes, etc. . . . I gave this to Liouville in a letter written principally at Knock, but posted at Camb., Oct. 8th. It was published in Liouville's *September* number. I had a great deal of correspondence about the *Math. Journal* all summer on acc^{t} of the change of name, etc. (Graves, Sir W. Hamilton, Townsend, Boole, besides the Cambridge men). My effected work till the present time is to be found in the *Math. Journal.* In the Christmas vacation (from the latter part of the October term) I was engaged a great deal in (as yet) illusory attempts to do anything for planes cutting at an angle not a sub-multiple of π. . . . Some results of the co-ordinates explained (Oct. 27th), including the actual solution by means of them of the problem for two spheres in contact, I sent to Liouville in a letter last week—I was writing him, at any rate, for a testimonial. I have resolved, if possible, to send him some more in acknowledging his answer, as I wish to get some of the results published at once, seeing that I have made many abortive attempts to commence my *treatise* on Electricity, which I have intended to publish in parts in the *Journal,* and the prospect of getting it ready is now rather more distant. I have also, since the beginning of the Lent term, been often trying to connect the theory of *propagation* of ele^{y} and magnetism with the solid transmission of force. I heard of the death of Dr. Meikleham on the afternoon of May 9th (the day of the second C.U.M.S. concert in the May term), and since that I have had an enormous amount of letters to write. In the examination I set four papers for the second year, viz. Newton and Dynamics,

Statics, Analyt. Geom. and Theory of Equations, and Diff. Calc. Since October I have been employed an hour a day (effectively a good deal more) except saints' days, lecturing the third year (Parnell, and Herringham, also Male till the middle of the Lent term ; and Rippingall occasionally from the 2nd year), for which Cookson gave me £80 before he started for Italy with Fuller. [Here follows a list of pupils: 6 in long vacation, 1845 ; 6 October term, 1845 ; 7 Lent term, 1846.]

So the chair of Natural Philosophy had at last fallen vacant, and the ambitions so long cherished by father and son were now to be put to the test.

CHAPTER IV

THE GLASGOW CHAIR

In May 1846 the chair of Natural Philosophy in the University of Glasgow became vacant by the death of Professor Meikleham, who had held it since 1803. For some years his health had been such that he had been unable to fulfil the duties of his office, and, as already narrated, they had been discharged at first by Professor J. P. Nichol, but for four years by David Thomson. Nichol now took the initiative in moving the Faculty of the University to consider the desirability of reorganising the teaching of Natural Philosophy, for which the University possessed little or no equipment. The Faculty recognised the need of bringing the teaching into conformity with the growing requirements of physical science and the advance of discovery, but postponed the rearrangements until such time as they should have appointed a new professor of Natural Philosophy, who should be able to advise them what steps to take. At the same time they recorded their intention to maintain the right of reopening the question of rearrangements at any future time, that they might not be debarred by any interposition of claims of vested rights, real or supposed, on the part of the new professor. Meanwhile information was circulated that the appointment was vacant, and that the Faculty would shortly proceed to election. David Thomson, who had discharged the duties from 1841 to 1846, had in the previous year succeeded to the chair at Aberdeen. Several candidates of experience announced themselves. But the members of the Faculty must have been well aware that in young Thomson, their own *alumnus*, and son of their honoured Professor of Mathematics, there was a man of no ordinary promise. His father, too, had been, as we have seen, preparing for the event, and at once took active steps. His first act, after sending a hasty note to his son, was to write to Forbes to ascertain his views and intentions. Forbes's reply was clear :—

CUMBERLAND, 11*th May* 1846.

MY DEAR SIR—Your note announcing the vacancy of the Glasgow chair of Natural Philosophy was forwarded to me here from Edinburgh. I have no intention of being a candidate. I hope that your son will, and that he will obtain it. I thank you for your mathematical papers, and remain, my dear sir, yours sincerely,

JAMES D. FORBES.

Dr. James Thomson, Glasgow.

The father sends advice to the son :—

GLASGOW, *May* 10, 1845[6].

MY DEAR WILLIAM—I have just had your note, which was written before you could have heard of Dr. Meikleham's death. In reference to the vacancy thus produced, you must do everything that honour and propriety will allow. . . . I have had no conversation, therefore, in reference to you with anyone except Dr. W. Thomson. He is quite friendly, but he evidently fears your instructions would be too deep for our students, and that you might fail in giving general interest. *Do all you possibly can* to obviate this impression. You cannot dictate to those who may give you testimonials, but you might, in many cases, be able to give a hint that the situation is half-mathematical and half-popular, and thus get persons to certify as to your clearness in writing, expression, etc. The testimonies of Hopkins, Cookson, Fuller, and others in your own college will obviously have the greatest effect. Get all you can, however, from other quarters. . . . A testimonial from Regnault (and perhaps others in Paris) as to your knowledge of experimental and the lighter subjects would do great good. If you do it in a cautious and judicious way, you can readily influence in considerable degree those to whom you apply. . . .

You should get, without delay, Kelland's new edition of Dr. Thos Young's *Natural Philosophy.* It is an interesting work, and is of a popular character. You should also, irrespective of your application here, write out some lectures of as simple and elementary a kind as possible; keeping in mind, what I find to be more and more the feeling, that Oxford and Cambridge men (Lushington, Kelland, Hitchens, etc.) have not given satisfaction here, and that you will have to contend against the feeling thus produced, and against the *handle* it will afford to the Dr. Flemings *et hoc genus omne.*—Yours affect^ly^

J. THOMSON.

Then there is much correspondence filling the rest of May and June; for both father and son busy themselves in collecting the necessary testimonials. William went to London to see his scientific friends and to consult Archibald Smith. He is dismayed to find Smith himself hankering after the post. He calls on Faraday, but knows that Faraday never gives any testimonials. He calls on the younger Lushington in London, and finds Tennyson in his rooms.

On May 26th he writes preliminary letters, announcing his candidature, to thirteen of the electors. A facsimile of one of these, to the professor of Moral Philosophy, is here given.

ST. PETER'S COLLEGE,
CAMBRIDGE, *May 26th*, 1846.

DEAR SIR — As the Chair of Natural Philosophy in the University of Glasgow has recently become vacant, and as you are one of the Electors, I take the liberty of announcing to you my intention of becoming a candidate for that situation; and, as soon as I shall have it in my power, I shall transmit to you testimonials in support of my application.

I have the honour to be, Dear Sir, Your most obedient servant, WILLIAM THOMSON,
Fellow and Mathematical Lecturer of St. Peter's College.

The testimonials produced by the candidate were circulated by his father, with the following covering letter:—

COLLEGE, *June* 20, 1846.

DEAR SIR—I take the liberty of sending you a *proof* of such of my son's testimonials as have reached me. He will have others, copies of which will be transmitted in due time to the electors. The number of his recommendations will be small, however, as he proceeds on what I consider to be a proper principle—that of procuring testimonials from no persons who are not able to speak of him from personal knowledge, or from having read his publications.

From the electors I wish no pledge or promise in his

St. Peter's College
Cambridge
May 26th 1846

Dear Sir As the Chair of Natural Philosophy in the University of Glasgow has recently become vacant, and as you are one of the Electors, I take the liberty of announcing to you my intention of becoming a candidate for that situation; and, as soon as I shall have it in my power, I shall transmit to you testimonials in support of my application.

I have the honour to be
Dear Sir
Your most obedient servant
William Thomson
Fellow and Mathematical
Lecturer of St. Peter's College

The Rev. Dr. Fleming

LETTER ANNOUNCING CANDIDATURE FOR THE CHAIR OF NATURAL PHILOSOPHY.

favour, and I have no doubt of their keeping themselves equally unpledged in reference to any other candidates that may come under their notice.—I remain, dear sir, yours faithfully, JAMES THOMSON.

On June 17 the younger Thomson wrote :—

ST. PETER'S COLLEGE,
June 17, 1846.

MY DEAR FATHER—I returned from London yesterday, and found your letter along with some others which had arrived since Saturday, waiting for me. I did not think that there would be any use in my being at home before the middle or end of July, when I engaged to go on with my pupil till the 15th, but from what you say I must endeavour to get him disposed of with some other tutor, and, if you still think it would be necessary, or at all useful, to be in Glasgow early in July, I think I can be quite sure of managing so as to get away as soon as the business about my Fellowship will allow me, which will be about the end of the first week in July. . . .

I called on Prof. Graham about Faraday's lecture, and again on the evening before I left London. He spoke to me a good deal about the prospects of the election at Glasgow. He does not think any of the London men will come forward, though there are some who would be very glad to have it, but are prevented by various reasons. If none of his friends there do come forward I think he might perhaps give me a testimonial if I were to ask him, but I doubt much whether I should, as he has such slight grounds for saying anything for me. He, as many others, especially of the best men here, spoke to me very strongly about the objectionable nature of the great multitude of testimonials which are commonly produced on such occasions. Arch^d Smith, as I think I told you before, is still in a state of indecision, and will probably remain so till the end, that is, till the electors offer him the situation, in wh^h case he would accept it, or give it to some other person. Graham says that Smith's father will probably make great exertions about the election, but still, unless Archibald Smith himself comes forward as candidate, I can hardly conceive of him being elected. Potter, the professor of Natural Philosophy at University College, would be a candidate if he was free from an engagement under which he is to remain for at least another year at University College. From what Graham says, I am sure he would have exceedingly high testimonials, but I have not heard that he is trying to get off with his engagement. All these things, in addition to the general uncertainty of the prospect, have been making me very anxious, and you may be quite sure that (although you think I am too cool on the subject) nothing which I can do to increase my chance will be left undone. Any slowness hitherto has been because I thought you would not want me to have the testimonials ready for circulation till the meeting on the 5th of August. I shall, however, think on nothing but getting any which I am to have ready as soon as possible. I have also been feeling very much a dislike of testimonials and want of faith in their power, and I think that if the *electors*, instead of the *candidates*, asked for the opinions of the people writing the testimonials, that much fairer results could be obtained. If, for instance, the electors would ask people here, who know both Smith and myself, for the opinion of our relative merits for the situation (which is a kind of request that a candidate could never make) something much more explicit would be known, than can be from general testimonials. Of course nothing of the kind will, I suppose, be done, and I must just take the usual plan of getting as many testimonials as I can. Cookson would have persuaded me if he could, not to get any testimonials except from Hopkins and Peacock, and even Hopkins recommended me to get very few.

I attended Faraday's lecture which, though on his new great discoveries, was very popular, and very interesting to a miscellaneous assemblage of fashionable people. I spoke to Faraday after the lecture, and told him that I am candidate for the vacant professorship. As I anticipated, he makes a rule not to give testimonials (and he gave the same opinion as others about the badness of testimonials in general), but he wished me success.

I enclose some more testimonials which I have already received.—Your affectionate son,

WILLIAM THOMSON.

The other candidates were :—

David Gray, M.A., then professor of Natural Philosophy in the Marischal College, Aberdeen.

Thomas Miller, A.M., rector of the Academy of Perth.

Thomas Aitken, M.D., formerly lecturer in Natural Philosophy in the Royal Institution of Liverpool.

Andrew Bell, mathematical master in Dollar Institution.

William Brydone Jack, A.M., professor of Mathematics and Natural Philosophy in King's College, Fredericton, New Brunswick.

A copy of the letters of application and testimonials presented by the several candidates has been preserved, and is in the possession of Dr. Hutchison, headmaster of the Glasgow High School.

The following is the complete list of the persons who furnished Thomson with the testimonials which he presented in support of his candidature :—

The Master and Fellows of St. Peter's College, Cambridge ; Rev. H. W. Cookson ; William Hopkins, Esq. ; Frederick Fuller, Esq. ; the Rev. W. Whewell ; the Very Rev. George Peacock ; the Rev. Professor Challis ; the Rev. S. Earnshaw ; the Rev. S. Blackall ; the Rev. Harvey Goodwin ; R. L. Ellis, Esq. ; Professor Forbes ; Dr. Thomas Thomson ; Dr. R. D. Thomson ; the Rev. Professor Lloyd ; Professor De Morgan ; Arthur Cayley, Esq. ; Richard Townsend, Esq. ; Dr. William Couper ;

Sir William R. Hamilton; William Walton, Esq.; George Boole, Esq.; J. J. Sylvester, Esq.; George G. Stokes, Esq.; James Armitage, Esq.; Professor David Thomson; George W. Hemming, Esq.; John Sykes, Esq.; M. Regnault; M. Liouville.

The more important of these are here reproduced.

From the Master and Resident Fellows *of St. Peter's College, Cambridge.*

Whereas, William Thomson, Bachelor of Arts, Fellow of this College, and Mathematical Lecturer, is a candidate for the Professorship of Natural Philosophy in the University of Glasgow, and hath requested of us letters testimonial of his scientific learning and good behaviour, We, the Master and Fellows whose names are underwritten, do testify that the said William Thomson has been resident amongst us nearly the whole period from his first entering the University in 1841, to the present; that he has been regular and exemplary in his conduct, and has diligently applied himself to his studies; that his talents and attainments in Mathematics and Natural Philosophy, as shown both in his own College and in this University, are of the highest order; and that we consider him, in all respects, eminently qualified for the situation which he seeks to obtain.

Given at St. Peter's College, under our hands and seal, this thirtieth day of May 1846.

W. Hodgson, *Master.*
H. W. Cookson, *Senior Tutor.*
Wm. Nind.
John Cocker.
Barnard Smith.
Philip Freeman, *Junior Tutor.*
Thos. Wollaston.
George West Piggott.
Frederick Fuller, *Assistant Tutor.*

From the Rev. HENRY WILKINSON COOKSON, M.A.,
Fellow and Tutor of St. Peter's College, Cambridge.

ST. PETER'S COLLEGE, CAMBRIDGE,
June 1st, 1846.

MY DEAR SIR—As I am about to leave England for a tour of several weeks, I shall not be in the way when the election takes place to the vacant professorship at Glasgow; there is nothing, however, which your son can require from the College, which we cannot provide at the present time. I sincerely trust that, both on his own account and for the cause of science, he may be successful in his application.

He is regarded here, by the most competent judges, as the first man of science, of the rising generation, in the country; and as he is devotedly fond of scientific pursuits, and ardent in his desire to inspire others with the love of them, he seems to be peculiarly suited for a professorship. It is a recommendation to him also, that he is extremely popular and much esteemed.

When he was an undergraduate, I was in the habit of examining him; and his expositions of principles and his proofs of ordinary propositions were so clear and satisfactory, that I frequently preserved them to give to my pupils as the best which could be met with on the subject; and on this account, and in consequence of the great interest which he was likely to take in encouraging the study of Mathematics, I proposed last year that he should be employed as College Lecturer, with the hope that private tutors might be rendered unnecessary. Happily he accepted the office, and he has discharged the duties of it with great ability and zeal, so that there is now no real necessity whatever for any of the students of his class resorting to the assistance of a private tutor.

Some persons may be inclined to suppose, that as he is much occupied with original investigations in the more abstruse parts of Natural Philosophy, he would often in his ordinary Lectures be above his audience, and be seldom understood; but those who know him can entirely rely upon his judgment to guard against this fault. He may be quite depended upon for adapting his instructions to his class. He has given us here too many proofs of his discretion in this particular to allow us to have any doubt of him. I might mention instances, but it is unnecessary to do so.

In the general character of his scientific investigations, as well as in the whole course of his study, he seems to preserve a proper medium between what is entirely speculative and what is exclusively practical; and he has a very high view of the objects of science. He may be expected, therefore, to give a good tone to the studies of a place; and it has already been often mentioned to me in Cambridge, how desirable it would be to retain him here, in order that he might have a good influence upon the studies of the University, and add to its scientific reputation.

Whatever he may do, I hope that, if he continues to be blessed with health and strength, he will soon make his name to be known in Europe. He is already blessed with a reputation which veterans in science might envy, but his friends look for still greater lustre. The very early development of his powers of accurate investigation in the most abstruse and difficult subjects is indicative of a mind of the first order, and raises the highest expectations of his future eminence. God grant that he may live and do honour to his country!—Very sincerely yours,

H. W. COOKSON.

Dr. James Thomson,
Professor of Mathematics, Glasgow.

From WILLIAM HOPKINS, Esq., M.A., *St Peter's College, Cambridge.*

To the Electors to the Chair of Natural Philosophy in the University of Glasgow.

CAMBRIDGE, *June* 26, 1846.

GENTLEMEN—My friend Mr. W. Thomson having offered himself as a candidate for the vacant Professorship of Natural Philosophy at Glasgow, I am anxious to bear testimony to his qualifications for the situation. I have known him intimately since the commencement of his residence at Cambridge, his mathematical studies there having been prosecuted principally under my direction.

It is scarcely necessary to offer individual testimony with respect to Mr. Thomson's mathematical talents and acquirements, which are best attested by the academic distinction which he has gained, and the original papers which he has written. But I would here remark, that his power as an *analyst* is not unaccompanied (as is not unfrequently the case) by a taste for physical investigations. I have known other young men as good analysts as himself; but I doubt whether, in the course of my long experience, I have ever met with any one of his own age who combines such a knowledge of abstract mathematics with such an almost intuitive perception of physical truths, so accurate a knowledge of physical principles, and such enlarged and matured views of the great physical problems which Nature presents to us.

I am aware that those higher intellectual qualities and acquirements may be regarded as having no very direct bearing on the elementary lectures which he would be called upon to give in his Professorial capacity; but I may be allowed, perhaps, to express the opinion, that the indirect influence of high attainments on even the most elementary lectures can hardly be too highly appreciated in their tendency to give to such lectures a precision and clearness which they will scarcely ever possess in the hands of a lecturer of inferior capacity; assuming always the existence of the requisite power of oral exposition. That Mr. Thomson will be found to possess this power I have myself no doubt. My conviction on this point rests, in the first place, on the style of his *written* expositions, which are singularly full and perspicuous, and possess a stamp of originality even on the simplest subjects; and, in the next place, I judge by the manner in which he discusses orally any point of scientific interest. Of this many of his friends can judge as well as myself, and I doubt not, all will draw the same inference. Of the general character of his written expositions no one can have had the same opportunity of judging as myself. I repeat my conviction that he will be found perfectly efficient as a *lecturer.*

I look to another cause of efficiency also in his lecturing. The amiableness of his character and the simplicity of his manners can hardly fail to render him as *popular* in Glasgow as he has been with all classes of his acquaintance in Cambridge; and should you do him the honour to elect him, I feel confident that he will be as much beloved by his class for his social and amiable qualities, as respected by them for his abilities and attainments.

I cannot, in conclusion, but express the most earnest hope that Mr. Thomson may prove the successful candidate for a situation for which he is so eminently qualified; and which will place him, should he obtain it, in so good a scientific position, and connect him with an Institution of which he cannot fail to become one of the most distinguished ornaments.—I remain, Gentlemen, Your obedient Servant, W. HOPKINS.

From FREDERICK FULLER, Esq., M.A., *Fellow and Assistant Tutor of St. Peter's College, Cambridge.*

ST. PETER'S COLLEGE, *June* 2, 1846.

As Mr. William Thomson is about to offer himself as a candidate for the vacant chair of Natural Philosophy in the University of Glasgow, I have great pleasure in bearing testimony to his high character and unusual endowments, both moral and intellectual.

With the evidence of his accumulated University and College honours, his published papers, and the testimony of the most distinguished mathematicians of Cambridge before me, it would hardly become me to speak of his extraordinary scientific acquirements; but in the capacity of College Lecturer, I have had an opportunity of observing his philosophical character more narrowly, and I can speak of these qualities which are more peculiarly requisite in an instructor. He combines the greatest clearness and precision with the most extended views in science, and he has always been as much distinguished for the simplicity and accuracy of his demonstrations of the more elementary propositions of Natural Philosophy, as for his talent in treating the abstrusest problems.

But what I conceive renders him particularly fitted for such a situation as that of a public professor, is the energy of his character, and his great enthusiasm for science. These he cannot fail to communicate, in some degree, at least, to his pupils; and it is needless to add that such qualifications form the best foundations for success in tuition. FREDERICK FULLER, M.A.

From the Rev. WILLIAM WHEWELL, D.D., *Master of Trinity College, Cambridge.*

TRINITY LODGE, *May* 11, 1846.

Having examined Mr. Thomson for the Smith's Prizes in January 1845, and having been one of the judges who assigned to him the *first* of these prizes, I beg leave to say that he appeared to me very decidedly to deserve this distinction, and that his superiority appeared principally in the most difficult of the mathematical problems proposed by me. W. WHEWELL.

From ROBERT LESLIE ELLIS, Esq., M.A., *Fellow of Trinity College, Cambridge.*

May 11, 1846.

In 1845, I was Senior Examiner, and I examined Mr. William Thomson for the degree of B.A.

I entertain the highest opinion of his genius and acquirements, and have no doubt whatever of his being admirably qualified to fulfil the duties of the office for which he is a candidate.

I may be allowed to express my belief, that Mr.

Thomson will hereafter occupy a very distinguished place among the scientific men of Europe.

R. L. ELLIS.

From JAMES D. FORBES, Esq., *Professor of Natural Philosophy in the University of Edinburgh.*

CARLISLE, 23*rd May*, 1846.

MY DEAR SIR—I am very glad to hear that your son is to be a candidate for the Chair of Natural Philosophy in Glasgow. From what I know of him personally, from all that I have heard by the report of competent and disinterested persons, and from what I have seen of his writings, I have no doubt whatever that his appointment will do honour to the College and to the Electors.—I remain, my dear Sir, Yours sincerely,

JAMES D. FORBES.

Professor James Thomson,
Glasgow College.

From THOMAS THOMSON, Esq., M.D., *Professor of Chemistry in the University of Glasgow.*

GLASGOW, *May* 28, 1846.

Mr. William Thomson attended the Chemistry class, and the regular Examinations during the Session 1838-9. His conduct was exemplary; and the figure which he made, when examined, was highly to his credit, and showed a good knowledge of the principles of science. He also attended the Laboratory, where he was a practical student for some time. His profound mathematical knowledge, together with his unremitting industry, give him peculiar claims to the Chair of Natural Philosophy, which, if properly taught, constitutes the most laborious chair (except Chemistry) in the College.

THOMAS THOMSON.

From AUGUSTUS DE MORGAN, Esq., *Professor of Mathematics in University College, London.*

To the Electors to the Professorship of Natural Philosophy in the University of Glasgow.

UNIVERSITY COLLEGE, LONDON,
June 19, 1846.

GENTLEMEN—On behalf of Mr. Thomson, of St. Peter's College, Cambridge, a candidate for the post above mentioned, I beg to state as follows :—

My knowledge of Mr. Thomson is limited to his writings, and some correspondence on matters connected with the *Cambridge Mathematical Journal*, of which he is editor.

A year or two ago, on inquiring the name of the author of some anonymous [1] papers of great ability in that *Journal*, I was told that they were by an undergraduate, Mr. Thomson of St. Peter's. From these, and from what Mr. Thomson has since written, there can be no doubt whatever that he is a mathematician of great promise.

With reference to his present application, there are two points particularly worth notice :

(1) That his mathematical speculations have turned very much in that line which is now so essential to a teacher of Physics—I mean the analysis by which the mathematical theories of electricity, heat, and light have been brought to their present form. That a Cambridge man of high honours has read in these subjects is now a matter of course; but it does not necessarily follow that he has made them one of his first subjects of interest. When the question comes on as to whether a professorship in Physics should be entrusted to a young man whose public character has been established by his mathematical writings, it should not be forgotten that, in the present instance, these writings are, in great

[1] These papers have since been acknowledged in the same *Journal* in an index.

part, not only connected with the most material questions of Physics, but indicate a strong turn that way.

(2) A Cambridge degree of the sort which Mr. Thomson has taken, though always a good test of knowledge and power, is a still better one, when the owner of it has been employed in original research during his undergraduateship. Very few men publish anything while undergraduates; and it is generally thought, and I believe, justly, that it is not wise, with reference to the degree alone, to distract the attention from the University course.

Confining my remarks strictly to what I know of Mr. Thomson, and to obvious and necessary inference, I shall say nothing further; but beg to remain, Gentlemen, Your most obedient Servant, A. DE MORGAN.

From GEORGE G. STOKES, Esq., M.A., *Fellow of Pembroke College, Cambridge.*

CAMBRIDGE, *July* 6, 1846.

Mr. William Thomson being a candidate for the Professorship of Natural Philosophy in the University of Glasgow, I am desirous, as a personal acquaintance of his, of expressing my opinion of his fitness for that office. The distinction which Mr. Thomson has already attained renders it almost unnecessary for me to say anything as to his mathematical abilities and acquirements. I would merely observe that, from my conversation with him, and from the perusal of his writings, I am fully satisfied of the original character of his mind. But besides this, I feel assured that Mr. Thomson possesses extremely clear views on physical subjects, together with a sound judgment, and a correct appreciation of the relative value of different hypotheses;—qualities indispensable to a philosopher, but which do not always fall to the lot of even eminent mathematicians. Should Mr. Thomson be elected, he will, I am persuaded, ably discharge the duties of his office, and reflect credit on the University which has chosen him. G. G. STOKES.

From M. REGNAULT, *Membre de l'Académie des Sciences, etc.*

PARIS, *le 1er juillet*, 1846.

J'ai l'honneur de recommander à la bienveillance de Messieurs les électeurs de l'Université de Glasgow, Mr. William Thomson qui se présente comme candidat à la Chaire de Physique vacante dans la dite Université.

Pendant son séjour à Paris, Mr. Thomson a travaillé pendant trois mois dans mon laboratoire, et m'a assisté dans mes expériences. J'ai été à même de reconnaître sa grande aptitude aux travaux de recherches et ses connaissances étendues dans les sciences physiques et mathématiques.

En nommant Mr. Thomson à la place qu'il postule, l'Université de Glasgow fera un choix dont elle se félicitera, et qui certainement lui fera honneur par la suite.

De mon côté, je m'estimerais heureux, si ma faible recommandation pouvait influer quelque peu sur la décision de Messieurs les électeurs.

V. REGNAULT.

Membre de l'Académie des Sciences,
Professeur de Physique au Collège de France.

The other testimonials not here quoted in full all speak in the same strain of Thomson's abilities. Challis, whose lectures on Practical Astronomy, Optics, Hydrostatics, and Pneumatics, Thomson had in 1844 attended, and who was one of the examiners for the Smith's Prizes in 1845, wrote that his answers to questions in Natural Philosophy exhibited a knowledge of the literature and a remarkable comprehension of principles. Blackall, one of the Moderators in 1845, spoke of his readiness in choosing the best mode of attacking unexpected problems, and his precision in arriving at an accurate result through the medium of complicated analysis. William Walton, the eminent mathematician, of Trinity Hall, wrote that he considered Thomson's published papers to place him in so high a position among the mathematicians of England, as to render entirely superfluous any testimonial to his ability as a mathematician or natural philosopher; and added that in relation to his powers of exposition, he considered him to possess, in an eminent degree, that facility of scientific elocution, and that energy of manner which are so essential to arrest the attention of a class of students. Sir William Rowan Hamilton emphasized his ability to deal with physical applications of mathematics, and Boole dwelt on the importance of his researches on the equilibrium of fluids. Sylvester testified to the influence of Thomson's editorship in raising the quality of the *Cambridge and Dublin Mathematical Journal.* Cayley praised his familiarity with mathematical physics, and the fullness and fertility of his mind, predicting that he would contribute largely to the future progress of science. Dr. R. D. Thomson attested that W. Thomson, having attended the Chemical Laboratory of Glasgow University during part of session 1843-44, exhibited a decided aptitude for manipulation; and that he handled his apparatus and performed his analyses in a neat and accurate manner. And David Thomson, now Professor in Aberdeen, stated that when, six years before, he had been suddenly called upon by the late Dr.

Meikleham to undertake the duties of his chair, he would certainly have sunk under the labours of his first session had not young William Thomson kindly undertaken to assist him in the preparation of the experiments by which the lectures were illustrated.

To the unique series of testimonials so presented, forming, as printed, an octavo pamphlet of twenty-eight pages, Thomson added an appendix enumerating his published original papers. Seventeen of these were from vols. ii., iii., and iv. of the *Cambridge Mathematical Journal*, and five more in the new *Journal*, of which he had become editor, together with four more from Liouville's *Journal*. He also presented the following addendum :—

The following communications from *M.* LIOUVILLE, member of the French Institute and Board of Longitude, and editor of the *Journal de mathématiques*, have been received since Mr. William Thomson's other testimonials were printed :—

TOUL, *24e juillet*, 1846.

MON CHER MONSIEUR ARMITAGE—J'apprends à l'instant même que la chaire de Physique à l'Université de Glasgow étant vacante, notre ami commun, *M.* William Thomson, se présente comme candidat. Vous avez pu voir, lors de votre dernier voyage à Paris, quelle haute opinion j'ai conçue des talents de ce jeune savant, quel cas je fais des mémoires déjà nombreux que nous lui devons, et combien d'espérances je fonde sur son avenir. C'est vous qui dans le temps avez bien voulu l'adresser à moi, et vous vous rappelez avec quel empressement j'ai accueilli quelques-uns de ses premiers travaux dans le *Journal de mathématiques*. On y voyait déjà briller les qualités principales qui le distinguent, la netteté, la précision. *M.* Thomson joint à l'esprit d'invention cette clarté d'exposition si nécessaire dans l'enseignement et que les inventeurs n'ont pas toujours. J'ai applaudi au choix qu'on a fait de lui pour diriger le *Cambridge and Dublin Mathematical Journal*, excellent recueil qui grandira encore entre ses mains. J'éprouverais une satisfaction égale en apprenant sa nomination à l'Université de Glasgow. *M.* William Thomson a parcouru presque toutes les parties des mathématiques. Mais c'est surtout vers les mathématiques appliquées que ses recherches se portent de préférence. Ses travaux sur la théorie de l'attraction, sur celle de la chaleur, et sur celle de l'Électricité, en particulier sa dissertation remarquable au sujet des objections opposées par *M.* Harris et par *M.* Faraday à la théorie de Coulomb, montrent un homme au courant de tout ce que l'on a écrit sur ces matières, et très digne de remplacer parmi vous ce géomètre éminent, si peu connu pendant sa vie, et si digne d'un sort meilleur, l'illustre George Green. Quant à la pratique des expériences, je ne suis pas compétent ; mais je sais que *M.* Thomson s'y est livré avec beaucoup de zèle pendant son séjour à Paris.

Vous jugez assez par ce qui précède combien j'aurais à cœur de pouvoir être utile à *M.* Thomson dans la circonstance importante où il se trouve. J'aime passionnément les sciences, et ceux qui les cultivent avec distinction dans quelque pays que ce soit. Prêtez-moi ici votre assistance si vous connaissez quelques-uns des Juges, si vous pensez que mon humble opinion puisse être près d'eux de quelque poids ; dites-leur, Monsieur, tout ce que je pense ; dites-leur que je regarde *M.* William Thomson comme destiné à figurer dans un rang élevé, au milieu de cette pléiade de savants dont l'Angleterre est fière à si juste titre. Faites ce que vous voudrez de cette lettre, et soyez assuré que notre jeune ami répondra de jour en jour par de nouveaux succès à mes éloges. Votre dévoué Serviteur,

J. LIOUVILLE,

Membre de l'Institut et du bureau des longitudes de France.

à *M.* Armitage,
Trin. Coll., Cambridge.

TOUL, *25e juillet*, 1846.

MON CHER MONSIEUR THOMSON—J'ai lu avec bien du plaisir les beaux théorèmes que vous m'avez envoyés relativement aux moments d'inertie, et qui font si bien suite à celui de *M.* Binet. Ces théorèmes nous montrent par un nouvel exemple comment des théories mécaniques et physiques, de la nature la plus diverse, peuvent en quelque sorte se rencontrer et se réunir sur un terrain géométrique commun. Voilà donc la surface des ondes de Fresnel que vous introduisez dans ces recherches ou figuraient déjà les surfaces isothermes de *M.* Lamé. De tels rapprochements ne sont pas seulement curieux. Ils sont pour la science du plus haut intérêt ; ils servent tout à la fois à l'agrandir et à la simplifier.

Quant au petit théorème dont je vous avais fait part, j'ai été bien content de la démonstration si courte et si nette que vous en donnez, et de l'extension que vous avez su lui faire prendre. Vous avez naturellement songé de suite aux moments d'inertie dont vous veniez de vous occuper ; pour moi, par une raison d'un genre semblable, je ne pensais qu'aux centres de gravité.

Continuez, Monsieur, à travailler comme vous l'avez fait depuis quelques années, et les plus brillants succès couronneront vos efforts. J'éprouverai une vive joie de vous voir ainsi justifier de jour en jour les espérances que j'avais conçues dès vos premiers travaux.

Je désire beaucoup vous voir réussir dans votre candidature à l'Université de Glasgow. La chaire de Physique vous convient de toutes manières ; car c'est surtout vers les mathématiques appliquées à la mécanique et à la Physique que vos goûts vous portent de préférence. Votre manière d'écrire, précise et nette, vos connaissances étendues, vos lectures variées, me prouvent d'ailleurs que vous serez un professeur excellent. Il y a des inventeurs qui ne savent pas être clairs, ou qui, contents d'avoir produit, se refusent à lire ce que font les autres. Vous, aussi, vous avez l'esprit d'invention ; mais vous y joignez les qualités du style, et vous savez vous tenir au courant de la science, même dans les parties sur lesquelles vous n'avez pas spécialement travaillé. Que ne suis-je en position de vous aider dans cette circonstance importante ! Que ne puis-je parler aux Juges et leur dire tout que je pense de vous ! J'ai du moins essayé de le faire en partie par l'intermédiaire d'un de mes amis. Mais sera-t-il à même de vous être utile ? Je l'ignore. Quoiqu'il arrive au reste, gardez-vous bien de vous décourager. Par la nature de votre talent, vous me semblez appelé à réparer la perte que les sciences ont faite dans la personne de votre compatriote, l'illustre Green, dont vous avez su, un des premiers, apprécier tout le mérite, et auquel je tâche en chaque occasion de faire rendre une justice qu'il n'a pas obtenue pendant sa vie. Vous serez plus heureux que lui ; votre avenir est beau, croyez-moi. Votre dévoué Serviteur, J. LIOUVILLE,

Membre de l'Institut et du bureau des longitudes de France.

à *M.* William Thomson,
St. Peter's College, Cambridge.

We have seen that Dr. James Thomson himself aided his son's candidature by sending advance copies of his testimonials to his friends. A side-light on the situation is thrown by a humorous reply sent by one of these, Professor Maconochie, who wrote on June 22 from Bath :—

MY DEAR DOCTOR—Your enclosure found me here this morning. It is plain from the testimonials that your son is, or is shortly to be, a 2nd Newton. He finds a warm advocate in Mrs. Maconochie, but I am much afraid of a father and son in our Faculty meetings. Pray, *what politics* does your William profess—in short, to what party in the State does he belong ? The question has been put to me, to which I can return no answer further than saying that I believe his father to have turned a most pestilent Whig, ready to do anything for Johnny Russell. Now, what party does your son favour ? Is he for Buckingham or Richmond? Or for Stanley, or for Peel, or for Russell, or for the Radicals? Is he a dutiful son, or will he sometimes oppose his father, who is in my opinion generally wrong? Reply at your leisure.

On July 3, his year of "probation" as Fellow having expired, Thomson went over to Ely to be "instituted" by the Bishop, who, learning of his candidature, expressed the hope that he "did not intend to fritter away his time taking pupils here."

The Faculty of the University of Glasgow had, as we have seen, already determined that in appointing their new professor they would reserve the right to make changes as might thereafter appear desirable in the arrangements of the department of Natural Philosophy. When they met on September 11, 1846, to make their choice, they were still in the same cautious mind. They knew that the spirit of advancing science required not only modern appliances, but a reform of methods. In Scotland, as in Germany at that date, the physical sciences were not yet emancipated from the shadow of the scholastics, who in the Middle Ages "rather wore away knowledge by their numerous treatises than increased its weight"; and deductive methods largely overshadowed the value and importance of direct experiment. Rhetoric and vague philosophic argumentation too often did duty in University lectures for sober exposition of reasoned facts; and the natural philosopher even shared the prejudice that startled Helmholtz, when a celebrated professor of Physiology announced during a discussion on vision that a physiologist had nothing to do with experiments. But all this was changing; in Glasgow the times were ripe, and Nichol's advocacy of radical changes, reinforced by his own enthusiastic pursuit of experimental research, met with a willing response in the Faculty. In the minute which they drew up in appointing the new incumbent of the chair, they began by asserting the right to originate or enforce such changes as might seem requisite during the incumbency of the new professor. After this preamble the resolution entered on their books runs thus :—

The Faculty having deliberated on the respective qualifications of the gentlemen who have announced themselves candidates for this chair, and the vote having been taken, it carried unanimously in favour of Mr. William Thomson, B.A., Fellow of St. Peter's College, Cambridge, and formerly a student of this University, who is accordingly declared to be duly elected; and Mr. Thomson being within call appeared in Faculty, and the whole of this minute having been read over to him, he agreed to the resolution of Faculty above recorded and accepted the office.

The next minute runs :—

The Faculty hereby prescribe Mr. Thomson an essay on the subject *De caloris distributione per terrae corpus*, and resolve that his admission be on Tuesday the 13th of October, provided that he shall be found qualified by the meeting, and shall have taken the oath and made the subscriptions which are required by law.

On October 13 the Faculty met and made the following resolution :—

The minute of last meeting having been read, Mr. William Thomson appeared and read a dissertation on the subject prescribed to him by the last meeting of Faculty, viz.: *De caloris distributione per terrae corpus.* The meeting having unanimously expressed their satisfaction with this trial essay, and Mr. Thomson having produced a certificate signed by Walter Crum, J.P., of his having taken the oaths to Government, and Mr. Thomson having further promised to subscribe the formula of the Church of Scotland as required by law on the first convenient opportunity, the following oath was then administered to him, which he then took and subscribed :

"Ego, Gulielmus Thomson, B.A., physicus professor in hac Academia designatus, promitto sancteque polliceor me in munere mihi demandato studiose fideliterque versaturum."

Upon which Mr. Thomson was solemnly admitted and received by all the members present, and took his seat as a member of Faculty.

As to the subject chosen by the Faculty for Thomson's essay on his induction to office—the movement of heat through the body of the earth—it was probably suggested either by Thomson himself, or by his father, as being one which he could readily write out as his dissertation. He had already, in February 1844, printed in the *Cambridge Mathematical Journal*, vol. iv. pp. 67-72, a "Note on some Points in the Theory of Heat." It begins by saying that "in problems relative to the motion of heat in solid bodies, the initial distribution, which is entirely arbitrary, is usually one of the data. When this is the case, and the circumstances in which this body is placed are known, the distribution at any subseqent period is fully determined, and if our analysis had sufficient power would become known in every case. It is in many cases an interesting investigation to examine what this expression becomes when negative values are assigned to the time." This will at once be recognised as the same problem that Thomson had struck in his earliest memoir (see p. 41 above). He went on to consider mathematically three classes of cases according to the convergency or otherwise of the series of terms in the expression. In one of these cases it is possible to assign a finite age to the given initial distribution of heat, in another it will not be possible to assign a limit, in a third the distribution cannot be any stage except a first or initial stage. He gave the mathematical criterion for any such essentially initial distribution. When, in 1882, Thomson included this paper in his volume of reprints (vol. i. art. xi. p. 39), he prefixed to it the following note :—

An application to terrestrial temperature of the principle set forth in the first part of this paper relating to the age of thermal distributions, was made the subject of the author's Inaugural Dissertation on the occasion of his induction to the professorship of Natural Philosophy in the University of Glasgow, in October 1846, *De Motu Caloris per Terrae Corpus*; which, more fully developed afterwards, gave a very decisive limitation to the possible age of the earth as a habitation for living creatures, and proved the untenability of the enormous claims for TIME which, uncurbed by physical science, geologists and biologists had begun to make and to regard as unchallengeable.

Of the Latin dissertation, which was burned by Thomson the same day, only one draft page exists. The handwriting shows that his father had gone over it to correct the composition. The text of it is as follows:—

De Distributione Caloris per Corpus Terrae.

In caloris theoria mathematica, problemata saepe proponuntur in quibus data temperaturae distributione in initio, distributio in quolibet temporis sequentis puncto invenienda est. In plurimis horum problematum quaestio oritur, an possibile sit, ad tempus aliquid datum, invenire distributionem antecedentem ex qua distributio data produceretur per motum liberum caloris. Si hoc fieri potest, inveniendum est maximum temporis intervallum inter distributionem datam et distributionem antecedentem ex qua data derivari potest. Veruntamen distributio data talis esse potest ut intervallum valorem maximum non habere possit. Haec considerationes suggerunt omnes distributiones quae possunt existere in tria genera redigi posse; sc.

i. Distributiones quae e distributione antecedente produci nequeunt.

ii. Distributiones quae produci possunt e distributionibus determinatis existentibus in temporibus datis ante tempus datarum distributionum, intervallis limites certos haud excedentibus. . . .

Thomson had but a month in which to compose his dissertation. Probably this was the first occasion on which he had been required to write a Latin composition since his student days in Ramsay's class. Probably also he found little difficulty in the task; for the training of his youth in *Literae Humaniores* had been thorough, after the Scottish manner. The illiterate science-graduate to whom the classical tongue is anathema is a product of a more recent age. The requirement of a Latin dissertation has been dispensed with—and wisely—in these days;

but, alas! with its disappearance there has come into the memoirs and theses of scientific writers, and even of occupants of university chairs, a neglect of precision in language and a slovenliness of composition that would have horrified the worthy members of Faculty of the year 1846. Of Lord Kelvin, at least, it can be said that his training in classics manifested itself throughout his life in a precision of diction, and a nice appreciation of accuracy in the use of terms, worthy of imitation by all who would follow him in science.

As to the theme of his dissertation: the "Age of the Earth and its Limitation as determined from the Distribution and Movement of Heat within it," this was to be more fully and emphatically set forth in after-years in his discourses on "Secular Cooling of the Earth," and on "Geological Time," which gave rise to the celebrated controversy with the geologists (Chapter XIII.). In the dissertation he suggested, as an application of mathematical principles, "that a perfectly complete geothermic survey would give us data for determining an initial epoch in the problem of terrestrial conduction." This suggestion he amplified at the meeting of the British Association in 1855.

The delight to Professor James Thomson in the fulfilment of his ambition by the election of his son to the chair of Natural Philosophy may well be imagined; and, happily, it has not been left unrecorded, for on the 12th September Dr. David King wrote[1] from Glasgow to his wife:—

[1] *Memoir of the Rev. David King, LL.D.*, p. 246.

> When I came up to Glasgow yesterday I was just in time to receive the joyful news of William's unanimous appointment to the chair of Natural Philosophy! The first announcement I had on the subject was your father's face as he came out of the hall where the election had been conducted. A face more expressive of delight was never witnessed. The emotion was so marked and strong that I only fear it may have done him injury.

After the election Thomson went over to Thornliebank to tell his cousins, Margaret Crum and her brother Alexander, of his success; and then father and son went down to join the family party at Knock, the scene of so many happy summers.

CHAPTER V

THE YOUNG PROFESSOR

On November 1st, 1846, being then aged twenty-two, William Thomson entered upon his duties as professor of Natural Philosophy by reading a carefully prepared introductory lecture on the scope and methods of physical science. The original manuscript of this lecture, much revised in later years, is still in existence, and from this the version now first printed as an Appendix to the present chapter has been taken. The lecturer began by stating a distinction, often emphasised by him in other connections, between the two stages of progress in science, the "Natural History" stage and the "Natural Philosophy" stage. The former is the descriptive or observational stage, in which the discoverer or narrator is busied with observing facts, cataloguing and classifying them, and filling up the gaps revealed in his natural classification. The latter stage is that in which, the facts being already gathered together, the investigator or teacher is concerned with discovering their mutual relations, and strives to bring them within the sweep of general laws or causes. General laws having been attained, they are then to be tested or further generalized by their application to discover fresh facts. The verification of such deductions by experiment and observation establishes the general laws or principles on a firm basis, and entitles them to be called theories. The validity of an induction is to be established by the pragmatical test whether, if worked deductively also, it will lead to new knowledge. These matters he illustrated by reference to Kepler's laws of motion of the planets, and to the verification afforded to the theory of planetary motion by the discovery of Neptune in consequence of the predictions of Adams and Le Verrier. In the subsequent part of the lecture the subject of dynamics was begun, and definitions laid down; stress being laid upon the distinction to be observed between kinematics, the geometry of motion as considered apart from the forces which produce it, and kinetics, or dynamics proper, where forces are considered also. Maxwell used to declare that when Thomson read this lecture its delivery took less than the allotted hour, and that the lecturer was greatly downhearted at its conclusion.

On the first day of every subsequent session, the Natural Philosophy class was opened by the delivery of the same lecture. Although the well-worn manuscript continued to do duty for over fifty years, the matter was not rigidly adhered to. Indeed, only twice, in 1871 and in 1880, was it read through to the end; and in the latter year there was much skipping of parts, with added digressions on the general problem of Dynamics and on Meteorology.

In 1856, and in some other years, he omitted the passage on Astronomy, including a now missing paragraph about the discovery of Neptune by Adams and Le Verrier. In 1862 he added a note about Geology and the Theory of Heat. In 1864 he read nothing after the fifth page, but extemporised about Meteorology and the use of Greek scientific terms. In 1866 nearly all was omitted, and he spoke on the Philosophy of Geology, on Materialism, and on the love of discovery in science. When he came to the passage on evidence of design he remarked that it was "written twenty years before, sometimes omitted, but never felt more strongly than now." In 1875 the address was read for him by Dr. Bottomley. In 1878 he substituted "Dynamics" for "Mechanics" wherever the word occurred. So late as 1895 he added a correction on the fourth page of the manuscript. But if he began by reading from his written discourse, he was sure to break away from it. Some train of thought evoked by one or other of its phrases led the lecturer away into a digression. The new discoveries in science, or some modernized way of stating the old and familiar truths, would send him off into an animated extempore discourse, and the manuscript was forgotten and put aside. There is no denying the fact that as a lecturer he was not methodical, as students expect a university professor to be; or rather, his method was to pour out before his students without restraint the immediate workings of his living thought, with an enthusiasm which, if it did not follow the strict and ordered lines of ordinary text-book teaching, was for those who were able to follow him[1] a veritable inspiration.

The youthful professor—fair-haired, slim, alert, filled with a boyish enthusiasm for experiment, and possessed with an almost feverish passion for submitting everything to calculation—must have presented an unusually attractive figure[2] in the dingy purlieus of the old College. He found the material adjuncts necessary to his teaching very inadequate and antiquated. "When I entered upon the professorship of Natural Philosophy at Glasgow," said Sir William Thomson[3] in 1885, "I found apparatus of a very old-fashioned kind, much of it was more than a hundred years old, little of it was less than fifty years old, and most of it was of worm-eaten mahogany. Still, with such appliances, year after year, students of Natural Philosophy had been brought together and taught. The principles of

[1] "It is, I believe, quite true," wrote Dr. Hutchison, who was a member of his class in early days, "that a vast amount of the abstruse teaching of Sir William never reached the brain of the average student; it might not perhaps be too much to say that some students left the class without picking up anything at all. One unfortunate, for example, who in my time sat on bench 10, has been heard to say, 'Well, I listened to the lectures on the pendulum for a month, and all I know about the pendulum yet is that it wags.' Of course sheer mental indolence far more than the thoroughness and abstruseness of the teaching was here to blame. Sir William unfortunately believed that everybody could learn mathematics."

[2] Writing in March 1909, the late Principal Lang, D.D., of Aberdeen, who entered Glasgow University in 1847, thus speaks of him:—"Nor can I honestly affirm that I greatly profited by the prelections of Professor William Thomson. They were on too high a level for the majority of his students. But William Thomson himself was an interesting study, and there could not fail to be interesting hours in *his* class. In 1850 he was in the vigour of youth, charming, with a face that shone, a figure lithe and graceful, distinction stamped on the personality."

[3] Address delivered at the opening of the Laboratories in University College, Bangor, N. Wales, February 2, 1885; see *Nature*, vol. xxxi. p. 409, Mar. 5, 1885; or *Popular Lectures and Addresses*, vol. ii. p. 475.

Dynamics and Electricity had been well illustrated and well taught, as well as lectures and imperfect apparatus—but apparatus merely of the lecture-illustration kind—could teach. But there was absolutely no provision of any kind for experimental investigation, still less idea, even, for anything like students' practical work." The young professor did not lose much time in making known to the Faculty the pressing needs of his department, and had an immediate grant of £100 to purchase instruments. A Committee was also appointed by the Faculty to see to this expenditure. Its report to the Faculty on November 26, 1847, is as follows:—

Natural Philosophy Class-Room and Instrument Committee.

This Committee having carefully investigated and considered the several applications made by the professor of Natural Philosophy, which formed the subject of the remit made to them at last meeting of Faculty, beg to report:

1. That the alterations on the platform of the class-room proposed by the professor of Natural Philosophy are essentially necessary for the proper exhibition of the various instruments and machines required for demonstration to the students.

2. The Faculty are aware that a purchase of instruments to a considerable amount was necessarily contemplated at the time of Mr. Thomson's election to the chair in question, seeing that this important department of the institution had, from the age and infirmity of the late professor, been almost if not entirely neglected for many years. Accordingly the sum of £100 was last year, and as a mere temporary measure, placed at the disposal of Professor Thomson and your Committee, of which amount £80 only were expended, showing the caution and cere-

mony with which the purchase and selection of the valuable instruments acquired was carried into effect. Your Committee, therefore, submit that a similar amount should be again placed at their disposal.

3. That the present presses and boxes used for keeping instruments pertaining to the Natural Philosophy class-room are, partly from age, and partly from faulty construction, wholly inadequate for their preservation, and, consequently, as the purchase of valuable and delicate instruments without the means of preserving them is a throwing away of money, your Committee submit that immediate steps should be taken towards procuring proper and sufficient accommodation, and of as portable a form as possible, so as to obviate future expense and inconvenience; and your Committee having taken estimates from several tradesmen, beg to report that if a sum of £50 be added to the £100 already mentioned they will be enabled to carry these additions, together with the abovementioned alterations on the class-room, properly into effect, and further, to make such purchases of instruments as may be advisable for this year; and your Committee beg to observe that this sum of £150 ought, in fact, to be considered only as £130, there being £20 of surplus from the amounts voted last year.

Further, your Committee think it proper to add to this report, that though they most emphatically deprecate all idea that such large annual expenditure is to be regularly contemplated for any one department of the College, yet they do not consider the present application by any means excessive considering the present state of the College funds and the very inadequate condition of the department in question, which will yet require further expenditure to put it in anything like fair and creditable order. Lastly, your Committee cannot permit this opportunity to pass without expressing to the Faculty their satisfaction with the reasonable manner in which the professor of Natural Philosophy has on all occasions readily modified his demands in accordance with the economical suggestions of the Committee, who view his ardour and anxiety in the prosecution of his profession with the greatest pleasure, and who heartily concur in those anticipations of his future celebrity which Monsr. Liouville, the French mathematician, has recently thought fit to publish to the scientific world.

ALLAN A. MACONOCHIE, *Convener*.
WILLIAM RAMSAY.

The consideration of the report was postponed till next meeting.

3rd December 1847.—The Faculty resumed consideration of the report of the Committee on the Natural Philosophy class-room and apparatus, and having approved of the terms of that report, agreed to place the sum of £150 at the disposal of the Committee for the purposes specified in their report; and direct the Committee to report from time to time the purchases they have made, in order that the Faculty may have an opportunity of forming a judgment with regard to the manner in which the above sum is expended.

As the months went on the activity of the professor and the exigencies of his department seem to have led to some slight difficulty; for a minute of April 2nd, 1852, reads thus:—

The Faculty agree to the recommendations of the Committee to defray by two instalments the sum of £137 : 6 : 1½d., as the price of purchases of philosophical apparatus already made. They approve of the suggestion of the Committee that without the special interposition of the Faculty's authority the expenditure on this behalf during the next twelve months shall not be allowed to exceed £50; and they desire that the purchases shall be made so far as possible with the previously obtained concurrence of the Committee.

This appointment to the chair did not lessen Thomson's powers of original thought; indeed, his note-book shows many developments to have been simmering in his brain. The following extract reveals the germ of ideas that haunted him to his dying day, and their occurrence at the very hour of his entrance upon formal duties is suggestive.

GLASGOW, *October* 31, 1846, 11.45 P.M.—I have this evening (in the middle of my work finishing an introd$^{y.}$ lecture) after thinking on Faraday's discovery of the effects of magnetism on transparent bodies and polarized light, been recurring to my idea (which occurred to me in the May term) which I had to give up, about magnetism and electricity being capable of representation by the straining of an elastic solid constituted in a peculiar way. I *think* the following must be true:—

If particles along a closed curve of any form be displaced equal ∞^{ly} small distances along the curve, the displacement produced, at any p$^{t.}$ of the medium, can be represented in some way by means of the diff$^{l.}$ coeff$^{ts.}$ of the *solid* angle whose vertex is the p$^{t.}$ and base the closed curve. This solid $\angle$ is the potential due to the action of a volt$^{c.}$ current, circulating in the closed curve, as is known.

Then a bar magnet would be represented by an axis turned round in the elastic solid so as to drag points of the solid round along with it, etc. . . . I am not at all sure of anything I have written just now, but I want to get it out of my head, as I have no time to spare during the session. . . .

November 28, 10.15 P.M.—I have at last succeeded in working out the *mechanico-cinematical* (!) representation of electric, magnetic, and galvanic f$^{ces.}$ I yesterday evening wrote to Cayley the two first, but I have only this moment got out the last case.

Incompressible elastic solid . . . § I. Electric Forces. . . . § II. Magnetic Forces. . . . § III. Galvanic Forces.

[Here follow two pages of equations ending with the equation of continuity.]

November 29, 2.45 A.M.—I have just completed a paper for the *Journal* on this subject.

This paper, which duly appeared in the *Journal* early in 1847, with the title "On a Mechanical Representation of Electric, Magnetic, and Galvanic Forces," was a remarkable attempt to formulate a mechanical theory in terms of the equations of equilibrium of an elastic solid, magnetic forces being represented by angular displacements. The paper concludes with these words:—

I should exceed my present limits were I to enter into a special examination of the states of a solid body representing various problems in electricity, magnetism, and galvanism, which must therefore be reserved for a future occasion.

It remains to add that the "future occasion" did not occur till 1889!

The diary from December 30, 1846, to March 24, 1847, is occupied with diamagnetic problems, the drift of which is to demonstrate the correctness of Faraday's views on the subject.

March 29, 1847.—I have found to-day that H. Jacobi (Taylor's *Scientific Memoirs*, vol. i.) has observed the circumstance which limits the power of an electromagnetic engine.

On December 2, 1846, Thomson was elected a member of the (Royal) Philosophical Society of Glasgow, to the proceedings of which he contributed down to its centenary celebration in 1902. His first communication was in April 1847, when he gave an account of Stirling's Air-Engine, exhibiting a model, and explaining the cycle of operations, concluding with the singular deduction that water at its freezing-point may be converted to ice without the expenditure of any work.

He was elected a Fellow of the Royal Society of Edinburgh on February 1, 1847.

In February, too, Thomson visited J. D. Forbes at Edinburgh University, to witness some experiments with electric light in his lecture-room. These he describes as very brilliant.

In those days the session of a Scottish University lasted for but six months, from November 1st to May 1st, thus leaving the students free in the summer to earn their living, and the professors at liberty for research or recreation as might seem best. For the first few years of his professorship Thomson used to repair to Cambridge for the first six or eight weeks of the summer half-year. To renew his colloquies with Hopkins or Stokes; to make acquaintance with the younger men—Steele, Maxwell, and Tait amongst them—who were working at mathematical physics; to row again in an eight-oar boat, or take his old place as second horn in the University orchestra, were attractions not to be resisted. There were also the meetings of the Cambridge Philosophical Society, and the high mid-summer pomps of the May week in the Colleges. Later in the season he would be preparing for, or attending, the meeting of the British Association. All of this might be varied with the pleasures of a foreign tour with some selected friend. The early autumn would find him back in Scotland, joining some family group in a country resort in Arran, or at Knock Castle, or the Gareloch, or he might be boating on the Clyde, until the closing days of October summoned him back to his duties at the University. To some degree this programme changed as his research and laboratory work at Glasgow grew and became more absorbing.

On his appointment to the professorship in Glasgow he had returned to reside in his father's house, "No. 2 the College," one of the best of the professors' residences in the dingy quadrangle of the old College, with back windows that looked out over the most horrible of slum properties[1] that can be imagined. Professor James Thomson was now sixty years of age, his widowed sister-in-law, Mrs. Gall, being his housekeeper. The eldest son, James, was still living under the parental roof; ill-health for some time proved a barrier to his pursuing his profession as a civil engineer, but he occupied his time profitably with scientific investigation and mechanical contrivances; the work then accomplished as a semi-invalid bore fruit in after-years, for it led up to some of his discoveries and inventions. The daughters, Elizabeth and Anna, now Mrs. King and Mrs. Bottomley, were in homes of their own with healthy,

[1] The late Principal Lang of Aberdeen thus wrote of it :—"In the 'forties of last century the environment was wretched—a congested mass of gloomy tenements, narrow lanes, squalid courts, in which day and night might be heard shrill cries and foul and blasphemous words. One of the worst of these lanes was called the Havannah, and I remember being horrified when I first encountered the miserable specimens of humanity that emerged from its purlieus. But, however bad the neighbourhood was, the buildings themselves had an impressive old-world dignity."

happy children, Elizabeth in Glasgow, Anna in Belfast. The two younger sons, John (who died, alas! too early after a brilliant course as a medical student) and Robert, were still living at home. William's return might well bring life and gaiety into that remarkable household. There were cousins, too, Grahams, Crums, and others connected by family ties, living near and within reach. Mr. Walter Crum, of Thornliebank, had a large family of young people. From their first removal to Glasgow from Belfast the Thomsons had visited at Thornliebank; and the young people as they grew up used to see much of one another at parties and dances, or taking rides together.

Thomson rejoiced greatly when the labours of his first session were ended. To his sister Anna he wrote:—"You need not go to infer that I am an unhappy person, I am really quite the reverse, and nothing but cutting another tooth could make me anything else when I have the prospect of six months' absolute and unmitigated enjoyment before me. I am looking forward even with greater pleasure to Cambridge than to Switzerland, and so you may imagine what a pleasant summer I am to have."

May of 1847 saw Thomson back at Peterhouse, revisiting his friends, examining for his College, playing in the C.U.M.S. concert, and delighted to occupy his old rooms looking out on the Grove. He wrote to his brother James of his doings:—

St. Peter's College,
June 11, 1847.

My dear James—I returned yesterday from London, where I spent a week very pleasantly, and I am now going to settle down as much as I can till after the installation, as I am really anxious to get some papers written that I have had on my mind for a long time. I shall, however, be sadly interrupted by the meeting of the British Association at Oxford which commences on the 23rd. Immediately after the installation, that is to say about 8th July, I shall start for Paris, and after remaining there for a few days I shall go to Switzerland, probably commencing along with Mr. and Mrs. Tom Shedden. On my way to or from Oxford I think I shall be able to see the apparatus of the Royal Institution (where Faraday lectures) in London along with the instrument maker who made it all, and as it seems to be on the best possible scale for a lecture room it will be of great use to me. I heard a lecture of Faraday's on Saturday, and had some conversation with him afterwards. I think I shall probably see a good deal more of him at Oxford, and afterwards, as his lectures end to-morrow, he will have some time to spare. Gordon [1] was away from town during my visit, but I hope to see him again soon. I think it is very likely he will remain in London and give up his professorship.

I heard Jenny Lind, and was as much delighted as possible in every way with her. I also was present at a rehearsal and concert of the Philharmonic Society along with some of my Cambridge friends who were in London at the same time, and I got several lessons on the French horn, and I have bought a horn with pistons, which I did not possess before. . . .—Your affectionate brother,

William Thomson.

P.S.—If possible persuade papa to come to Oxford. I have ! ! ! just heard to-day that Fischer has been unanimously elected ! ! ! at St. Andrews : that is old news, however, I suppose.

[1] Lewis Gordon, first professor of Engineering in the University of Glasgow.

More is told of the trip to London in a letter from J. B. Dykes to one of his sisters.

Last Thursday, Wray, Thomson, and I left Cambridge for London, and in the evening, of course, all went to hear Jenny Lind. . . . The next day we went to the Exhibition, and very much I was pleased. We saw divers other things, and dined at Wray's rooms. . . . After that I went with Thomson to the Royal Institution, Albemarle Street, and he got Faraday to introduce us to his lecture, which took place a little after three. There was a most fashionable attendance, and a most interesting lecture on divers branches of Natural Philosophy—chemistry, to wit, acoustics, electricity, etc. That done we went again to Wray's to dinner, and afterwards went out to divers mathematical instrument makers to look over all the new instruments which are being invented, and to get some for the Glasgow College.

This lecture was the last of a course on Physico-Chemical Philosophy.

On returning to Cambridge, Thomson wrote to Faraday:—

ST. PETER'S COLLEGE, CAMBRIDGE,
June 11, 1847.

MY DEAR SIR—I enclose the paper which I mentioned to you as giving an analogy for the electric and magnetic forces by means of the *strain*, propagated through an elastic solid. What I have written is merely a sketch of the mathematical analogy. I did not venture even to hint at the possibility of making it the foundation of a physical theory of the propagation of electric and magnetic forces, which, if established at all, would express as a necessary result the connection between electrical and magnetic forces, and would show how the purely *statical* phenomena of magnetism may originate either from electricity in motion, or from an inert mass such as a magnet. If such a theory could be discovered, it would also, when taken in connection with the undulatory theory of light, in all probability explain the effect of magnetism on polarized light.

I have only just returned to Cambridge, and so I have not had time to look for the paper to which you kindly sent me a reference. I have lately received a number of a Pisa Journal, containing a paper by Matteucci on electrostatical induction, in which the author seems to endeavour to resolve your results into effects of conduction; and I suppose this is the paper you alluded to in speaking to me. I have not quite been able to learn its contents yet, as I do not understand the language, and it is not enlivened by any x's or y's.

I should consider it a great favour if you would allow your assistant to show me some of the lecture room apparatus belonging to the Royal Institution, when I am in London on my way to or from Oxford, or at any other convenient time. As I am now Professor of Natural Philosophy at Glasgow, and have had a session's experience of the inadequacy of our apparatus, I am anxious to learn as much as I can about good apparatus, and how to procure it.—I remain, yours very truly,

WILLIAM THOMSON.

Prof. Faraday.

On June 20 he writes from Peterhouse to his father about the approaching meeting of the British Association:—

I shall be at Pembroke College, Oxford (care of the Rev. Bartholomew Price, who is one of the secretaries of Section A, and has kindly offered me hospitality), from Wed. the 23rd till Tuesday the 29th, and after that at Peterhouse till about the 10th of July. I am enjoying myself very much here at present as it is remarkably quiet, and I really have for a week found a little time for myself. I have been getting out various interesting pieces of work, along with Stokes, connected with some problems in electricity, fluid motion, etc., that I have been thinking on for years, and I am now seeing my way better than I could ever have done by myself, or with any other person than Stokes.

In July 1847 Thomson attended the meeting of the British Association at Oxford, memorable, as is to be subsequently narrated (p. 264), for his meeting with Joule. At Oxford also he again met Faraday, who gave the second of the evening discourses, his subject being "Magnetic and Diamagnetic Phenomena." After Oxford Thomson went to London on his way to the Continent. From Paris he wrote to his brother James, who was then busy with the water-turbine that he had invented.

PARIS, *July* 22, 1847.

MY DEAR JAMES—[The letter begins with a long and interesting description of water wheels, and reference to M. Poncelet.]

Since Monday I have been as busy as possible (before that I could find nobody and got nothing done), commencing by calling on Le Verrier, to whom I had been introduced at Cambridge by Mrs. Hopkins. He engaged me to go to Mr. Milne-Edwards in the evening, with whom he and Struve were going to dine. I breakfasted with him next morning, and he took me and Struve and a little son of his own to Versailles, where we were for the day; but before I breakfasted with him I saw part of the Cabinet de physique of the Polytechnic, to which he got me admitted, and made engagements with Regnault the préparateur at the Polytechnic. I have been all day, to-day, and a good part of yesterday at Marloye's shop, where I ordered a quantity of acoustical apparatus. M$^{\text{dme.}}$ Dubret is away for two months. I start for Geneva to-morrow at 10.30. W. T.

He stayed a month at Geneva, making acquaintance with the university professors, taking lessons in French, and "trying to get on" with a memoir on electrical images. On August 22 he sent word home that he would leave on the morrow *via* Sallenches for Cluses, from which he would walk to Sixte, there to meet Auguste Balmat, who was to take him on to Chamounix over the Col d'Anterne. On August 29 he was at Chamounix, where he met Joule, with whom he experimented at the cascade of Sallenches. He visited the Great St. Bernard, then up the Valais, walking from Sion to Sierre, and over the Gemmi to the Oberland. On September 12 he wrote from the Faulhorn to Liouville a letter containing some "first ideas and physical especially hydrodynamical demonstrations" of certain theorems on magnetism. These were unfortunately lost, and were not written out again until 1871 for the article entitled "Inverse Problems" (p. 453 of the reprint of Papers on Electrostatics and Magnetism).

In this year he had two mathematical papers in Liouville's *Journal*, and four other short papers of great importance in the *Cambridge and Dublin Mathematical Journal*, besides two which he read at the British Association, on "Electric Images" and "On the Electric Currents by which the Phenomena of Terrestrial Magnetism may be Produced." The latter was a discussion of an ingenious "means of constructing an electro-magnetic model of the earth."

On September 27 he was back at Cambridge, and visited Ireland before November brought the winter session at Glasgow. His diary of November 20 shows him at work :—

To-day (having commenced extra lectures on Theory of Attrn.), in thinking on the best way of putting all investigations on attrn. of spheres into a geometrical form (banishing the notation of the diff. calc., and the *potential* method), I found the following theorem. [There follows a draft of propositions published in the *Journal*.]

In March 1848 he was working at a new absolute thermometric scale with the help of his pupil Steele. After Easter he was again at Cambridge lecturing, examining, and playing in the C.U.M.S. orchestra. In May he wrote to his sister that he was now a Master of Arts and wore the Master's gown. To the Cambridge Philosophical Society he communicated also a paper on an Absolute Thermometric Scale founded on Carnot's Theory, of which more in the next chapter. He also contributed several papers to the *Cambridge and Dublin Mathematical Journal*, including two "Notes on Hydrodynamics," and some articles on the "Theory of Electricity in Equilibrium," already mentioned on p. 142 above. In June he again visited Faraday in London, and on leaving wrote him the following letter :—

WILLIAM THOMSON to FARADAY

BORLEY RECTORY, SUDBURY,
SUFFOLK, *June* 27, 1848.

MY DEAR SIR—Since I had the conversation with you last week I have been reconsidering the subject with some care, and I am quite satisfied that the theoretical views which I then mentioned are correct. I should not, however, expect that it would be at all easy, or perhaps possible, to verify experimentally the result that a needle of a diamagnetic substance tends to arrange itself in the direction of the lines of force, when in a situation where there is no sensible variation of magnetic intensity through the space in which it is free to move, since this tendency arises from the mutual influence of the different portions of the diamagnetic substance itself, and is consequently excessively slight. An experiment showing that a diamagnetic needle will be sensibly astatic when the intensity in its neighbourhood is nearly constant would be extremely interesting, as so far confirming the conclusions deduced from theory; but I fear that a complete verification would be unattainable on account of the excessive feebleness of the force of which the existence is to be tested.

The same process of mathematical reasoning enables us to infer that a needle of soft iron and a needle of a diamagnetic substance would both rest stably in the direction of the lines of force, or unstably in a perpendicular direction; but in the former case the directive tendency is extremely sensible, so much so that it may be easily verified by observing the position which a delicately suspended needle of soft iron will assume when acted upon only by the earth.

I have not yet had an opportunity of referring to Poggendorff to find the account of the remarkable researches which you mentioned to me, but as soon as I return to Cambridge I shall read it with great interest. . . .—I remain, dear sir, very truly yours,

WILLIAM THOMSON.

From Scotland, in August, he sent to the British Association meeting at Swansea two papers. One of these was on the "Equilibrium of Magnetic or Diamagnetic Bodies under the Influence of Magnetic Force," the other on the "Theory of Electro-Magnetic Induction." The first of these was a brief abstract of a longer research afterwards published by the Royal Society, the second an expansion of Neumann's mathematical statement of Faraday's law.

In October Thomson's elder sister Mrs. King, being in weak health, was ordered to Jamaica. William Thomson accompanied her and Dr. King on the steamer as far as Ailsa, returning with the tug to Greenock.

The winter of 1848-49 was marked by a visitation of cholera, from which Glasgow with its insanitary slums did not escape. Its ravages spread far, and amongst its victims was Professor James Thomson. "I have sad intelligence to give you," William wrote to his brother-in-law Dr. King on January 12, 1849. "We have lost our father. He died this afternoon at two o'clock. . . . On the Sunday night he became delirious, and since that time he has been gradually sinking in strength. I could not believe last night at this time that we were to lose him. . . . But God has willed it for the best, and has tried us with a heavy affliction. . . . It is a terrible and irreparable loss, and a sad void is now left."

After this sorrowful event Thomson continued to reside at No. 2, The College, his aunt Mrs. Gall keeping house for him.

A glimpse of the young professor is afforded by the autobiographical notes of Professor John Nichol in a passage written in 1861:—

During my first session (1848-49) I attended the experimental course of the Natural Philosophy, taught by the young Mr., now the famous Dr. William Thomson.

The lectures I heard were on electricity and magnetism. I took careful notes, read, thought, and made experiments on subjects which interested me intensely. I was regularly examined during the course, and gave in my name for a series of special examinations at the end. The result was that I got one of the two prizes which were offered at the close of the year, and made my first appearance on the platform on the 1st of May, under the auspices of my old playfellow's brother. I remember his handing me the prize with the phrase, "A very young, but very ardent, natural philosopher."

Thomson had big things in hand. On January 2, 1849, he had read to the Royal Society of Edinburgh a paper on "Carnot's Theory of Heat," the significance of which is discussed in the chapter on Thermodynamics (p. 269). He was struggling to clear Carnot's acute reasoning from the assumptions which underlay it, assumptions which had been undermined by the investigations of Joule. To-day this paper is of interest principally so far as it reveals Thomson's mind in this transition stage of the doctrines as to the nature of heat.

In March he was with Forbes in Edinburgh discussing underground temperatures. In April he filled six pages of his mathematical note-book with a discussion of the action of pointed conductors in promoting electric discharges, intending to communicate it to the British Association in July. It has never been published, however.

In June Thomson presented to the Royal Society of London (his first communication to that august body), a very elaborate memoir, an abstract of which appeared in the Proceedings for June 1849, entitled "A Mathematical Theory of Magnetism." Important as his earlier papers in mathematical physics had been, they were for the most part brief and disconnected. The most methodical of them were those on the mathematical theory of electricity which he had written apparently because he could find no published work in which the principles were stated (to use his own words) "in a sufficiently concise and correct form, independently of any hypothesis, to be altogether satisfactory in the present state of science." He now set himself to perform a like service for the theory of magnetism. Poisson had, indeed, following the lines of Lagrange's *Mécanique analytique*, published in 1821 an analytical theory of magnetism; but it was based on the hypothetical existence of magnetic fluids. Moreover, during the years 1845-48 Faraday had announced new magnetic discoveries of the most fundamental kind—the action of magnetism on polarized light, and the diamagnetic repulsion of bismuth and other metals. Also Plücker had investigated the relations between magnetism and the crystalline structure of bodies. Thomson had, therefore, plenty of experimental material on which to exercise his thoughts. He had listened to Faraday discoursing at Oxford in 1847 on magnetic and diamagnetic phenomena. Later, when on visits to London, he seldom failed to call at the Royal Institution to see Faraday in his laboratory, and discuss with him[1] the latest

[1] See pp. 203 and 207 above, for the letters of this date.

investigations. He was now to reduce to precise mathematical form—in terms of the analysis which he had learned from Green and from the great French mathematicians—the theory of magnetism freed from all unnecessary hypothetical assumptions. So with an obvious sense of the importance of the task he was undertaking, he set himself to formulate the theory with a systematic and methodical exposition hardly attempted in any previous paper, and certainly never afterwards repeated. Whereas in his later life most of his original communications to the learned societies were rough drafts, usually in pencil, sometimes indeed copied out by an amanuensis, but often consisting of a few leaves torn from his note-book, in the present instance the manuscript was written in a bold and almost stately hand on large quarto pages, bearing a sense of importance. As Thomson was not yet a Fellow, it was formally communicated to the Society by Lieut.-Col. Sabine, then Foreign Secretary (later President) of the Society.

The memoir begins by explaining how, since the notion of magnetic fluids had become excessively improbable in the light of recent discoveries, the theory of magnetism may now be established upon the sole foundation of facts generally known, and the experimental researches of Coulomb in especial. The first chapter was devoted to preliminary definitions and explanations. In the second chapter a method of specifying the intensity and direction of the magnetization was formulated. In the third chapter, by a method of investigation akin to that used by Fourier in constructing the equation of motion of heat in a conducting body, a plan was found for representing the polarity of a magnet by the distribution of imaginary magnetic matter, leading to formulae which agree with Poisson's, and to expressions in chapter four, analogous to those of Laplace and of Green, for the mutual actions between any two finite portions of magnetized matter. This opened out the prospect of a further investigation[1] of the "mechanical value" of any given distribution of magnetism. The fifth chapter, communicated in June of 1850, dealt with two ways of regarding magnetic distributions. In the first of them the magnet was considered as made up of filamentary magnets laid side by side, so that the poles of all the filaments constituted the surface-distribution of the magnetism. In the second it was regarded as represented by equivalent electrical currents circulating around the polar areas. Thomson assigned the names of "solenoidal" and "lamellar" respectively to these two distributions, the properties of which were then investigated. A sixth chapter, "On Electromagnets" (using that term in its most general sense as including any circuit traversed by an electric current), was written in 1849, but not published[2] till 1871, when also chapters seven ("On Mechanical Values"), eight ("Hydrokinetic Analogy"), and nine ("Inverse Problems"), of

[1] See the article, dated January 1872, in *Electrostatics and Magnetism*, p. 432.

[2] *Electrostatics and Magnetism*, §§ 524-554.

which last a sketch had been given at Oxford in 1847. Chapter ten ("Magnetic Induction"), which had been originally planned as Part II. of this memoir, was for some reason not communicated to the Royal Society, but published in the *Philosophical Magazine* in March 1851. Here, again, Thomson's aim was to reconstruct the theory of the magnetism received by bodies during induction, without resorting to any hypotheses such as that of fluids, and to apply it to cases where the crystalline nature of the body might give it different magnetic susceptibilities in different directions. This article is notable for its precision of definition and its announcement of the principle of superposition. In it Thomson brought out the notion of a magnetic inductive capacity in the material. The name "susceptibility," which is now preferred, and the other term "permeability," for the "conducting power for lines of magnetic force," were not coined by him till 1872.

The importance of this memoir was at once recognized. It was accorded full publication in the *Philosophical Transactions*, vol. cxli., 1851, Part I., pp. 243-286.

Thomson wrote to Faraday supporting his views on diamagnetic forces:—

32 DUKE STREET, ST. JAMES',
Saturday, June 19.

MY DEAR SIR—After our conversation to-day I have been thinking again on the subject of a bar of diamagnetic non-crystalline substance, in a field of magnetic force which is naturally uniform, and I believe I can now show you that your views lead to the conclusion I had arrived at otherwise, that such a bar, capable of turning round an axis, would be set stably with its length along the lines of force. As I may not have another opportunity of seeing you again before you leave town, I hope you will excuse my continuing our conversation by writing a few lines on the subject.

Let the diagram represent a field of force naturally uniform, but influenced by the presence of a ball of diamagnetic substance. It is clear that in the localities A and B the lines of force will be less densely arranged, and in the localities D and C they will be more densely arranged than in the undisturbed field. Hence a second ball placed at A or at B would meet and disturb fewer lines than if the first ball were removed; but a second ball placed at D or C would meet and disturb more lines of force than if the first ball were removed. It follows that two equal balls of diamagnetic substance would produce more disturbance on the lines of force of the field if the line joining their centres is perpendicular to the lines of force than if it is parallel to them. But the disturbance produced by a diamagnetic substance is an effect of worse "conducting power," and the less of such an effect the better. Hence two balls of diamagnetic substance, fixed to one another by an unmagnetic framework, would, if placed obliquely and allowed to turn freely round an axis, set with the line joining their centres *along* the lines of force.

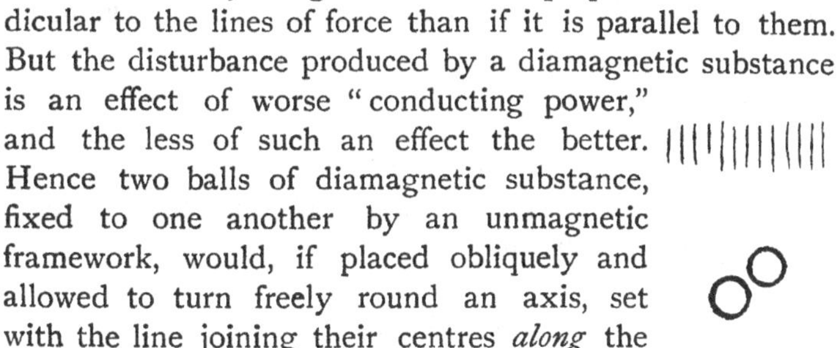

The same argument, for a contrary reason, shows that two balls of soft iron similarly arranged would set with the line joining their centres also *along* the lines of force. For this position more disturbance is produced on the lines of force than in any other, but now the *more* disturbance (being of better "conduction") the better. Hence the conclusion. Of course similar conclusions follow for bars, or elongated masses, of the substances.

I can never, however, make any assertion regarding the tendency of a diamagnetic bar in a uniform field without repeating that I believe no experiments can make it sensible. I doubt even whether the corresponding tendency in the case of a neutral bar in as strong a solution of sulphate of iron as could be got could be rendered sensible by actual experiments, as excessively slight deviations from uniformity in the field would entirely mask the results of this tendency, even if by themselves they might produce appreciable effects.—I remain, my dear sir, yours very truly, WILLIAM THOMSON.

M. Faraday, Esq.

He followed this with a second letter:—

9 BARTON STREET, WESTMINSTER,
July 24, 1849.

MY DEAR SIR—In the conversation which we had about the beginning of this month I mentioned several objects of experimental research which occurred to me as of much importance with reference to a theory of Diamagnetic, and, still more, of Magnecrystallic, action. I now take the liberty of addressing to you a few memoranda on the subject.

1. If a ball, cut out of a crystal of bismuth, be placed so as to be repelled by a magnet, will the repulsion not be stronger when the magnecrystallic axis is held perpendicular to the lines of force than when it is held in the direction of these lines. (Reference to §2552 of your Researches.)

2. It would be a valuable acquisition to our experimental elements if a ball cut from a crystal of bismuth were suspended in the manner described by you in §2551, and experiments were made by varying the length of the lever, and altering the general disposition so as to perceive cases in w[h] the tendency to move, due to the

repulsive action, might be exactly balanced by a tendency to move in the contrary direction arising from the magne-crystallic action. A sketch, with dimensions, of the arrangements in any such case of equilibrium would be most valuable.

3. In such a case as the preceding, if the strength of the magnet (a pure electro-magnet without soft iron would be the most satisfactory kind for such an experiment) be increased or diminished, will the equilibrium remain undisturbed?

4. Is the repulsion on a non-crystalline or crystalline diamagnetic ball or the attraction on a ferro-magnetic ball exactly proportional to the square of the strength of the magnet? Thus in any case of pure repulsion, or of pure attraction, if the strength of the magnet be doubled, would the force be quadrupled; if the strength of the magnet be increased threefold would the force be increased ninefold? In this investigation, as in the preceding, a pure electro-magnet would be the best, since in such a magnet the strength may be altered in any ratio, which ratio may be measured with much precision by a torsion galvanometer, while the character and form of the lines of force remains absolutely invariable.

5. How are crystals of magnetic iron ore related to other crystals in their magnetic properties? Are they intrinsically polar, or are they merely axial? For example, if, supposing that to be possible, a crystal of magnetic iron ore have its polarity reversed, will it remain permanently magnetized in this reverse way? or, if a crystal of magnetic iron be demagnetized, will it remain non-magnetic? Will it not, in virtue of an intrinsic tendency to magnetization, gradually become magnetized in its original way?

I have a small ball of loadstone from the island of Elba which I have employed in place of a needle in a torsion galvanometer, and which appears to be susceptible of inductive action like soft iron (returning apparently to its primitive magnetic state when the inducing magnet is removed), and to be susceptible of this action to a greater degree when its axis is along the lines of inducing force than when it is perpendicular to them. My means of experimenting are, however, so very limited that I cannot be confident with reference to any such conclusions.

My intended departure for Norway, of w[h] I spoke to you, has been necessarily delayed for a fortnight by important and unexpected business. I hope to-morrow, however, to be on my way to Copenhagen by steamer.

I hope you are at present enjoying a pleasant and refreshing tour, as I heard to-day at the Royal Institution that you are travelling.

Believe me, my dear sir, yours very truly,

WILLIAM THOMSON.

After the reading of this memoir to the Royal Society on June 21st, 1849, Thomson remained for some time in England. He went down in July to Plymouth to visit Sir William Snow Harris, and to examine Cavendish's then unpublished manuscripts on electricity. Apparently he did not attend the meeting of the British Association, held that year at Birmingham, but took a tour in Scandinavia with his former tutor Frederick Fuller, and Norman Macpherson, afterwards Professor in Edinburgh. They reached Christiania about the middle of August. Thomson writes that he has met Ole Bull, who had played his violin to them in his own rooms: he has also met Langberg, Professor of Natural Philosophy, who, being also a director of the theatre, has to go every night for a week to superintend the lighting of the lamps! They bought three carioles, with which they travelled through Southern Norway, visiting Drammen and Konksberg, and eventually crossing into Sweden. They met Professors Malmstén and Svenberg in Upsala. From Stockholm, which he describes as "by far the most beautiful city I have ever seen," on September 20th, Thomson dispatched to the *Mathematical Journal* one of his geometrical investigations on the equilibrium of electricity in a sphere electrified by influence. On returning he visited his sister Mrs. Bottomley, at Belfast, before settling down for the winter session in Glasgow. An account of his doings was sent by him thence to Mrs. King.

FORTBREDA, BELFAST, *Oct.* 15, 1849.

I shall give you a short history of myself from the 1st of May to the present time.

I suppose you know that I went to Cambridge almost immediately after the distribution of prizes. The only delay which I suffered was a short visit of about two days to Meadowbank, which I enjoyed very much, although I was impatient to get to Cambridge. I saw James and got dinner from him in his lodgings on my way through London, and got to Cambridge about the 4th of May. I found all my friends there very well, and met with a very kind reception. The "May term" is usually very gay at Cambridge, and my time was a good deal taken up with the University Musical Society, for which I played, as of old, in two concerts, and with other similar engagements. I had my friend and former pupil Pollock staying with me in my rooms for about a fortnight. I had also a short visit from William Ker, who spent a Saturday afternoon, Sunday, and Monday morning with me. . . . Mr. and Mrs. Maconochie[1] also visited Cambridge and spent ten days there in lodgings which I had taken for them in Fitzwilliam Street. They were constantly with me and Mr. and Mrs. Hopkins, and the whole family were much

[1] Professor Allan A. Maconochie, Professor of Jurisprudence in Glasgow University, 1842-1855.

pleased with them, and were very glad to show them a great deal of attention.

Cambridge, as you know, always becomes deserted immediately after the College examinations which take place about the end of May. I had undertaken a good deal of work in the way of "setting" examination papers and conducting the examinations, and I had besides, in the first half of the month, to complete a paper on "The Motive Power of Heat" for the Edinburgh Royal Society, so that my time was not entirely spent on music and gaiety. When the quiet time commenced I began my paper on "Magnetism," and worked hard at it till the 20th of June at Cambridge. . . .

After this digression I must return to London on the 20th of June, where I arrived and took up my quarters in James's lodgings. I was very glad to be with him again, as I had seen him very little since our father's death on account of his leaving Glasgow so soon. I remained there till the 25th of July, with the exception of three or four days spent very pleasantly at Plymouth on a visit to Sir W. Snow Harris, the "electrician," and two very brief visits to Cambridge. One of these was to make arrangements with Fuller about the Scandinavian tour which he intended to make, and in which he persuaded me to join him. He went away early in July, but I could not go with him on account of a piece of business connected with a small Government grant of money for my apparatus room, with which, along with Mr. Maconochie, I was occupied in London. . . . I left London on the morning of Wednesday the 25th in the steamer *Camilla*, for Copenhagen, overtook Fuller at Christiania, had a refreshing and in every way delightful tour with him in Norway and Sweden, returned by Hamburg, where we stayed three days. . . . I arrived in London on Monday morning October 1st, got all I had to do over as quickly as possible, and went to Cambridge on Tuesday night. There I saw my friends, and was as usual most kindly received. I also had several College friends to see, and a great deal to say to all. I got my business executed with the utmost diligence, and went on Friday at 11 h. 13 min. by the night train to Hull. Next day I went to Goodmanham Rectory to see Blow, who is curate to his father. . . .

I arrived here last Friday morning and found all well. . . .—Your ever affectionate brother,

WILLIAM THOMSON.

In this year he contributed two more Notes to the *Cambridge and Dublin Mathematical Journal.*

The year 1850 witnessed James Thomson's singular prediction of the lowering by pressure of the melting-point of ice, and William Thomson's experimental verification of his brother's prediction. William wrote to Faraday, on January 10th and January 14th, two letters describing his experiments. On January 31st he wrote to Mrs. King:—

My class is keeping me very busy, and besides, for some weeks back I have been taking advantage of the abundance of ice (it is now over, however, I hope) to make some very curious experiments, by which I have been able to verify a discovery made by James last year, and published by the Royal Society of Edinburgh, that the freezing temperature of water is lowered by pressure. The effect, however, is very slight, so that about ten times the atmospheric pressure upon the ice and water makes the mixture only about a quarter of a degree colder.

The letter went on to speak of his devising with James a new kind of heating apparatus for houses. He had also been giving musical parties at his house at the College. In June he read to the Royal Society the second portion of his great memoir on Magnetism.

He had more Notes in the *Cambridge and*

Dublin Mathematical Journal, two of them electrical, and two in the *Philosophical Magazine*, one of which was a letter to Joule,[1] announcing his adhesion to Joule's views on the mutual convertibility of heat and work. The other was on Magnetism in Non-crystalline Substances, and was dated from the Gareloch in August. To the British Association at Edinburgh he communicated a paper dealing with magnetism in crystalline substances. The substance of this was embodied in the later part of the "Theory of Magnetism." The diary returns to the subject:—

ROW, GARELOCH, *Wed. Aug.* 14, 1850, 11.15 P.M.—I have just finished a letter (of 18 pp. note paper) to Tyndall (whom I met at the British Assoc. meeting at Edinb. last Thursday and Friday week). . . . In the letter (by way of answer to Tyndall's of Aug. 7) I gave a hasty account of Poisson's conjecture regarding crystals, and of his general theory, stating my conviction that his hypothesis is false; I made some statements of my own (positive) views regarding the distinctive action that a row of balls of strong or weak ferromag^{c} subs^{ce}, or of diamag^{c} subs^{ce} would experience if placed in a uniform field of force; mentioned my conjecture regarding the quasi-crystalline structure induced in the dough itself, or the gum water in drying in Mess$^{rs.}$ Tyndall and Knoblauch's experiments on the cakes containing powder of bismuth, etc.; and that this occurred to me in consequence of having had my attention directed to young Clerk Maxwell's (now of Peterhouse) optical experiments on isinglass dried under constraint, which analogy would lead us to suppose should have magnecrystallic properties; that I have asked C. M. to make some preparations; that I do not anticipate obtaining any results by my own experi-

1 See p. 278 below.

menting, but would be glad if Messrs. T. & K. would make some experiments according to suggestions which I promise to make if wanted.

The next entry has an interest of another kind.

Thursday, Aug. 15, 5 P.M.—I have just written to Rankine telling him of Clausius' paper in Poggendorff. . . .

Then he began, on the receipt of some data about underground temperatures, to formulate the equations of heat conduction, a work which he still continued in October when he revisited Paris. Thence he wrote to his sister, Mrs. King, then in Torquay:—

HOTEL CORNEILLE, PLACE DE L'ODÉON,
PARIS, *Oct.* 6, 1850.

MY DEAR ELIZABETH—I suppose by this time you are settled, and I hope comfortably and pleasantly at Torquay. Not knowing your address, I am going to let this take its chance of finding you at Torquay, which I daresay I am quite safe in doing.

I arrived here on Thursday evening, having been dreadfully detained on the road by the difficulty of getting a steamer from Brighton to France. I went direct to Brighton from Glasgow, only stopping for an hour or two in London, and I spent the Sunday with Ellis, intending to leave on Monday for France. I found, however, there was no steamer from any tolerably near port for France till Tuesday morning. So I had to wait, and on Tuesday I set out by rail at 8.30 for Newhaven, and when I arrived there the steamer had not made her appearance, having been unable to leave Dieppe the day before on account of the storm. So I had a whole day to wait, and at last was on the point of demanding back my fare, when the steamer appeared about 4 P.M.—contrary to the expectation of the nautical wise men about the harbour. She could not sail again, or at least need not, on account of the tide at Dieppe, till 9 P.M., so I took a walk and saw Seaford Cliff, which had recently been blown up by gunpowder and galvanism (see the newspapers). I had rather a rough but not disagreeable passage to Dieppe, during which I slept very well till we arrived at the mouth of the harbour, where we had to lie for an hour and a half for water, and I was a little sea-sick. I spent twenty-five hours at Rouen, and refreshed my memory with a sight of St. Ouen, the Cathedral, and St. Vincent, and saw besides many other churches which we had not seen before, one of which, dating from the third century, was very interesting.

I find none of my scientific friends in Paris, but Regnault comes in from time to time, and I shall see him to-morrow. I have got a good deal done regarding my instruments, and have been enjoying myself very much. I was at the English church to-day, where the sacrament was administered to a very large congregation.—Your affectionate brother, WILLIAM THOMSON.

On this visit he went with the Abbé Moigno to the workshop of Duboscq to see some striking and beautiful experiments, due to Foucault[1] on the production of spectra of the metals by the electric arc.

A prism and lenses were arranged to throw upon a screen an approximately pure spectrum of a vertical electric arc between charcoal poles of a powerful battery, the lower one of which was hollowed like a cup. When pieces of copper and pieces of zinc were separately thrown into the cup, the spectrum exhibited, in perfectly definite positions, magnificent well-marked bands of different colours characteristic of two metals. When a piece of brass, compounded of copper and zinc, was put into the cup, the spectrum showed all the bands, each precisely in the place in which it had been seen when one metal or the other had been used separately.

[1] See *Popular Lectures and Addresses*, ii. p. 172.

After this there is a long gap in the mathematical diary, until April 29, 1851, when he is excited by a communication from Joule about the dryness of steam issuing from an orifice.

On March 2, 1851, he writes to Mrs. King from Glasgow :—

I have been riding a good deal of late, and going some long distances on days when I have had time. I yesterday had a ride of twenty-four miles with Alexander Crum, his two sisters Margaret and Mary Gray, and Miss Couper (Dr. John Couper's daughter). A. Crum and I rode out with Miss Couper, and we took up the others at Thornliebank, and rode about six miles further till we got to a very wild pastoral sort of district, high, and giving splendid views of the valley of the Clyde and the hills beyond it, the air being very clear. I have engaged a horse for two months (only one, if I please to break at the end of the first), a very good one both for long distances, speed, and general zeal in jumping or anything that may be wanted.

Such relaxations did not, however, interfere with Thomson's activities. He was now in full cry in the subject of Thermodynamics.

Great work was done by Thomson in this year. In March he read to the Edinburgh Royal Society the classical memoir "On the Dynamical Theory of Heat," as developed from the standpoint of Joule's experiments, which is analysed in the chapter on Thermodynamics (see p. 280, below).

In April he gave to the Royal Society of Edinburgh a paper in which he suggested a method of investigating the mechanical equivalent of heat by forcing air through a porous plug.

In 1851 Thomson was proposed as a candidate for the Fellowship of the Royal Society. The certificate of candidature, which was signed by Faraday, Hopkins, Whewell, Sedgwick, Willis, Adams, Booth, Lyon Playfair, Wheatstone, Graves, and Baden Powell, describes him as the "author of several Mathematical papers in the *Cambridge and Dublin Mathematical Journal*, of which Journal he is the editor, and of two papers on the Mathematical Theory of Terrestrial Magnetism ordered to be printed in the *Philosophical Transactions*." The certificate is in the handwriting of Hopkins. The proposal of candidature was read to the Society on January 9, 1851, and he was elected a Fellow on June 6, of the same year, Huxley and Stokes being elected Fellows on the same day. His formal admission did not take place till June 10, 1852, when he signed the Charter Book.

He was absent from England in the summer of 1851 for nearly two months, visiting Regnault in Paris, and then staying with the Blackburns in Florence, where he spent much of his time in writing out for Regnault an account of the dynamical theory of heat. He went on to Venice and Verona, returning to England on August 11.

Ipswich was the meeting-place of the British Association in 1851, but Thomson does not appear to have been present; in the Presidential Address of Sir George Airy there is, however, a reference to Thomson's magnetic investigations.

From the end of August to the end of October,

some twenty pages of the note-book are devoted to Laplace's functions, to a polemic with Clausius, to the working out for Joule of the values, in absolute measure, of electro-chemical equivalents, and to the "mechanical values" of transient electric currents and of currents in a coil.

In November he communicated two papers to the *Philosophical Magazine*. One of these was on the application of the principle of "Mechanical Effect to the Measurement of electromotive Forces, and of Galvanic Resistance in absolute Units." He is here applying thermodynamic principles to the actions in an electric circuit, and to the measurement in terms of mechanical units of the chemical actions of the battery, but, in particular, he is advocating and extending the use of "absolute units" as introduced by Weber, which he now restates in British measure, with the foot, grain, and second as the fundamental units of length, mass, and time. He gives a table of measurements of resistances by different workers, showing considerable discrepancies, and concludes by advocating determinations of the absolute resistance of the same conductor by the direct method of Weber, and by the indirect method of calculating it by Joule's equivalent from the heat developed in it by a current. His other article, "On the Mechanical Theory of Electrolysis." This is an extraordinary discussion, by the light of Joule's discoveries, of the absolute value of the electromotive forces concerned in electrolytic processes. The argument is based upon a hypothetical machine or model generating electromotive forces by the revolution of a copper disc in a magnetic field, on Faraday's plan. The electromotive force generated would be, as he shows, proportional in the angular speed, and to the square of the radius. He even gives the expression for its efficiency. At the end he considers the globe of the earth,—a revolving magnet,—and calculates its inductive effect on a hypothetical conductor held above it, which he estimates as about 204,000 times as great as that of one Daniell's cell. He concluded that if the space surrounding the earth were capable of conducting electricity, and were affected as a fixed conductor by the motion of a magnet in its neighbourhood, there would be electric currents induced; and he added, "It is, I think, far from improbable that the phenomena of aurora borealis and australis are so produced." It is not known whether in later life Lord Kelvin still held the same view.

Later in the year he sent to the Edinburgh Royal Society an "Appendix on a Mechanical Theory of Thermo-electric Currents." Seebeck had discovered in 1822 that a current can be created and maintained by heating that point in an electric circuit where there is a junction between two different metals. Peltier, a few years later, had discovered that if a current from a battery is caused to traverse such a junction, the surface of contact between the two metals is the seat of an evolution or of an absorption of heat according to the direction of the current through the junction. Applying to these known facts the thermodynamic principles of Carnot and of Joule, Thomson came to the remarkable conclusion that there must be a third, and hitherto unknown phenomenon; that an electric current must produce in one and the same metal, different thermal effects according as it passes from cold to hot or from hot to cold. He pointed out several lines of experimental research, in which he suggests measurements of electric quantities in absolute units, and incidentally mentions that he has found the specific electric resistance of different specimens of copper to differ considerably from one another. Incidentally, also, he suggests an experiment of driving an electric motor ("a galvanic engine") from a thermal battery, stating the problem in terms which show that he already had a correct and adequate grasp of the theory of the efficiency of electric motors.

In the spring of 1852 Thomson sent to the Royal Society a note on a paper by Joule on the Air-engine. To the Royal Society of Edinburgh he read a paper on the "Mechanical Action of radiant Heat or Light, on the Power of animated Creatures over Matter, and on the Sources available to Man for the production of Mechanical Effects." This was in part a commentary on Helmholtz's *Die Erhaltung der Kraft*, which he had but just read. The published text is only an abstract. It shows that Thomson was thinking of the philosophical question of the origin of vital forces. He answers in the negative the question whether animated creatures can set matter in motion in virtue of an inherent power of producing a mechanical effect. Both the heat they generate and the work they perform demand an equivalent consumption of energy in the chemical processes of the animal body. The body does not act as a thermodynamic engine; and very probably the chemical forces produce the external mechanical effects through electrical means. But, whatever the nature of these means, they are to some extent subject in every individual to the direction of his will. Will power can direct motions which are due to chemical forces. Of the sources of mechanical effect available to man the principle is the heat radiated from the sun.

On April 19, Thomson communicated to the Edinburgh Royal Society his famous discovery of a Universal Tendency in Nature to the Dissipation of Mechanical Energy, the complement to the Theory of the Conservation of Energy enunciated by Helmholtz. This brief but pregnant paper is described below in the chapter on Thermodynamics (see p. 287).

ST. PETER'S COLLEGE,
CAMBRIDGE, *June* 7, 1852.

MY DEAR ELIZABETH—I have been anxious to hear from you for a long time, but I could scarcely expect a letter without having first done my duty in writing to you. I have been on the point of writing for a long time, and on Saturday I had actually commenced a letter, but was not able to go on on account of interruptions. I wrote an account of our gaieties in Peterhouse, including ladies dining in Hall, an evening party, and an extempore dance and a great deal of good music in the rooms of one of our junior fellows, besides a regular concert of vocal music on another evening in the same rooms, to Aunt, which shows that I have been idle enough since last Tuesday when the College examination was concluded. But one has not always most time to write letters, or to read or do anything requiring a little quiet time, when one is most idle, and so I have got more and more behind in writing letters and other good things. I am going to London to-day when I do hope I shall get a little regular work done every day.

I spent ten days at Manchester very pleasantly and profitably with Joule making experiments, and I came after that to Cambridge where I have been since, that is a fortnight ago to-morrow. I found Blow here, to my surprise, when I arrived, he having been persuaded to come up to play at the 50th concert of the University Musical Society, which was established in Peterhouse when we were undergraduates, and is now a most flourishing and stable institution.

I have had altogether an extremely pleasant time here, seeing many old friends, and enjoying this pleasantest time for Cambridge in general and our Peterhouse Grove and gardens most particularly. I had once more a pull in an eight-oar, for the first time since seven years ago, and I have been bathing at "Sheep's Green" again, generally early in the morning.

Peterhouse has earned great credit by two students from Scotland, one[1] of them, Steele, being a former pupil of Papa's and mine, who were senior and second wrangler last February. They were both here for some time after I arrived, and one will in a few weeks be a Fellow of the College.—I remain, your affectionate brother,

WILLIAM THOMSON.

After London he returned to Glasgow.

Allusion has been made to the relations between the families of the Thomsons and of the Crums of Thornliebank; and that relation was now to

[1] The other—the senior—was Peter Guthrie Tait, afterwards for so many years associated with Thomson.

be strengthened by a much more intimate tie. Thomson was now twenty-eight years old; he was becoming famous, and he held an assured position as University professor. Margaret Crum, whom he had known from boyhood, was an accomplished and beautiful girl of twenty-two. Well educated, well read, possessed of a lively imagination and a poetic fancy, she was also of a deeply religious nature and of great earnestness of purpose. Thomson might well be attracted by so rare and sweet a character as hers; and the attraction, now that he had so many opportunities of meeting her, ripened into a deep attachment. The summer of 1852 witnessed the culmination of his hopes.

2, COLLEGE,
Tuesday night, July 13, 1852.

MY DEAR ELIZABETH—I have a piece of news to tell you which I think will please you as much as it will surprise you. Margaret Crum has consented to be mine, and from to-day we are engaged to be married. I came up from Largs in the afternoon, and rode out to Thornliebank to tell Mr. and Mrs. Crum and ask their consent, which was given with the greatest goodwill. I cannot tell you any more at present, as I am writing this and two or three other letters in the middle of the night, to get them posted immediately, so that they may leave by the early mails. I shall tell you more when I see you, which I hope will be on Friday. If I can I shall let you know my plans beforehand, but whether you hear from me or not, I shall probably be with you on Friday evening. I am going down to Largs to-morrow, and I think I shall be there a good deal till Anna[1] comes. Write to me addressing to Mrs. Arthur's, White Hart Inn, Largs.

[1] Mrs. William Bottomley of Belfast, second daughter of Dr. James Thomson.

Emery Walker Ph. sc.

[1852]

Yours very truly
William Thomson

Margaret and I despatched a joint letter to Anna by to-day's post, so that she will hear of it at the same time as you. I am now going to write to James, although I suppose he will hear of it from Anna before my intimation can reach him.—I remain always your affectionate brother,
WILLIAM THOMSON.

A week after their engagement Margaret Crum wrote to Mrs. King: . . . "We have one interest in common that can never fail, and as I told Mrs. Gall, I feel that in William's love for his sisters and her, lies my best security for the continuation to me of those feelings on which the happiness of my life must now depend."

The marriage took place, in accordance with Scottish custom, at the home of the bride's father at Thornliebank, on September 15, 1852, the officiating minister being the Rev. Dr. Brown of Edinburgh, father of the author of *Rab and His Friends*, Dr. John Brown. The young couple went to Wales for their wedding tour.

By his marriage he vacated the Fellowship which he had held in St. Peter's College since 1846.

BANGOR, *Saturday* [*Sept.* 19, 1852].

MY DEAR ELIZABETH—You have probably heard from Aunt, to whom I wrote on Thursday at Carlisle, of our doings after we drove away on Tuesday. On Thursday afternoon we had a most agreeable railway journey to Chester. Yesterday we walked quite round the walls, which you may remember of old, and came on in the afternoon to this place. We are in the Penrhyn Arms Hotel, which is beautifully situated, out of the town, with a garden running down to the sea, and a view of Anglesea, over the straits, from our windows. We set out this morning in a car to see the Britannia Tubes and the old Suspension Bridge ; but rain coming on slightly, but rather obstinately to appearance, we turned, and are now sitting in our parlour writing letters. The day is somewhat dark and cold, and some people might say dreary, but it does not seem so at all to me. I scarcely think it does so to Margaret either, although she has just been saying that it is, and what is more, laying particular emphasis on the most dismal parts. Perhaps she is only joking, but whether or not, she looks cheerful, and has quite got rid of her cold. In fact, I do not think either of us are going to apply to Dr. Brown to undo what he did on Tuesday.

We are not going to travel about a great deal, as we are both most inclined to get to some quiet seaside place and settle there, taking little excursions up the hills and on the water from it. We shall, however, see *something* of Wales in general before we leave it, if it were only in our search for the little Elysium that we have been imagining when we spoke to one another of "what was to be done in Wales." . . .

If we could get just such a little house as the one you are in, and similarly situated (at Clinder), somewhere on the coast hereabouts, I think we should be contented without much travelling about till the end of October.—I remain, your affectionate brother,

William Thomson.

The British Association of 1852 had been held in Belfast at the beginning of September. Thomson was this year chosen President of section A, and himself read five papers there. One of these was on the Sources of Heat generated by the Galvanic Battery. Another on Certain Magnetic Curves, being the Field of Force, accurately calculated, as perturbed by the introduction of Spheres of Substances of either higher or lower Magnetic Permeability than the surrounding Medium. Another on the Equilibrium of Elongated Masses of Ferro-magnetic Substance in uniform and varied Fields of Force. This narrated experiments in confirmation of Faraday's doctrine that magnetic bodies in such fields tended to move into positions where their magnetization would be greatest. Another was on the Mutual Attraction of two Electrified Spherical Conductors. Then lastly, a joint paper by Thomson and Joule on the Thermal Effects of Air rushing through Small Apertures. This narrated experiments begun jointly in May 1851 at Manchester, and continued by Joule, to test a point which had been assumed by Mayer without proof.

In December Thomson discoursed to the Glasgow Philosophical Society on the economy of the heating or cooling of buildings by means of currents of air. This suggested the mechanical method of refrigeration by the expansion of cooled compressed air, afterwards commercialised on a large scale in the Bell-Coleman process. In 1853 two papers were communicated to the Glasgow Philosophical Society on the Mechanical Values of Distributions of Electricity, Magnetism, and Galvanism, and on Transient Electric Currents. The former of these was a concise statement of the potential energy of an electric or magnetic system. The latter, which was given in fuller mathematical form in the *Philosophical Magazine* for June, was an investigation into the rushes of current which occur when a condenser is suddenly discharged through a conducting circuit. Riess had observed that in attempting to magnetize steel needles by such discharges, they are sometimes magnetized in one direction, sometimes in the reverse. Faraday had also found some anomalies in the amounts of electrolytic gases produced in water by such discharges, and from these Helmholtz had conjectured that they might be due to the discharge being oscillatory in character. Moreover, the existence of double, triple, and even quadruple flashes of lightning which had occasionally been observed, suggested that in certain cases the lightning flash might itself be an oscillatory discharge. Starting from energy-principles, Thomson determined to test mathematically what was the motion of electricity, at any instant after making contact, in a circuit containing capacity, resistance, and self-induction. He knew that capacity such as that of a Leyden jar tended to operate as if the dielectric of glass possessed a sort of electric elasticity. He knew that resistance in the circuit tended to fritter away the energy of the current as mere heat, and played therefore the same part as friction does in destroying the motion of mechanical bodies or of moving fluids. He also knew from Faraday's work that the energy of a current flowing in a circuit depended on the form of a circuit, being great in one that consists of coils; and that it was proportional to the square of the current, the circuit acting, if coiled, as though it gave inertia to the motion of electricity through its convolutions. So, then, starting from energy-principles, he ingeniously built up the differential equation for the motion of electricity in such a circuit, and deduced the integral. The result was very remarkable. He discovered that a critical relation occurred if the capacity in the circuit was equal to four times the coefficient of self-induction divided by the square of the resistance. If the capacity was less than this the discharge was oscillatory, passing through a series of alternate maxima in opposite directions before dying out. If the capacity was greater than this, the discharge was non-oscillatory, the charge of the condenser dying out without reversing. He suggested that it might be possible by discharging a Leyden jar, or other condenser of small capacity, through a circuit having large self-induction and small resistance, to produce artificially such oscillatory discharges. If the oscillations followed one another too rapidly for the eye to distinguish them, Wheatstone's method of observing them in a rotating mirror might be employed, and would show the sparks as several points or short lines of light separated by dark intervals, instead of a single point of light, or an unbroken line of light, as it would be if the spark were instantaneous, or were continuous and of appreciable duration.

This beautiful bit of mathematical analysis, and the prediction, which passed almost unnoticed at the time, laid the foundation of the theory of electric oscillations subsequently studied by Oberbeck, Schiller, Hertz, and Lodge. It forms to-day the basis of wireless telegraphy. Feddersen, in 1859, succeeded in photographing these oscillatory sparks, and sent photographs to Thomson, who with great delight gave an account of them to the Glasgow Philosophical Society.

In May 1853, Thomson, accompanied by his wife, made a trip to the Mediterranean, visiting Gibraltar, Malta, and Sicily. The exertions of travel, and of walks in Sicily, overtaxed Mrs. Thomson's strength, and on their return to England her health was still suffering. At the end of the year she was obliged to go to Edinburgh for surgical nursing, and for many weeks could not return to Glasgow. All through the ensuing spring and summer, though slowly advancing toward recovery, she was still far from well, and her sufferings cast a cloud upon the life of the home.

In 1853 Thomson published one note in the *Cambridge and Dublin Mathematical Journal*, and two papers in the *Philosophical Magazine*, one of these being electrical; the other on the restoration of mechanical energy from an unequally heated space.

The British Association Meeting of 1853 was held at Hull on September 3rd. Thomson was not present, and he made no communication of scientific work to its proceedings. His mathematical notebook shows no further entries till 1855.

APPENDIX TO CHAPTER V

INTRODUCTORY LECTURE TO THE COURSE ON NATURAL PHILOSOPHY

In entering on a new branch of study it is natural to look for a distinct statement of its subject. But nothing in science is more difficult than definitions. Attempts to give sharp and complete definitions, especially to define branches of science, have generally proved failures. Where definition and logical subdivision become practically valuable is to give *method*, and to promote order and regularity in the prosecution of a study. I do not propose in this introductory lecture to lay down with logical precision any definite and sharp line round our province: I shall rather attempt to explain in general terms the relation which Natural Philosophy bears to other branches of human inquiry, observation, science, and philosophy, and to divide our treatment of it in the manner which we find most convenient for our work in the Natural Philosophy Class and Laboratory of the University.

Mind and Matter, the two great provinces of Nature, are very remarkably and wonderfully distinct subjects of investigation. Our knowledge of the first is called *mental science*; and although *nature*, in the strict sense of the word, includes all things created, the knowledge of the *material* part alone is usually understood by the expression *natural science*.

In the progressive study of natural phenomena, that is, the phenomena of the external world, the first work is to observe and classify facts; the process of inductive generalisation follows, in which the laws of nature are the objects of research. These two stages of science are designated by the expressions of *natural history* and *natural philosophy*.

In mental science, that other principal branch of human knowledge, there are corresponding gradations which we may consider for a moment, as they serve to illustrate the division with which we are particularly occupied. Thus, what by the analogy of terms might be called *mental history*, that is to say, a combination of personal experience and a knowledge of men and of manners, with the study of politics and history, leads us to reason upon the abstract properties of mind, and to investigate that system of general laws on which *mental philosophy* is founded. So in the study of external nature, the first stage is the description and classification of facts observed with reference to the various kinds of matter of which the properties are to be investigated; and this is the legitimate work of *Natural History*. The establishment of general laws in any province of the material world, by induction from the facts collected in natural history, may with like propriety be called *Natural Philosophy*.

In the ordinary use of the terms this distinction is not strictly observed; for, although, according to it, every part of the material world should have its place in the first stage of science, the expression *natural history* is commonly restricted to the description and classification of the various natural products in the mineral, vegetable, and animal kingdoms of the earth; while the vicissitudes of temperature and climate and the flowing of the tides, as well as the general appearance and motions of the heavenly bodies, are considered as forming subjects for distinct branches of natural science, and are known by the names of Meteorology and Descriptive Astronomy. The reason of this apparently arbitrary limitation is sufficiently well founded. In the study of that class of objects which belong to the restricted province of Natural History, according to the specific meaning of the term, description and classification are of remarkable importance; while in the other branches of natural sciences we enter more immediately upon those speculations which belong properly to Natural Philosophy, or at least upon that series of systematic observations and experiments which have for their object the establishment of laws and the formation of theories; so that in these the primitive study which might strictly be called Natural History forms a much less prominent part of the whole science than it does in that province to which the name has been peculiarly assigned.

The term Natural Philosophy once detached from the study of the characteristic properties of minerals, plants, and animals, has not been subsequently bestowed upon those of the more advanced stages of this branch of science in which inductive generalisations lead to general laws and theoretical speculations; but distinct names such as Geology,[1] Chemistry, and Vegetable and Animal Physiology, have been applied to its different divisions; although in the strict logical sense these might be considered as belonging to Natural Philosophy.

Considering the sciences I have just mentioned as formally excluded from Natural Philosophy, I shall now proceed to give some explanation connected with the great province that remains.

The fundamental subject of Natural Philosophy is Dynamics, or the *science of force*; mechanics, kinetics, statics. Every phenomenon in nature is a manifestation of force. There is no phenomenon in nature which takes place independently of *force*, or which cannot be influenced in some way by its action;[2] hence dynamics has application in all the natural sciences; and before any considerable progress can be made in a philosophical study of nature a thorough knowledge of dynamical principles is absolutely necessary. It is on this account that dynamics is placed by universal consent at the head of physical sciences. It deserves this position, no less for its completeness as a science, than for its

[1] [Note added 1862:—

1. Geology; 2. Theory of heat and mechanical effect: (1) shows us the successive introduction of the different races of animals on the earth; (2) shows that the inanimate world must have had a beginning, and that all motion except that of heat must have an end, unless it please God to restore by an act of new creative power the dissipation of mechanical effect which always goes on. Chemistry occupies an intermediate position; electricity also, according to Sir H. Davy, Berzelius, Faraday, and Joule, etc. etc.].

[2] Here explain light, heat, chemical changes, taste and smell; every modification of matter; every object of sense.

general importance. From a few simple, almost axiomatic principles, founded on our common experience of the effects of force, the general laws which regulate all the phenomena, presented in any conceivable dynamical action, are established; and it is thus put within our power by a strict process of deductive reasoning to go back from these general laws to the actual results in particular cases of the operation of force; the instrument, or form, by which this deductive process is conducted being mathematical analysis. Hence it is that dynamics is said to be a branch of mixed or applied mathematics.

Statics and Kinetics are the two great divisions of mechanical philosophy; statics treats of the balancing of forces, while kinetics deals with the effects of unbalanced forces in producing motion or altering the motion of bodies. Some investigations in kinetics have reference to the forces which must be applied to communicate a given motion to a body; others again to the resistance necessary to reduce a moving body to a state of rest; or, lastly, the principles of kinetics may be employed to determine the forces called into action in the connecting pieces or links joining parts of any system of bodies in motion, such as the effect of centrifugal force in stretching the cord of a sling or the strain on the different parts of a steam-engine at work. But the circumstances of mere motion, considered without reference to the matter of the bodies moved and the forces producing the motion, or by it called into action, do not belong to the province of kinetics, but constitute the subject of a distinct science, to which, although it might be regarded as a branch of pure geometry, a distinguished French author has given the name *cinematics*, a term derived from the Greek word *κίνημα*, signifying a *motion*.

There are also subdivisions of statics and kinetics depending on the nature of the substance considered as subject to the action of force. Thus hydrostatics and hydrokinetics are the subordinate branches in which fluid bodies are treated of; and still more specifically, pneumatics is that part of hydrostatics in which the equilibrium of aeriform fluids is considered. Again, acoustics is that portion of general kinetics in which are investigated the laws of the small vibratory movements that give origin to the various phenomena of sound. The special characters of these branches of Natural Philosophy are best illustrated by considering examples of the subjects treated. Thus all the knowledge we possess regarding the mechanical condition of the atmosphere with which we are surrounded and of its relation to solid bodies or to other fluids, belongs to the science of pneumatics. And the investigations concerning the vibrations of stretched strings or membranes, of air in tubes, as in the organ or flute, or of solid laminæ, such as the reeds used in certain musical instruments, are instances of the researches which occur in the science of acoustics.

The two last-mentioned branches of dynamics, pneumatics and acoustics—on account of the experiments and observations necessary, in the first place, to enable us to rank them in their proper position in dynamical science, and after that to furnish the data necessary for deducing the solution of any practical question connected with them —may with propriety be still considered as distinct branches of Natural Philosophy. This remark is applicable in even a greater degree to Astronomy, which, although the oldest of the physical sciences, did not attain to its present rank as a branch of Natural Philosophy (all attempts to reduce to a common principle the various complicated motions which had been observed in the heavenly bodies having proved entirely futile) until the genius of Newton penetrated the mystery of their connection with dynamical principles, and discovered the simple law of Universal Gravitation by which they are all regulated. It may be thought that this great step in the theory of astronomy, at once its beginning and its end, must have superseded the further labours of the practical astronomer. Instead of having such an effect, however, it only showed the true object of those labours, and proved that they can never come to a termination, the vastness of their field being without limit. It is now no longer the dimensions of crystal spheres, or the number and arrangement of cycles and epicycles; it is not the finite mechanism of an orrery that the astronomer has to investigate, but it is the elements by which the positions of the planets and countless stars may be determined at any assigned instant of time past or to come. The immensity of the task might be appalling were it not that progress is sure, and the brilliant discoveries by which the diligent observer and the bold theoretical adventurer is occasionally rewarded give that deep interest to the work of the astronomer which encourages and excites him through all his toil. [Here followed in the original lecture three pages of MS. relating to the discovery of Neptune, predicted by Adams and Le Verrier. They were omitted after 1862 on the advice of Tait that Neptune had been "ridden to death." In later years there used to be about here a reference to the *Mécanique Céleste* of Laplace, and a remark on the *romance* of that title.]

Another great division of our subject is Optics, or that part of Natural Philosophy in which the properties of light and the laws of vision are investigated. Whether we consider the infinitely varied optical phenomena of nature or the brilliant experiments which may be made in this branch of science, the study of optics presents attractions of the highest order; and when in addition we reflect that it is through the assistance of instruments constructed according to our knowledge of optical principles that all the observations are made on which the science of astronomy depends, and that such instruments are used more or less in nearly every accurate research in any branch of natural science, then truly the practical importance of the study seems if possible to exceed its attractive interest.

(Most wonderful of instruments, the eye. Demonstration of design sometimes by fanciful or too zealous advocates discredited.)

The thoughtful student must, besides, be deeply interested in tracing the proofs of design in the adaptation of our organs of vision in accordance with the physical laws of light to receive the impressions by which we see external objects. There is indeed no department of the study of nature in which we meet with more remarkable instances of beautiful and wonderful phenomena, or of perfect mechanism designed for rendering them sensible to animated creatures.

The other subjects which belong to the Natural Philosophy course are included in three divisions,—Heat, Electricity, and Magnetism. Our knowledge of these branches of the science is not so far advanced as to enable us to reduce all the various phenomena to a few simple laws from which, as in mechanics, by means of mathematical reasoning every particular result may be obtained; but observation and experiment are the principal means by which our knowledge in this department can be enlarged. Hence what is called the experimental or physical course includes these three subjects; while the more perfect sciences of mechanics and optics, being really mathematical subjects, form a distinct division of the studies prescribed by the University for the complete course of Natural Philosophy. I should here explain that although to advance to a profound knowledge of dynamics or optics, or to make discoveries in these sciences, an extensive and accurate knowledge of mathematical analysis is almost absolutely necessary, yet a very useful and interesting study may be made of all the branches of mechanics (astronomy and the theory of heat and light, as I have already remarked, being included), and of optics, by persons moderately acquainted with the more elementary parts of mathematics. It will therefore be my object to limit as much as possible the merely mathematical part of the treatment of these subjects in the general lectures, and to separate off the higher mathematical treatment, so that students qualified by the amount of mathematical preparation prescribed by the University regulations before entering publicly on the Natural Philosophy course, may by careful attention and diligent private study be able to follow out the full business of the class.

When we look back upon knowledge of nature from

earliest times, and on the progress which has been made by the human mind in the discovery of truth, we feel that the power of investigating the laws established by the Creator for maintaining the harmony and permanence of His works is the noblest privilege which He has granted to our intellectual state. If we neglect cultivating to the utmost of our opportunities the faculties which for this end He has bestowed upon us, we reject His gifts, and are unworthy of His beneficence. But while power is given to us to learn, the exercise of this power has been made to be a source of happiness. "Wisdom's ways are ways of pleasantness" (surely Natural Philosophy was included), is the saying of one who spoke from experience, who, as Bacon says, was enabled by the gift of God, "not only to write those excellent parables or aphorisms concerning divine and moral philosophy," but also "to compile a natural history of all verdure from the cedar tree that is in Lebanon to the hyssop that springeth out of the wall, as well as of all things that breathe or move."

Whether in the active investigations by which we arrive at truth, or in the gratification felt in the possession of knowledge, the intellectual value of science is conspicuous, and the adaptation of the human mind for such enjoyment is a manifestation not less remarkable of the divine goodness than the various arrangements by which the physical powers of the animal world are produced and maintained. The contentment felt in the knowledge of truth is well illustrated by Lucretius when he says, "it is a view of delight to stand or walk in safety on the shore, and to see a ship tossed with tempest on the sea; or to be safe in a fortified tower and to see two armies joining in battle on a plain; but it is a pleasure incomparable for the mind of man to be settled, landed, fortified in the certainty of truth; and from thence to descry and behold the errors, perturbations, labours, and wanderings up and down of other men."

Bacon places the delights of knowledge and learning above all other in nature. "Shall," he asks, "the pleasures of the affections so exceed the senses as much as the obtaining of desire or victory exceedeth a song or a dinner; and must not, of consequence, the pleasures of the intellect or understanding exceed the pleasures of the affections? We see in all other pleasures there is satiety, and after they be used their verdure departeth; which sheweth well they be but deceits of pleasure, and not pleasures; and that it was the novelty which pleased, not the quality. But of knowledge there is no satiety, but satisfaction and appetite are perpetually interchangeable; and therefore it appeareth to be good in itself simply without fallacy or accident."

The deep interest of scientific research cannot be entirely appreciated by those to whom such inquiries are strange; but it is felt in some degree by all who apply themselves in earnest to the study of Natural Philosophy. All of you, who are now commencing it, will I hope bestow enough time and thought upon the work to enable you to surmount the initial difficulties, such as necessarily occur when ideas of an entirely novel kind are brought before the mind. By sufficient perseverance, and by continual reference to the examples which you see in the world around you of the various physical actions and effects you will have to study, the habit of reasoning on the phenomena of nature will be acquired, and you will *then* begin to feel the enthusiasm which the subject inspires. Each one of you when he attempts the solution of a problem of difficulty, or struggles to comprehend a new principle, will have a share of that spirit of enterprise which led Newton on to his investigations; and when the problem is solved, when the doubts have vanished, a feeling of satisfaction will be the reward, similar to that which Newton himself must have felt after some of his great discoveries.

A strong recommendation of the study of Natural Philosophy arises from the importance of its results in improving the physical condition of mankind.

At no period of the world's history have the benefits of this kind conferred by science been more remarkable than during the present age; still we must not consider that the practical benefits to be derived from Natural Philosophy are only now recognised. A popular writer toward the end of the last century compares in the following terms the primitive state of mankind with that which existed at the time he wrote, "Man in a rude and savage state, with a precarious subsistence, exposed to the inclemencies of the seasons and the fury of wild beasts, is an object of pity when compared to man enlightened and assisted by philosophy. Ignorant of architecture, of agriculture, of commerce, and of all the numerous arts which depend on the mechanic powers, he exists in the desert, comfortless and unsocial, little superior in enjoyment to the lion or the tyger, but much their inferior in strength and safety." "If it be true that man ever existed in this state, it could not have lasted long; the exertion of his mental strength must have given rise to the arts. Aided by these the wilderness became a garden, embellished with temples, palaces, and populous cities; and he beholds himself removed to an immense distance from the animals to which in his original ignorance he seemed nearly allied." Since the time when this estimate was made of the advantages society has derived from science, some of the highest principles of Natural Philosophy and Chemistry have been applied to produce effects which at that time would have been deemed incredible. Who would have believed that we should at present consider twenty-five or thirty miles an hour a slow average rate of travelling? or that our messages should now be communicated for thousands of miles by sea or land, literally with the speed of lightning? These are only single instances of the vast resources which we derive from direct applications of modern science; but I need not enumerate more, as every one is convinced of the immense practical importance of the principles of Natural Philosophy at present known. We must not, however, by considerations of this kind be led to regard applications to the ordinary purposes of life as the proper object and end of science. Nothing could more effectually stop the advancement of knowledge than the prevalence of such views; even the desired *practically useful* discoveries would not be made if researches obnoxious to the fatal question *cui bono* were to be uniformly avoided. To take one example of a very great discovery; the foundation of a new branch of Natural Philosophy; which will form one of our subjects in the elementary course of this session. Oersted would never have made his great discovery of the action of galvanic currents on magnets had he stopped in his researches to consider in what manner they could possibly be turned to practical account; and so we should not now be able to boast of the wonders done by the electric telegraphs. Indeed, no great law in Natural Philosophy has ever been discovered *for* its practical applications, but the instances are innumerable of investigations apparently quite *useless* in this narrow sense of the word which have led to the most valuable results. Besides, we must never forget that the true end of philosophy is the investigations of laws of nature, and towards this object we must go on, turning neither to the right nor to the left, working steadily and honestly, and at the same time thoughtfully. The true spirit in which science ought to be cultivated is well distinguished from the base motives which sometimes actuate its votaries in the following passage which I have great pleasure in quoting to you from Bacon's *Advancement of Learning*: "This is that which will indeed dignify and exalt knowledge if contemplation and action be more nearly and straitly conjoined and united together than they have been; for men have entered into a desire of learning and knowledge sometimes upon a natural curiosity and inquisitive appetite; sometimes to entertain their minds with variety and delight; sometimes for ornament and reputation; and sometimes to enable them to victory of wit and contradiction; and most times for lucre and profession; and seldom sincerely to give a true account of their gift of reason to the benefit of man; as if there were sought in knowledge a couch whereupon to repose a searching and restless spirit; or a tarasse for a wandering and variable mind to walk up and down with a fair prospect; or a tower of state for a proud mind to raise itself upon; or a fort or commanding ground for strife and contention; or a shop for profit or sale; and not a rich storehouse for the glory of the Creator and the relief of man's estate."

"A little learning" has been found fault with as "a dangerous thing," but profound knowledge, or the earnest attempt to attain profound knowledge, is liable to no such objection. It has been well said by Sir John Herschel, in his eloquent *Discourse on the Study of Natural Philosophy*, that "the character of the true philosopher is to hope all things not impossible, and to believe all things not unreasonable. He who has seen obscurities which appeared impenetrable in physical and mathematical science suddenly dispelled, and the most barren and unpromising fields of inquiry converted, as if by inspiration, into rich and inexhaustible springs of knowledge and power, on a simple change of our point of view, or by merely bringing to bear on them some principle which it had never occurred before to try, will surely be the very last to acquiesce in any dispiriting prospects of either the present or future destinies of mankind; while on the other hand the boundless views of intellectual and moral as well as material relations which open on him on all hands in the course of these pursuits, the knowledge of the trivial place he occupies in the scale of creation, and the sense continually pressed upon him of his own weakness and incapacity to suspend or modify the slightest movement of the vast machinery he sees in action around him, must effectually convince him that humility of pretension, no less than confidence of hope, is what best becomes his character."

But while we exult in the progress we have made in the cultivation of the natural sciences, we must not omit to consider the impressions which the true study of nature ought to make on every mind, and we must remember that as the depth of our insight into the wonderful works of God increases, the stronger are our feelings of awe and veneration in contemplating them and in endeavouring to approach their Author. At a time when astronomy could hardly have been said to exist as a science, the Psalmist exclaimed, "When I consider thy heavens, the work of thy fingers, the moon and the stars which thou hast ordained; What is man that thou art mindful of him, and the son of man that thou visitest him?"

How much must this spirit of humble gratitude be enhanced when we have been allowed to trace even remotely the laws according to which the heavens have been created.

When we comprehend the vastness of the dimensions of that part of creation of which we know a little, and yet consider what an infinitesimal portion this is of the whole universe, how insignificant a being we must feel that man is, and how grateful ought we to be that God should still be mindful of him and visit him, and for the gifts and the constant care bestowed on him by the Creator of all. By such feelings the earnest student of philosophy must always be impressed; so will he by his studies and successive acquirements be led "through nature up to nature's God."

[Added, 1848.] In conclusion, gentlemen, let me express the hope that you will do everything in your power to second whatever exertions I may be able to make for your improvement, devoting yourselves with energy and cordiality to the business of the class; and allow me to anticipate for you as the result a pleasant and profitable session.

CHAPTER VI

THERMODYNAMICS

Crescunt disciplinae lente tardeque; per varios errores sero pervenitur ad veritatem. Omnia praeparata esse debent, diuturno et assiduo labore, ad introitum veritatis. Iam illa certo temporis momento, divina quadam necessitate coacta, emerget.

C. G. J. JACOBI.

IT has been told how Thomson, when studying in Paris in 1845, sought for the treatise of Carnot on the motive power of heat, of which he had learned through the memoir of Clapeyron. In the laboratory of Regnault he had taken part in experimental researches involving severely accurate measurements of temperatures, of coefficients of expansion, and of the specific and latent heats of materials.

In the year that he had spent as mathematical lecturer at Peterhouse, and in the next twelve months, crowded with new duties and preoccupations of his Glasgow chair, he had found scant time to put into shape the mathematical and physical conceptions which formed themselves with embarrassing rapidity of succession in his fertile brain. Magnetism, hydrodynamics, and electricity were all furnishing him with problems for solution. Amid such pressing demands upon his activities the study of the problems of the motive power of heat might well have been left aside. And yet, owing to circumstances presently to be narrated, his thoughts were destined to be directed to this subject with an absorbing interest that dominated his work for at least a decade, and in which some of his highest achievements were reached. If his work in Thermodynamics stood alone it would suffice to place his name as a natural philosopher beside that of Newton in its grasp of principles and generality of outlook; yet in this subject, in which his peculiar genius found scope in the knitting together and unification of the fundamental laws, there will ever be associated with his name that of another worker, James Prescott Joule. It would be difficult to conceive of a man better fitted than Thomson was by training and habit of mind to grasp the ultimate importance of Joule's patient and individual investigations, and to guide, counsel, and co-operate with him, as well as to expound, develop, formulate, and systematize the new philosophy for which they afforded a basis.

Here again it will be necessary to make some digression in order to elucidate the part which Thomson took in this great forward step in physical science.

Down to the middle of the eighteenth century the term *heat* was used indiscriminately for the invisible agent which when communicated to a body warms it, and for the degree of sensible warmth attained, which we now call *temperature*. Toward the end of the eighteenth century the chemists Black and Irvine, in Scotland, and Lavoisier, in France, by their investigations into latent and specific heat, compelled a clear distinction between the cause and the effect. When heat is given to a body the usual effect is that its temperature is raised. This is not always so; for in the cases where a body is changing its state, as when a solid is melting, or when a liquid is evaporating, the communication of heat may not produce a rise of temperature. Thus when heat is given to a melting lump of ice, the resulting liquid is not hotter than the ice was. The heat has disappeared *as heat*, and in the language of Black, it has become *latent* in the liquid. So also when boiling water is converted into steam, the steam is no hotter than the liquid was, a very large quantity of heat having been rendered latent in the process. In fact, during change of state the heat has been employed, not in warming the substance, but in changing its state of molecular aggregation. Lavoisier gave the name *caloric* to that which we should now call the *heat* per se. Though he did not bind himself[1] to the

[1] "Cette substance, quelle qu'elle soit, étant la cause de la chaleur . . . on ne peut pas, dans un langage rigoureux la désigner par le nom de 'chaleur'; parce que la même dénomination ne peut pas exprimer la cause et l'effet. . . . Nous avons en conséquence désigné la cause de la chaleur, le fluide éminemment élastique qui la produit, par le nom de *calorique*. Indépendamment de ce que cette expression remplit notre objet dans le système que nous avons adopté, elle a encore un autre avantage; c'est de pouvoir s'adapter à toutes sortes d'opinions, puisque, rigoureusement parlant, nous ne sommes pas même obligés de supposer que le calorique soit une matière réelle; il suffit, comme on le sentira mieux par la lecture de ce qui va suivre, que ce soit une cause répulsive quelconque qui écarte les molécules de la matière et on peut ainsi en envisager les effets d'une manière abstraite et mathématique."—LAVOISIER, *Traité élémentaire de chimie*, 1789, p. 4.

notion that "caloric" was a kind of imponderable matter, which could pass from one substance to another, or become latent amongst molecules of a liquid or of a vapour, that notion was generally received and came to be associated with the use of the term. To this doctrine of the material nature of heat was superadded the further notion that caloric itself was indestructible, and could neither be created nor destroyed,—the evolution of heat in combustion being regarded as merely the liberation of caloric previously latent, and the cooling of a hot body by conduction or radiation being considered merely as a dilution or dissipation of the caloric through a larger quantity of matter.

On the other hand Bacon, and after him Boyle, Hooke, and apparently Newton, had held that heat, in a body, consisted in a species of motion of the minute particles,—a view which was revived with experimental demonstrations just at the beginning of the nineteenth century by Davy, and by Count Rumford, who produced heat by friction, and concluded that the quantity was limited only by the amount of labour expended mechanically in maintaining the motion against the resistance of friction. Though British philosophers generally held this view, the continental doctrine of *caloric* was usually taught, or at least its language was adopted, with the implication that caloric was a subtle and imponderable fluid that could neither be created nor destroyed, but which could pass readily from a hot body to a cold one; could be conducted from the hot parts of an unequally heated body to a cold one; could be absorbed or rendered latent in changes of state, as from solid to liquid, or liquid to vapour; and could even be radiated across space, as from the sun to the earth.

Carnot's wonderful little treatise of 1824—"an epoch-making gift to science," as Lord Kelvin styled it—had discussed the question how it was that work could be done in an engine by the motive power of heat. He had grasped the fundamental point that the steam that drives the engine always comes out of the cylinder colder than it was when it entered; that heat tends to pass from a hotter to a colder body. Assuming the view of the indestructibility of heat, he was led to explain the action of the steam-engine (and indeed of all heat-engines) by the analogy of water-engines. In these none of the water disappears or is destroyed; the work is done by the water descending from a higher level to a lower level. The amount of work that a water-wheel can do, by the passage of any given quantity of water, depends upon the "head," that is, upon the difference of level from the supply-flume above to the tail-race below. So, according to Carnot, the work done by a heat-engine, by the passage through it of a given quantity of heat, depends upon the difference of temperature between the source (boiler) and the refrigerator (condenser or exhaust). He supposed that exactly as much heat left the engine and entered the refrigerator as the engine itself received from the source. As a final result, he says, the cold water of the condenser takes possession of the caloric developed by combustion in the boiler. It is heated by the intervention of the steam as if it —the water of the condenser—had been placed directly over the furnace. The steam is here only a means of transporting the caloric. The production of motive power in steam-engines is due, not to any actual consumption of caloric, but to its transportation from a warm body to a cold body.

It is true that Carnot before he died had outgrown these notions, though his maturer reflections were not published till long after his death. But there can be no doubt of the assumption from which he started.

He had, however, meditated upon the puzzling circumstance, that whereas in the engine the transmission of heat from a hot body to a cold one effects motive power, the communication of heat from a hot body to a cold one by mere conduction or radiation produces no work. He saw that in the cases where motion is produced the heat operated only by virtue of the changes of volume or of form which it causes. He saw that alternations of heating and cooling are requisite, and that corresponding alternations of change of volume are equally necessary. So he conceived—and the conception remains as a permanent contribution of the utmost value to science—the notion of a *cycle of operations*. In this cycle heat should be given to a prescribed quantity of air or steam (or other working fluid); then the hot fluid should expand, doing work, but becoming cooler as it expanded, and giving out its heat to the refrigerator; then it should be compressed back to its original volume, pressure, and temperature to recommence the cycle. The cycle consists of four operations, which, as originally described, are these:—Heat having been previously imported from the source at high temperature, and the piston having risen to a certain point, and the source of heat having been cut off, the first operation (1) consists in permitting the fluid still further to expand while, as no heat can enter or leave, the temperature falls, the expansion being continued till the lower temperature of the refrigerator is reached; (2) the piston is pushed back while the fluid is compressed isothermally, giving up heat to the refrigerator; (3) the piston is pushed farther back under adiabatic conditions, the temperature rising till it reaches that of the source; when (4) heat is allowed to enter from the source, the fluid expanding isothermally until it reaches the original volume and pressure; and then the cycle recurs. In this round of operations—famous as *Carnot's cycle*—heat is transferred from the source to the refrigerator, and a certain nett amount of work is performed. Carnot, with the clearest reasoning, saw that cycle to be reversible; that is, that if the operations were all reversed and performed in reverse order by the application of external work, the heat would be pumped back from the refrigerator to the source. He also perceived clearly that the "motive power" in the cycle was independent of the particular kind of fluid employed; and that for a given quantity of the fluid the quantity of motive power was determined solely by the difference between the two temperatures between which the transfer of caloric (as he expressed it) was effected. He also observed that the fall of caloric through a given difference of temperatures produces more motive power at a lower average temperature than it does at a higher. He also acutely concluded that the superiority of high-pressure engines over low-pressure engines lay essentially in their power of utilizing a greater "fall" of caloric.

This most remarkable conception was further developed by Clapeyron, who gave precision to its views by constructing an indicator diagram of the four operations of the cycle, and emphasising the point as to the efficiency of the heat concerned in the cycle being dependent solely on the temperatures. Taking Carnot at his word he, however, insisted, in his description of the cycle, that in the second operation the same quantity of heat was given up to the refrigerator as was received from the source in the fourth operation. He thus made the indestructibility of heat an essential part of the cycle theory, which Carnot had not explicitly done. Clapeyron's methodical exposition, published in 1834, was translated into English, and appeared in the third series of Taylor's *Scientific Memoirs* in 1839.

Quite independently of Carnot or Clapeyron, a still more potent investigator arose a few years later in the person of James P. Joule, the young Manchester brewer, whose work was destined profoundly to influence physical science. In 1838, at the age of nineteen, Joule began his scientific labours with the simple aim of improving the electro-magnetic motor of Sturgeon. He built for himself such an engine, of which he recorded: "It weighs $7\frac{1}{2}$ lbs., and the greatest power I have been able to develop with a battery of forty-eight Wollaston four-inch plates was to raise 15 lbs. a foot high in a minute." He was thinking as an engineer, and measuring the power in true engineering fashion. After a year or more spent on investigating the production of heat by electric currents, he turned to the magneto-electric machine, which, he observed, enabled him "to convert mechanical power into heat by means of the currents which are induced in it." But being at first uncertain whether the heat was *generated* or merely *transferred* by this apparatus, he set himself to clear up the difficulty. After a close investigation of singular experimental difficulty, the result was, in his own words—"We have, therefore, in magneto-electricity an agent capable, by simple mechanical means, of destroying or generating heat." And he proceeded with astonishing insight to deduce the numerical equivalence, which he found (at first) to be 838 foot-pounds of work as the mechanical equivalent of the quantity of heat capable of increasing by one Fahrenheit degree the temperature of one pound of water. [The correct result of after-years is 778 foot-pounds.] This research, communicated by Joule to the British Association meeting at Cork, in 1843, was received with general silence. His conclusion, all too sweeping in its generality, of "the convertibility of heat and mechanical power into one another, according to the above numerical relations," was received with entire incredulity. He had proved that work can be converted into heat. He had not yet proved that heat can be converted into an equivalent quantity of work. He had indeed raised the question *how much* of the heat produced by the combustion of coal is converted into mechanical work by the steam-engine; and on his assumed equivalence had shown that the steam-engine utilizes less than 10 per cent of the assumed equivalent. Yet in the question "*how much*," he had touched the vital point. Before the paper was published he had already taken another step. Forcing water to flow with friction through minute nozzles in a perforated piston, he had found the mechanical work expended to be converted into heat, with a numerical equivalent of 770 as against his previous 838. In 1844 Joule, unaware of a similar suggestion by Mayer—a suggestion, however, unsupported by any fresh experimental evidence—made an inquiry into the changes of temperature due to the rarefaction and condensation of air, and from these deduced a fresh determination of the mechanical equivalence of heat. In this paper, which was rejected by the Royal Society, he combated the views of Carnot and Clapeyron, and reiterated his own conclusion that the steam in the cylinder of an engine loses heat while it is expanding and doing work, and that on condensation of the steam the heat thus converted into power is not given back. Joule, then, had read Carnot and had rejected his theorem; but he had read Carnot without realizing that Carnot's cycle, apart from the erroneous notion that the heat given out to the condenser was equal to the heat received from the source, was essentially true. Nor did he perceive that if the word "caloric" had been used in the extended sense laid down by Lavoisier—in whose mind it clearly connoted "energy" rather than "heat"—then Carnot's theorem became a mere truism, that the work done in the cycle was equivalent to the difference between the heat received from the boiler and the heat passed on to the condenser. He failed, as Osborne Reynolds points out, to observe that his own discoveries cleared up the only thing that was obscure in Carnot's. But, like many others, Joule lived too near to his own work to be able even to do justice to the significance of his own achievements.

At the British Association Meeting at Cambridge in June 1845, Joule, undaunted by the chilling reception accorded to his earlier papers, read a communication[1] before the section of Chemistry "On the Mechanical Equivalent of Heat." In this he described an early form of his apparatus with paddles for measuring the heat disengaged by

[1] *Brit. Assoc. Report*, 1845, p. 31; reprinted in Joule's *Scientific Papers*, p. 202. This is an abstract only; the full paper seems to have been re-written for the *Philosophical Magazine*, ser. 3, vol. xxvii. p. 205; or *Scientific Papers*, pp. 202-205.

friction of water. This gave 819 foot-pounds as the equivalent, or, averaging with previous experiments, 817. He went on to suggest that the water of waterfalls should be found correspondingly warmer at the bottom than at the top of the cascade. Niagara, for example, 160 feet high, should show about one-fifth of a Fahrenheit degree of difference. The paper ended with a deduction from the expansion of gases that there should be a "zero of temperature" at 480° F. below the freezing-point of water, the first suggestion of an absolute zero. Whether Thomson was present at the reading of this paper is unknown; but there arose no discussion upon it.

Two years passed, and yet once more Joule, in 1847, brought his work before the British Association[1] meeting in Oxford in more perfected form. In the new apparatus brass paddles revolving in a fluid were propelled by the descent of weights. And now for the first time he succeeded in attracting notice to his discoveries.

Joule's account of the circumstance, written[2] in 1885, is worth recording:—

It was in the year 1843 that I read a paper "On the Calorific Effects of Magneto-Electricity and the Mechanical Value of Heat" to the chemical section of the British Association assembled at Cork. With the exception of some eminent men, among whom I recollect with pride

[1] *Brit. Assoc. Report*, 1847, p. 55 (abstract); printed in full *Philos. Magazine*, ser. 3, vol. xxxi. p. 173; reprinted in *Scientific Papers*, vol. i. pp. 276-81. See also *Comptes rendus*, August 23, 1847, in an article reprinted in *Scientific Papers*, vol. i. pp. 283-86.

[2] *Scientific Papers*, vol. ii. p. 215.

Dr. Apjohn, the president of the section, the Earl of Rosse, Mr. Eaton Hodgkinson, and others, the subject did not excite much general attention; so that when I brought it forward again at the [Oxford] meeting in 1847 the chairman suggested that, as the business of the section pressed, I should not read my paper, but confine myself to a short verbal description of my experiments. This I endeavoured to do, and discussion not being invited, the communication would have passed without comment if a young man had not risen in the section, and by his intelligent observations created a lively interest in the new theory. The young man was William Thomson, who had two years previously passed the University of Cambridge with the highest honour, and is now probably the foremost scientific authority of the age.

Thomson, in 1882, gave his version[1] of this notable episode:—

I made Joule's acquaintance at the Oxford meeting, and it quickly ripened into a life-long friendship. I heard his paper read at the section, and felt strongly impelled to rise and say that it must be wrong, because the true mechanical value of heat given, suppose to warm water, must for small differences of temperature be proportional to the square of its quantity. I knew from Carnot's law that this must be true (and it *is* true; only now I call it "motivity" in order not to clash with Joule's "mechanical value"). But as I listened on and on I saw that (though Carnot had vitally important truth not to be abandoned) Joule had certainly a great truth and a great discovery, and a most important measurement to bring forward. So instead of rising with my objection to the meeting, I waited till it was over, and said my say to Joule himself at the end of the meeting. This made my first introduction to him. After that I had a long talk over the whole matter at one of the conversaziones of the Association, and we became friends from thenceforward. However,

[1] *Nature*, vol. xxvi. p. 618.

he did not tell me that he was to be married in a week or so; but about a fortnight later I was walking down from Chamounix to commence the tour of Mont Blanc, and whom should I meet walking up but Joule, with a long thermometer in his hand, and a carriage with a lady in it not far off. He told me that he had been married since we parted at Oxford! and he was going to try for elevation of temperature in waterfalls. We trysted to meet a few days later at Martigny, and look at the Cascade de Sallenches to see if it might answer. We found it too much broken into spray. . . .

Joule's paper at the Oxford meeting made a great sensation. Faraday was there and was much struck with it, but did not enter fully into the new views. It was many years after that before any of the scientific chiefs began to give their adhesion. It was not long after when Stokes told me that he was inclined to be a Joulite.

Miller and Graham, or both, were for many years quite incredulous as to Joule's results, because they all depended on fractions of a degree of temperature, sometimes very small fractions. His boldness in making such large conclusions from such very small observational effects is almost as noteworthy and admirable as his skill in extorting accuracy from them. I remember distinctly at the Royal Society, I think it was either Graham or Miller saying simply he did not believe in Joule because he had nothing but hundredths of a degree to prove his case by.

In another account of the same incident given by Lord Kelvin in 1893, at the unveiling of the Joule statue in Manchester,[1] he declared that he was "tremendously struck with the paper," and added, "This is one of the most valuable recollections of my life, and is indeed as valuable a recollection as I can conceive in the possession of any man interested in science."

[1] *Nature*, vol. xlix. p. 164; and *Pop. Lect.* ii. p. 558.

The meeting of Thomson and Joule proved to be a notable event for both. Thomson was obviously impressed by the sincerity and scientific merit of Joule's investigations. Yet he found himself at first, and for many months to come, unable to accept the views of Joule as to the mutual equivalence of heat and work. To his brother James he wrote after the meeting :—

> I enclose Joule's papers, which will astonish you. I have only had time to glance through them as yet. I think at present that some great flaws must be found. Look especially to the rarefaction and condensation of air, where something is decidedly neglected in estimating the total change effected in some of the cases.

Assuredly Joule had shown that in the circumstances of his varied experiments work could be turned into heat. But the converse proposition that heat could be turned into its equivalent of work, Thomson could by no means accept as proven. He still clung to the reasoning of Carnot that heat could furnish motive power when being let down from a higher to a lower temperature, or when passing from a hotter to a colder body. And as he had not yet perceived that this would still be true even if during the transference some portion of the heat disappeared, as heat, to be converted into its equivalent in work, he held back from assenting to Joule's generalization. He had indeed on April 21, 1847—that is two months prior to his meeting with Joule—given to the Glasgow Philosophical Society an account of Stirling's Hot-Air Engine, with the theory of it deduced on Carnot's principles; and at the end of this paper he had propounded the conclusion that water at 32° F. may be converted to ice at 32° F. without the expenditure of any work. Carnot's theory clearly held his mind. Confronted now on the one hand by the apparently irrefragable arguments of Carnot, on the other by the indisputable experiments of Joule, Thomson refused to accept the new doctrine until he could see how it could be reconciled with the old. The apparent conflict took strong possession of Thomson's mind and dominated his thoughts.

Eleven months later Thomson read to the Cambridge Philosophical Society a paper "On an Absolute Thermometric Scale Founded on Carnot's Theory of the Motive Power of Heat, and Calculated from Regnault's Observations." If he had been considering Joule's suggestion that an absolute zero of temperature might be deduced from the expansion of gases, he put it from him on the ground that any scale based on the expansion of any gas, even under standard conditions, though it might be a definite, would not be an *absolute* scale. But perceiving that in the Carnot cycle the work performed by a unit of heat in being let down (to retain Carnot's phrase) through a degree of temperature is independent not only of the working substance, but also of the particular part of the scale at which the operation occurs, he suggested this as basis of an absolute scale of temperature, and calculated a set of tables from Regnault's experimental data of latent heats and pressures of saturated vapour. This paper[1] is interesting for the view it incidentally affords of Thomson's ideas at that date. Throughout it he uses the term "mechanical effect" for work done. And he declares that "the conversion of heat (or caloric) into mechanical effect is probably impossible, certainly undiscovered. In actual engines for obtaining mechanical effect through the agency of heat, we must consequently look for the source of power, not in any absorption and conversion, but merely in a transmission of heat." But he at once qualified this sweeping statement by appending at the word "impossible" in the sentence above quoted a footnote, which must be repeated in full :—

> This opinion seems to be nearly universally held by those who have written on the subject. A contrary opinion, however, has been advocated by Mr. Joule of Manchester ; some very remarkable discoveries which he has made with reference to the generation of heat by the friction of fluids in motion, and some known experiments with magneto-electric machines, seeming to indicate an actual conversion of mechanical effect into caloric. No experiment, however, is adduced in which the converse operation is exhibited ; but it must be confessed that, as yet, much is involved in mystery with reference to these fundamental questions of Natural Philosophy.

The fact is that Thomson was still living in the pre-Joule order of ideas, and had been prompted to write this paper by the publication earlier in the year of the first volume of Regnault's memoirs, giving the data of latent heats and pressure

[1] *Mathematical and Physical Papers*, vol. i. p. 102.

of vapour on which Thomson's calculations were founded.

The footnote wherein Thomson had noted Joule's heterodox views on the nature of heat—views which, in the text of his paper, he dismissed as "probably impossible, certainly undiscovered"—is the first trace of an impending admission. But incidentally it proves that Thomson had either not read the evidence afforded by Joule's earlier researches on magneto-electric machines, or else that he doubted their relevancy. Thomson was unconvinced.

At the Swansea meeting of the British Association in August 1848, Joule recapitulated his former experiments, described improvements which corrected the value of the mechanical equivalent to 771 foot-pounds (per degree Fahrenheit), and claimed to have established: (1) proof of the mechanical nature of heat, (2) proof from work done by condensing and rarefying air that the heat of gases consists simply in the *vis viva* of their particles, (3) that the zero of temperature, determined by the expansion of gases, is at 491° F. below the freezing-point.

Seven months passed, and Thomson communicated to the Royal Society of Edinburgh[1] a second paper, "An Account of Carnot's Theory of the Motive Power of Heat, with Numerical Results deduced from Regnault's Experiments on Steam."

[1] *Trans. R. Soc. Edinb.* xvi. p. 541; read Jan. 2, 1849; reprinted in *Mathematical and Physical Papers*, vol. i. p. 113; see also *Ann. de chimie*, xxxv. p. 248, 1852.

In this paper, which is valuable as being an exposition in detail of the Carnot cycle, drawn up after he had succeeded in getting hold of an actual copy of Carnot's treatise, he recalculated the tables of his former paper. But he had clearly been studying Joule's writings, and the references to them are weighty and numerous. He still uses the term "mechanical effect" for work performed, and still accepts as axiomatic "the ordinarily received and almost universally acknowledged" principle that in the cycle of operations as much heat must leave the body as entered it. After quoting Carnot, and admitting that a most careful re-examination of the entire experimental basis of the theory of heat has become urgent, he turns to Joule and to the evidence which previously he had ignored.

The extremely important discoveries recently made by Mr. Joule of Manchester, that heat is evolved in every part of a closed electric conductor, moving in the neighbourhood of a magnet, and that heat is *generated* by the friction of fluids in motion, seem to overturn the opinion commonly held that heat cannot be *generated*, but only produced from a source where it has previously existed either in a sensible or in a latent condition.

All this is favourable, but he immediately followed it by a denial :—

In the present state of science, however, no operation is known by which heat can be absorbed into a body without either elevating its temperature, or becoming latent, and producing some alteration in its physical condition; and the fundamental axiom adopted by Carnot may be considered as still the most probable basis for an investigation of the motive power of heat.

So, still unconvinced, he denied any conversion of heat or caloric into mechanical effects, declaring that the quantity of heat "discharged during a complete revolution (or double stroke) of the engine must be precisely equal to that which enters the water of the boiler." He even added in a note that this application of Carnot's principle, tacitly admitted as an axiom, had "never been questioned by practical engineers." He therefore enunciated the following principle :—

> The thermal agency by which mechanical effects may be obtained is the transference of heat from one body to another at a lower temperature.

Then he is struck, as Carnot was, by the difficulty, that in the ordinary conduction of heat from a hot body to a cold one there is no mechanical effect ; and he gives expression to his perplexity in a footnote :—

> When "thermal agency" is thus spent in conducting heat through a solid, what becomes of the mechanical effect which it might produce? Nothing can be lost in the operations of nature—no energy can be destroyed. What effect then is produced in place of the mechanical effect which is lost?

No "energy"! The very word, then unknown in its modern signification, here bursts forth in the intuitive utterance. The note continues :—

> A perfect theory of heat imperatively demands an answer to this question, yet no answer can be given in

the present state of science. A few years ago a similar confession must have been made with reference to the mechanical effect lost in a fluid set in motion in the interior of a rigid closed vessel, and allowed to come to rest by its own internal friction; but in this case the foundation of a solution of the difficulty has been actually found, in Mr. Joule's discovery of the generation of heat by the internal friction of a fluid in motion. Encouraged by this example, we may hope that the very perplexing question in the theory of heat by which we are at present arrested, will, before long, be cleared up. It might appear that the difficulty would be entirely avoided by abandoning Carnot's fundamental axiom. . . . If we do so, however, we meet with innumerable other difficulties—insuperable without farther experimental investigation, and an entire reconstruction of the theory of heat from its foundation. It is in reality to experiment [1] that we must look—either for a verification of Carnot's axiom, and an explanation of the difficulty we have been considering; or for an entirely new basis of the theory of heat.

Having thus shown how his mind was working, he proceeds on the basis of the Carnot axiom to expand the theory of the Cycle of operations, demonstrating anew the reversibility of the Cycle, the independence of any particular fluid—air or steam—as the working substance, and the dependence solely on the two temperatures, high and low, between which the cyclic operation is conducted. So far, this is merely Carnot restated with precision. Then he notes that the amount of mechanical effects produced in the Cycle by letting down one unit of heat between

[1] Here Thomson overlooked the existence of several of Joule's series of experiments which in reality went far to supply the need. It is, however, curious that no one before Hirn in 1863 seems to have thought it worth while to measure the actual amount of heat received by the condenser from the engine, or to compare it with the amount given to the engine by the boiler.—S.P.T.

two given temperatures depends upon a certain coefficient or function, which he calls "Carnot's Coefficient," which (as Carnot himself observed) has high values for low temperatures, and low values for high temperatures; and he grasps the facts that the complete investigation of the motive power of heat turns on the experimental determination of the numerical values of this coefficient at different temperatures. Resorting, then, again to Regnault's experimental data for air, he calculates these values and tabulates them for use in the Carnot cycle.

In an Appendix, read April 30th, 1849, he continues this line of thought. He is still worshipping Carnot. "Nothing," he says, "in the whole range of Natural Philosophy is more remarkable than the establishment of general laws by such a process of reasoning." But, with the doubt as to the fundamental axiom always haunting him, he tries to deduce further inferences that can be put to the test of experiment. He compares the values of the Carnot's Coefficient as deduced from Clapeyron's experiments. He discusses the quantities of heat disengaged by compressing air in Joule's research. He compares the relative advantages of air-engine and steam-engine, and the economy of actual steam-engines. And here, face to face with the engineer's problem, he is compelled to think about the number of foot-pounds of work, and to compare the actual output of a good Cornish engine with the theoretical output calculated on the basis of Carnot's Coefficient from the temperatures of boiler and condenser, and from the data already tabulated by himself. He is already approximating to Joule's point of view, but is still not convinced.

Joule did not publish anything further for a time, but his mind was busy with the same problem, how to reconcile Carnot's principle with his own discoveries. Indeed he held the key in his own hands. He had discovered the law of equivalence in those operations where transformation takes place. Carnot had found that the amount of work done was determined by the temperatures, and depended in particular upon an unknown function or coefficient ("Carnot's Coefficient," see above), the values of which were smaller with higher temperatures. Writing to Thomson on December 9, 1848, Joule suggested[1] that the value of this Carnot's Coefficient varied inversely as the temperature at which the heat was received, if that temperature were measured from the absolute zero calculated by himself from the expansion of gases. Thomson's reply is unknown.

In June 1849 Faraday communicated to the Royal Society for Joule, the memoir "On the Mechanical Equivalent of Heat," which, having been printed in the *Philosophical Transactions*, is most often referred to. It summarised experiments on the friction of water, mercury, and cast-iron, and gave the final numerical value of 772 foot-pounds as the equivalent of the heat required to warm 1 lb. of water 1 degree of the Fahrenheit scale.

[1] Osborne Reynolds. *Memoir of James Prescott Joule*, *Manchester Literary and Philosophical Society*, series 4, vol. vi. p. 128, Manchester, 1892. See also Thomson's *Math. and Phys. Papers*, vol. ii. p. 199.

Meantime Thomson had been talking to his brother James[1] upon a curious paradox that arose from the application of Carnot's reasoning to the case of ice, leading him to the conclusion that water at its freezing point may be converted into ice by a process wholly mechanical, and yet without the final expenditure of any mechanical work. Professor James Thomson, on consideration, acutely reasoned that, unless the absurdity of a perpetual motion is to be admitted, it is necessary to conclude that *the freezing point becomes lower as the pressure to which the water is subjected is increased.* He supported this view by imagining a corresponding cycle of operations in compressing ice. This paper was read[2] on January 2, 1849. A few months later William Thomson submitted the question to experiment in which a pressure of 16.8 atmospheres was applied. The temperature of the freezing point was found to have fallen to about 0.232 of one Fahrenheit degree; the theoretical value, calculated from the known coefficient of expansion of water in freezing being, for this pressure, 0.227. This verification was published[3] in January 1850.

Success in the prediction of a new and unknown

[1] A letter written by James to his brother on August 4th, 1844, containing a curious piece of primitive thermodynamics, shows that his mind had long previously been directed to such problems.

[2] *Trans. Roy. Soc. Edinb.* xvi. p. 541, Jan. 2, 1849; "Theoretical Considerations on the Effect of Pressure in Lowering the Freezing Point of Water," by James Thomson; reprinted in *Cambridge and Dublin Mathematical Journal*, vol. v. p. 248, Nov. 1850; also reprinted in Lord Kelvin's *Mathematical and Physical Papers*, vol. i. p. 156.

[3] *Proc. Roy. Soc. Edinb.* ii. p. 372, Jan. 1850; reprinted in *Philos. Mag.* xxxvii. p. 123, 1850; and in *Pogg. Annalen*, lxxxi. p. 163, 1850; also in *Mathematical and Physical Papers*, vol. i. p. 165.

fact that is afterwards experimentally verified has ever been held to be a test of the truth of any theoretical principle. Again and again in later years Thomson would revert to his brother's brilliant prediction as an illustration of the power of a sound theory to anticipate undiscovered truth. Coming as it did at a moment when Joule's discoveries seemed to throw doubt on the validity of Carnot's theory, it confirmed him in holding fast to it, even though there seemed no way out of the existing perplexity save by new experiments. It confirmed Carnot, but did not establish the thermodynamic basis for an absolute zero, nor reveal the law of transformation.

Thomson was a strong man, and, like many other strong men, found himself incapable of accepting from others views which, though he could not refute them, he had found himself unable to justify. For more than two years he had had Joule's demonstrations before him, and had enjoyed the personal friendship of their author. He, alone, had insisted publicly on the importance of these discoveries, even though they seemed to overturn the views he held. He resisted the temptation to throw over Carnot's theory when he could not reconcile it with the new views, clinging even more tenaciously to it while he recognized that it could not express the whole truth. And the very strength of his suspended judgment, discouraging as it must have been to Joule, wrought its own reward as the solution dawned upon him. He began to perceive, as his Appendix of April 1849 shows, that Carnot's principle in itself did not deny the possibility of the transformation of heat into work; it dealt only with the proportion which the amount of work accomplished by the engine bore to the amount of heat that had entered from the boiler. As stated by Carnot, it proved nothing as to the amount of heat that passed into the condenser, and which Carnot has *assumed* to be the same. If transformation took place, then Carnot's principle still governs the transformation by determining what amount of the heat that entered should be transformed into work. It showed that the efficiency of such transformation depended solely on the temperatures. It provided a basis for calculating the answer to the question *how much* of the heat that entered was available for conversion into work. Even when he doubted most, Thomson had affirmed that "nothing can be lost in the operations of nature—no energy is lost." He now saw that, though not lost, it might not be available for transformation.

In the middle of this time of suspense two new thinkers, Rankine and Clausius, stepped into the arena. Rankine, a rising engineer, only three years senior to Thomson (and later his colleague at Glasgow), in December 1849 sent to the Edinburgh Royal Society a remarkable paper[1] "On the Mechanical Action of Heat," in which he deliberately developed a theory of heat from the basal hypothesis that in an elastic fluid the quantity of heat is the *vis viva*

[1] *Trans. R. S. Edinb.* xx. p. 147.

of the molecular movements of the molecules. He imagined a special hypothesis of molecular vortices with rotational or oscillatory movements. On this basis he developed with great mathematical skill the equations for the relations between pressure, volume, temperature, and heat, for air and for steam. He accepted, with certain reserve, Joule's experimental results; and in this and another paper two months later he deduced general laws which in effect included Carnot's principle, and defined a certain "thermodynamic function" of great importance.

Incidentally Rankine discussed the phenomenon well known to engineers, that steam issuing through an orifice from a high-pressure boiler does not scald, as low-pressure steam does. Now any saturated steam, if slightly cooled so as to permit of the condensation of minute liquid spherules, does scald; hence the inference that the friction through the orifice actually superheated the steam and prevented saturation, even though the steam in expanding must absorb the heat necessary for its expansion. This circumstance seems to have arrested Thomson's attention, for on October 15th of the same year he wrote to Joule in the following terms:—"This conclusion can, I think, be reconciled with known facts only by means of your discovery, that heat is evolved by the friction of fluids in motion." Joule, overjoyed, at once sent Thomson's letter for publication in the *Philosophical Magazine*.

Clausius, of Zürich, communicated to the Berlin

Academy of Sciences, in February 1850, a paper[1] in which he had followed out almost the same line of thought. Starting with Joule's principle of equivalence, and from a general proposition, somewhat like Rankine's but less restricted, as to the dynamical properties of gases, he deduced, as Joule had done, an absolute zero of temperatures; and, as Joule had already suggested in his (unpublished) letter to Thomson fourteen months earlier, he adopted the reciprocal of the absolute temperature (assumed from the properties of supposed perfect gases) as the Carnot Coefficient, so enabling him to state the Carnot theory in terms consistent with Joule's discoveries. He thus arrived at a vague statement of the law of transformation as embodying the principle that heat tends to pass from hot bodies to colder ones. The best thing in Clausius's paper is the following remark:—

> It is not even necessary to cast the theory of Carnot overboard. . . . On a nearer view of the case we find that the new theory is opposed, not to the real fundamental principle of Carnot, but to his addition "no heat is lost"; for it is quite possible that in the production of work both may take place at the same time, a certain portion of heat may be consumed, and a further portion transmitted from a warm body to a cold; and both portions may stand in a certain definite relation to the quantity of work produced.

This, so far as it goes, is admirably clear. If it still needed to be formulated in terms that would enable the proportions of heat utilized and wasted

[1] *Pogg. Ann.* It should be remembered that Clausius restated his theory much more clearly in 1854, and again in 1864, on its republication.

to be calculated, and to be freed from undemonstrated assumptions, this conclusion was yet important enough to secure for Clausius a place amongst the founders of thermodynamics.

Thomson was therefore forestalled in two directions in his quest for a consistent basis for the theory of heat: by Rankine, in the application of the principle of equivalence to explain some difficulties in the known behaviour of steam and other vapours under change of pressure; by Clausius, in the adoption of a hypothesis which would enable him to reconcile the theories of Carnot and Joule.

Thomson's mind was never satisfied by reasoning that started from insecure hypotheses or unwarranted assumptions. He could be satisfied neither with Clausius's doctrine that heat always tends to pass from the hotter body to the colder (an assumption incompatible with Prévost's law of exchanges), nor with Rankine's hypothesis of molecular vortices. He must have some more valid grounds for generalization. Nor was he long in finding such. He drew Joule's attention to Clausius's work, and then continued to think the thing out in his own way. In March 1851 he read to the Royal Society of Edinburgh a memoir "On the Dynamical Theory of Heat," which at once put matters in a new light. This memoir was an example of exquisite codification of isolated points, and a masterly exposition, freed from unnecessary hypotheses, of the new developments of the theory. Beginning with Davy, proceeding then to the experiments of Mayer and

Joule, whose work he now takes as fundamental, and glancing at the mathematical arguments of Rankine and Clausius, he states that the objects of the present paper are: (1) To show how Carnot's theory must be modified; (2) to point out the significance of his own former deductions from Carnot's theory and Regnault's observations (viz. the calculation of an absolute scale of temperatures), and to show that when taken in connection with Joule's mechanical equivalent of heat a complete theory of the motive power of heat is obtained; (3) to deduce some remarkable relations connecting the physical properties of all substances. In the first part he considers the "work done," or "mechanical effect" of an engine in a cycle of operations. Then he asserts that "the whole theory of the motive power of heat is founded on the two following propositions":—

PROP. I. (Joule).—When equal quantities of mechanical effects are produced by any means whatever from purely thermal sources, or lost in purely thermal effects, equal quantities of heat are put out of existence or are generated.

PROP. II. (Carnot and Clausius).—If an engine be such that when it is worked backwards, the physical and mechanical agencies in every part of its motions are all reversed, it produces as much mechanical effect as can be produced by any thermodynamic engine, with the same temperatures of source and refrigerator, from a given quantity of heat.

The first of these propositions is *the law of equivalence*, merely needing the numerical value of Joule's equivalent to be stated.

The second proposition, though not expressed in a form that would serve as a basis for calculations, is the great *law of transformation.*

Thomson generously assigns it—or so much of it as is not Carnot's—to Clausius, who had given a demonstration based upon the axiom: *It is impossible for a self-acting machine, unaided by external agency, to convey heat from one hot body to another at a higher temperature.* But Thomson himself gave a different demonstration derived from the much more fundamental axiom:—*It is impossible, by means of inanimate material agency, to derive mechanical effect from any portion of matter by cooling it below the temperature of the coldest of the surrounding objects.* The limitation to *inanimate* agency is characteristic of Thomson, and significant. He then went on to apply these laws in a rigorous way to the discussion of the known methods of producing mechanical effect from thermal agency: first, considering Joule's investigation of the heat-values of the mechanical and chemical effects produced by electric currents; secondly, dealing with the work done by the expansion of air when heated. In this latter investigation, which is in strict mathematical form, he assigned the symbol J, in honour of Joule, to the number expressing the mechanical equivalent of heat. He retained the old form of "Carnot's function," making no assumptions. And he pointed out that if the temperature-range in a Carnot's cycle be small, the amount of the realizable effect will be only a small fraction of the equivalent of the heat supplied; "the remainder being irrevocably lost to man, and therefore 'wasted,' although not *annihilated*." The bearings of this pregnant sentence[1] demand consideration later. He then proceeded to reconstruct his thermodynamic scale of absolute temperatures, and to deduce afresh the formulæ for the efficiency of the perfect engine working between two given temperatures, and to recalculate the tables for efficiencies at various temperatures of boiler and condenser. In a third part of the memoir he extended the application of the theory to other relations between the physical properties of all substances, particularly changes of state, such as freezing and evaporation. In a fourth part, dated April 17th, 1851, he is testing the laws which govern the expansion and pressure of gases, the work spent in compressing it, and the heat it produces by friction when forced through an orifice. Still working at the subject, he published, in December 1851, a fifth part of this Dynamical theory; and now he had definitely adopted the term *energy*, and is discussing the mechanical energy of a body in a given state, and its investigation by experiment. With these extensions the Memoir of 1851, as reprinted in Thomson's *Mathematical and Physical Papers*, occupies fifty-eight pages. A sixth part, which, with notes, occupies one hundred pages more, and which deals chiefly with thermo-electric currents, was not added till May 1854.

[1] In 1875 Sir W. Thomson wrote to the Duke of Argyll: "The *wear and tear*, as it were, which I have shown to accompany every dynamical action is *not* an annihilation of energy."

Before considering further developments, we may pause on the significance of the law of transformation—or second law of thermodynamics—now established. Thomson had derived it, while faithful to the Carnot conceptions of a cycle of operations and an ideal engine of perfect reversibility, by postulating a certain negation: that it is impossible by means of inanimate material agency to derive mechanical effect from any portion of matter by cooling it below the temperature of the coldest of the surrounding objects. Though called an axiom it is by no means self-evident: the utmost that we can say about it is that while it is true (for cyclical processes), its truth is not rigorously demonstrable from first principles, but is consonant with the facts of experience, no contradiction of it having ever been observed. It is, in one sense, a mere pragmatism. To the end of his life Lord Kelvin would admit that the second law of thermodynamics was a law of "natural history," rather than of natural philosophy. Further, in the formulation of the law Thomson held back from accepting the suggestion of substituting for the "Carnot's function" the reciprocal of the absolute temperature, in terms of the gas scale, as proposed independently by Joule and Clausius. Having thought out the correct principles of an absolute thermodynamic scale, he was not prepared to accept in lieu of it a scale derived from the expansion of any known "permanent" gas, whether air, hydrogen, or other, until he should have satisfied himself that the physical properties of the gas in question justified such acceptance. The test of perfection of such a gas was to be found by trying whether, if driven under high pressure through a fine nozzle or porous plug, its temperature was either raised or lowered. If it showed no change of temperature it might for this purpose be regarded as perfect, and its scale of expansion might be taken as accordant with the absolute scale. Already Thomson had arranged with Joule to conduct a conjoint research on this very point. As a matter of fact a slight cooling effect — called the Joule-Thomson effect—is observed with most gases, and is nowadays utilized in the commercial processes for the liquefaction of air and hydrogen.

The law of transformation governs the answer to the question: what fraction of the heat that enters is utilized by being transformed into work? And, when numerically stated, the answer which it gives is, the same fraction[1] as the fall of temperature bears to the absolute temperature of supply, as measured from the true zero of temperatures. Thomson saw, however, much deeper than the formulation of a numerical value. If heat, in a gas, consists in a

[1] As is well known, the fraction which states how much of the heat supplied is transformed by a perfect engine into work is expressed by the formula—

$$\frac{T'-T}{T'};$$

where T′ and T are the respective temperatures, on the thermodynamic or absolute scale, of the boiler and condenser. It was immortalized by James Napier, in 1877, by one verse of a poem read at the Glasgow Philosophical Society, addressed to Thomson:—

When you yourself once taught me
Heat's greatest work to know,
Wasn't it T dash minus T,
With a T dash down below?

diffused motion of the particles, jostling and colliding with one another as they fly about, then it will be intelligible that to obtain work from them by causing the expansion to produce a mechanical movement in some definite direction, there must be some directing or guiding of the movements of the particles. Man's power to recover thus from the miscellaneous molecular movements their inherent energy clearly depends upon his being able to take advantage of the tendency to equilibrium between different parts of the system that are in different states. When any portion of hot gas has cooled down to the temperature of its surroundings, it is no longer available as a source of power, whatever the degree of temperature of itself and its surroundings may then be. But as the movements of the particles in a gas at a given temperature are not all equal amongst themselves, some moving quicker and some slower than the average, the temperature is after all only an average statistically. And if we could sort out the quick-moving ones from the slow-moving ones, we could avail ourselves of the greater pressure of the quick-moving crowd to produce further energy from the gas. But our means are too clumsy to deal with such minute things as individual molecules; and the very magnitude of the scale on which we have to work renders such further energy unavailable to man; unavailable, that is, by any inanimate or unintelligent material agency. Maxwell, indeed, indulged in the imagination that there might be conceived little intelligent beings—"demons"—who by opening frictionless sliding doors might let through any quick-flying molecules, and close them against slow-moving ones. It is even conceivable that there might be bacteria or other organisms of minute enough scale to conduct similar operations. From the law of transformation as formulated by Thomson all such supposititious cases are excluded.

Now it happened that in the very year when Joule and Thomson first met, Helmholtz, then a young army surgeon at Potsdam, published his celebrated paper *On the Conservation of Force*, which though rejected by the leading physicists of Germany, won its way to permanent recognition. Helmholtz, starting like Carnot from the denial of the possibility of perpetual motion, conceived the great generalization that the sum of the energies in the universe (supposed to be itself finite) is constant, and that whenever force (meaning thereby *vis-viva*, or energy of motion) disappears it is not lost, but is converted into an equivalent of some other kind of energy. Helmholtz had already for several years abandoned the idea of the material nature of heat, and had accepted Joule's earliest determinations of the equivalent. His physiological studies under Müller had led him to consider the origin of the heat of animals. In this new work, which proved his powers as a mathematical physicist, he went the round of the branches of physics, showing instance after instance of the conservation of energy during its transformation from one species to another, in mechanics, heat, electrostatics, and magnetism.

For some reason Thomson never saw Helmholtz's memoir until January 20th, 1852, when he found that Helmholtz had already touched some of the matters with which he had been occupied. But Helmholtz, in accepting Joule's position as to the conversion of heat into work, had thrown over Carnot's principle, and had not attempted to formulate any law as to the proportion of heat available for transformation.

If the doctrine of the Conservation of Energy thus enunciated by Helmholtz won its way to acceptance by all physicists, for whom indeed the earlier work of Faraday and of Grove had made its comprehension easy, there yet remained another general principle, of equal importance, to be discovered. By the very suspense of judgment which Thomson had exercised in the middle of his perplexity as to the Carnot principle, he had in reality won a new insight. If only a fraction of the heat that entered the engine was available to be transformed into mechanical work, what became of the remainder? He had already (p. 283) found the answer. It was "irrevocably lost to man, and therefore 'wasted,' though not annihilated." Having been "let down" (in Carnot's phrase) to the cool temperature of the surrounding objects, it had lost its availability to be transformed into useful work. And, clearly, as heat tended always to equilibrium —the hot body to cool down to its surroundings— the diffusion of heat by conduction (which had puzzled both Carnot and himself at an earlier stage), where none of the heat is utilized, could only end in a loss of available energy. What was lost in transformation processes was not energy, but availability of energy. And this loss was always going on in Nature wherever hot bodies were cooling down; and there was no compensating regeneration, for that would have required a fresh expenditure of mechanical energy.

All this dawned on Thomson early in 1852. First he sent to the Royal Society of Edinburgh a discussion[1] of the source of animal power, and of the sources available to man for the production of mechanical effect. He sees clearly that animals can do work at the expense of their food, through the chemical oxidation going on in digestion and respiration. He denies that the animal's body acts as a thermodynamic engine, converting heat into work between definite temperatures, and regards it as more probable that the chemical forces produce the external mechanical effects through electrical means. Of the sources available to man the chief one was the heat radiated from the sun—the source (if we include luminous radiations) of our coal-fields and our timber as well as of our water-power; the second source being the motions and attractions of the earth, sun, and moon, to which the tides are due.

In April 1852 he followed this up by a short

[1] *Proc. Roy. Soc. Edin.* iii. p. 108, Feb. 1852: "On the Mechanical Action of Radiant Heat or Light; On the Power of Animated Creatures over Matter; On the Sources available to Man for the production of Mechanical Effect"; also *Philos. Mag.* iv. p. 256, Oct. 1852; reprinted *Math. and Phys. Papers*, vol. i. p. 505.

paper[1] bearing the pregnant title, "On a Universal Tendency in Nature to the Dissipation of Mechanical Energy." This paper begins by classifying stores of energy into two categories, statical and dynamical, or, as we should now say (using the adjectives introduced by Rankine and Thomson respectively), potential and kinetic. Then recurring to his own axiom as to the impossibility of deriving mechanical effect by cooling below the temperature of the coldest surroundings, he points out that in the case only of a reversible process can the heat energy be restored to its primitive high-temperature condition; that in any irreversible process, such as friction, there is a dissipation of mechanical energy, and perfect restoration is impossible; and that there is also dissipation when light or radiant heat are absorbed otherwise than in vegetation or chemical action. Then in the steam-engine, the remainder of the heat that is not converted into work is "absolutely and irrevocably wasted," unless some use is made of the heat discharged from the condenser. The paper concludes with the following three propositions:—

1. There is at present in the material world a universal tendency to the dissipation of mechanical energy.

2. Any *restoration* of mechanical energy, without more than an equivalent of dissipation, is impossible in inanimate material processes, and is probably never effected by means of organised matter, either endowed with vegetable life or subjected to the will of an animated creature.

[1] *Proc. Roy. Soc. Edin.* iii. p. 139, April 19, 1852; also *Philos. Mag.* iv. p. 304, Oct. 1852; reprinted *Math. and Phys. Papers*, vol. i. p. 511.

3. Within a finite period of time past the earth must have been, and within a finite period of time to come the earth must again be, unfit for the habitation of man as at present constituted, unless operations have been or are to be performed which are impossible under the laws to which the known operations going on at present in the material world are subject.

With this momentous pronouncement, the truth of which has never been seriously challenged, one might well conclude a chapter on Thomson's work in thermodynamics, were it not that his contributions to that science did not end there.

We have seen how Clausius and Rankine were on his heels, and during subsequent years there arose much needless controversy as to individual claims to particular points. Tyndall, on the one hand, recklessly advocated a priority for Mayer over Joule, while Tait with unsparing vigour denounced the claims of Mayer and of Clausius. Thomson stood as far as possible aloof from all this. In matters of priority in scientific discovery he was always generous; and he had shown remarkable generosity[1] toward Clausius in coupling his name with that of Carnot as to the law of transformation, since Clausius had then done little more than restate in mathematical language the equation of the Carnot cycle, improved by the arbitrary substitution of the reciprocal of the absolute

[1] In the *Revue des Sciences Scientifiques* for Feb. 8, 1868, there appeared an article, on the second law of thermodynamics, by Clausius, from which all mention of the name of Carnot was absent. Thomson, on receiving this article, sent it on to Tait, after writing on it this comment :—

"!! With no mention of the name of Carnot. O T′ return—*But don't forget to pray for Clausius.*"

temperature reckoned on a gas-scale not yet proven to be thermodynamically valid. But Thomson never was grudging of the fame of independent discoverers. "Questions of personal priority," he wrote, "however interesting they may be to the persons concerned, sink into insignificance in the prospect of any gain of deeper insight into the secrets of nature." In March 1855 Clausius wrote[1] to the *Philosophical Magazine* grumbling that Thomson, by speaking of "Mr. Joule's conjecture" that Carnot's function might prove to be reciprocal of the absolute temperature, had ascribed the discovery of this proposition to Joule instead of to himself. To this claim for priority Thomson replied[2] by quoting exactly what he had said in his paper of 1851, where, discussing Joule's suggestion of 1848 he had referred not only to Clausius's deduction, but to Mayer's still earlier assumption which Clausius had adopted without adducing any reason from experiment. To Thomson's quiet and unassertive reply no rejoinder was made. Thomson's dignified refusal, in May 1862, to bandy recriminations with Tyndall over Mayer's claims is another instance of his attitude.

To perfect the available data for further developments, Thomson embarked upon a large number of experimental investigations which occupied much of his time for some years. Conjointly with Joule he published several important papers on the thermal effects experienced by air in rushing through small

[1] Published in the May number, p. 388. [2] *Ibid.* p. 447.

apertures, and on the thermal effects of fluids in motion. They also investigated the density of saturated steam, and compared the scale of the air-thermometer with the thermodynamic scale deduced from Carnot's principle. They measured the heating effect experienced by a body whirled rapidly through the air. These researches were conducted at Joule's house near Manchester.

Other researches, principally on thermo-electric phenomena and on the thermal relations of elasticity, and on thermometry, were carried out in the laboratory of Glasgow University by Thomson and his enthusiastic handful of volunteer students.

Both Thomson and Clausius continued to work at the theoretical aspects of the subject.

Clausius confined himself rather to the purely thermal questions, handling them with great mathematical skill. Coming in 1855 upon the physical conception previously considered by Rankine under the name of "thermodynamic function," which is the quotient of any quantity of heat by the absolute temperature at which the quantity enters (or leaves) the cycle of operations, he gave to it the name of *entropy*, and proceeded to lay down the doctrine that the entropy of the universe tends to a maximum. This is indeed only another way of viewing the doctrine of the dissipation of energy, and is much less easy to grasp, because of the inherent difficulty of framing a conception as to what entropy itself is. Even so acute a mind as that of Maxwell failed to comprehend it; and the conception of entropy does not seem to have appealed to Thomson. He makes no use of it in any of his later writings. For, indeed, he had his own way of thinking of the matter, and that way was of much wider generality than that of Clausius. His conception of the availability, or non-availability, of energy is of much greater sweep, not being confined to thermodynamics, but running through the whole of energetics. The later developments of gas theory due to Maxwell, Boltzmann, and others came mostly after 1871; and such part as Thomson took in them belongs chiefly to a later period of his life. But in 1855 he had already travelled far ahead. In a paper[1] which appeared in April 1855 in the first number of the *Quarterly Journal of Mathematics*, under the title, "On the Thermo-elastic and Thermomagnetic Properties of Matter, Part I.," occurs a very important departure. In an exquisite bit of argument after the manner of Carnot, based on the consideration of the work done on a system in a cycle of operations, in changing its physical state at a constant temperature, Thomson deduced the conclusion that the heat taken in is equal to the difference between the increment of the total intrinsic energy and the potential energy acquired by the system. Or, conversely, the available energy that the system can utilise in any such operation is the difference between the decrement of its total intrinsic energy

[1] Reprinted with certain additions in *Philos. Mag.* v. p. 4, Jan. 1878; also *Math. and Phys. Papers*, vol. i. p. 291

and the heat given up by the system. This is in effect the starting-point for the later extensions of thermodynamics, which in the hands of Helmholtz, Gibbs, Van't Hoff, and Planck have carried energy-principles into every corner of physics.

Helmholtz cherished an unconcealed admiration for Thomson's deductions. In his last Königsberg report, on matters bearing on the Theory of Heat, in the year 1852, he referred to them[1] in the following terms:—"These consequences of the Law of Carnot are, of course, only valid provided that the law, when sufficiently tested, proves to be universally correct. In the meantime there is little prospect of the law being proved incorrect. At all events, we must admire the sagacity of Thomson, who, in the letters of a long-known little mathematical formula, which only speaks of the heat, volume, and pressure of bodies, was able to discern consequences which threatened the universe, though certainly after an infinite period of time, with eternal death."

[1] Helmholtz, *Popular Lectures*, translated by E. Atkinson, 1873, vol. i. p. 172.

CHAPTER VII

THE LABORATORY

Nata est ars ab experimento.—QUINTILIAN.

IN the first years of Thomson's professorship at Glasgow he had reorganized the teaching of Natural Philosophy, and, as we have seen, had secured many improvements in the material equipment for the illustration of the lectures by experiments. But as his investigations of physics proceeded, he found himself hampered by the lack of accurate data upon which to base his theoretical investigations. From Regnault in Paris he had learned the importance of minute and accurate measurement of physical quantities, and he had seen how, in the patient and skilful hands of Joule, a magical accuracy of measurement could be attained in quantitative experiment. If data were not forthcoming he must ascertain them for himself. There was no laboratory attached to the department of Natural Philosophy, still less was there any idea of instructing students in laboratory exercises. The altogether inadequate state of the apparatus at the disposal of the Professor has been detailed on p. 193, above. Up to the year 1847 the University class had met at 9 A.M. for Natural Philosophy lectures, and at 8 P.M. for lectures in Experimental Physics. In 1847 the latter course was altered to 11 A.M. By 1863 the work of the second hour had been changed to the more mathematical part of the work and to exercises, the 9 o'clock lecture being more experimental.

Driven, then, by force of circumstances Thomson sought for the opportunity to follow out experimental research. About the year 1850 an old disused wine-cellar[1] in the College basement, adjoining the natural philosophy class-room, was taken possession of. To this was joined, eighteen years later, the abandoned Blackstone examination room; an unauthorised annexation. In these inconvenient surroundings Thomson set to work, with such appliances as he could lay hands upon, to supply the data of which he stood in need. Some of his earliest determinations were concerned with

[1] "In my time," said the late Professor Ayrton, who was a student in the 'sixties, "Thomson's laboratory consisted of one room and the adjoining coal-cellar, the latter being the birth-place of the siphon recorder. . . . There was no special apparatus for students' use in the laboratory, no contrivances such as would to-day be found in any polytechnic, no laboratory course, no special hours for students to attend, no assistants to advise or explain, no marks given for laboratory work, no workshop, and even no fee to be paid. But the six or eight students who worked in that laboratory felt that the *entrée* was a great privilege. . . . Thomson's students experimented in his one room and the adjoining coal-cellar, in spite of the atmosphere of coal dust, which settled on everything, produced by a boy coming periodically to shovel up coal for the fires. If for some test a student wanted a resistance coil, or a Wheatstone's bridge, he had to find some wire, wind the coil, and adjust it for himself. It is difficult to make the electrical student of to-day realise what were the difficulties, but what were also the splendid compensating advantages of the electrical students under Thomson in the 'sixties. . . . But oh! the delight of those days! Would we have exchanged them, had the choice been given us, for days passed in the most perfectly designed laboratory of the twentieth century without him? No! for the inspiration of our lives would have been wanting."

the thermo-electric properties of the metals. Others related to elasticity and other properties of matter under stress. To assist him, he organized a body of voluntary helpers from amongst his students, who, excited by his infectious enthusiasm, worked with him and for him.

The Bangor address (p. 845) tells of their work:—

> I found as many as three-quarters of the students were destined for service in the religious denominations in after-life. I have frequently met some of those old students who had entered upon their profession as ministers, and have found that they always recollected with interest their experimental work at the University. They felt that the time they had spent in making definite and accurate measurements had not been thrown away, because it educated them in accuracy—it educated them in perseverance if they acquired such education. . . . There is one thing I feel strongly in respect to investigation in physical or chemical laboratories—it leaves no room for shady, doubtful distinctions between truth, half-truth, whole falsehood. In the laboratory everything tested or tried is found either true or not true. Every result is *true*. Nothing not proved true is a *result*; there is no such thing as doubtfulness.

Fitful as the work was at first, and erratic as were the varying requirements of the time, the results which followed from this primitive laboratory were of enduring worth. It was the first physical laboratory to be put at the disposal of students in any of the universities. Years afterwards, at the jubilee of Sir George Stokes, in June 1899, Lord Kelvin declared that Stokes's room at Cambridge was the first physical laboratory to be formed in any university in Great Britain. But Stokes was essentially an individual worker, making his own experiments, unaided by assistants or students, so that the Glasgow laboratory must really be accounted the first working laboratory for physical science. Thomson does not seem to have made the acquaintance of Stokes till after his return from Paris in May 1845; at least Stokes's name never occurs in Thomson's earlier letters or in his diary. Yet he was settled in Pembroke as a junior fellow, having been Senior Wrangler in 1841. It seems that it was the task of editing the *Mathematical Journal* which brought them together; and Stokes's penchant for experimenting led Thomson often to seek his advice. Stokes was indeed guide, philosopher, and friend to his eager and enthusiastic disciple. While Thomson was daring in speculation, moving swiftly, almost erratically, to some intuitive result, Stokes was methodical, conservative, cautious. When as young men they paced the quadrangles of the ancient colleges, discussing keenly the deep problems of mathematical physics, each supplied something to the other's mental equipment, and each in his own mode was unrivalled in attacking the unsolved problems of physics. Whenever the letters between Stokes and Thomson come to be published, it will be seen how invaluable to both was the friendship so formed. Stimulus on the one hand, measured judgment on the other, went ever with the deepest and most scrupulous regard for scientific truth. "Consult Stokes" was Thomson's continual word—and action—when any doubtful proposition presented itself for decision. "What would Thomson think?" was equally the mental attitude of Stokes. For more than fifty years each was in the habit of communicating to the other the progress of his ideas—so much so that often they could not be certain with which of them the original suggestion arose. A notable example occurs in relation to the subject of Spectrum Analysis. Thomson had, in 1850, witnessed the brilliant experiments of Foucault on the spectra given by the use of the electric arc lamp (see p. 224), the bright lines in the spectra of the metals sodium, copper, silver, etc., and of the reversal of the sodium line, giving a black line of absorption in place of the characteristic yellow. Talking these things over with Stokes, the suggestion of a physical explanation was perceived by them. In the hot vapours each metallic particle is free to vibrate with its own natural mode or modes of vibration, and emits its own characteristic kind of light as a bright line or lines; and the absorption indicated by the dark lines in the spectrum can be accounted for by admitting that the vibrating particles can take up, in their own natural modes of vibration, all (or most) of the energy of those constituents of mixed light trying to pass through the vapour, which have the same periods as those modes. Thomson often said that this, the elementary theory of spectrum analysis, he learned from Stokes. "He taught it to me at a time that I can fix in one way indisputably. I never was at Cambridge once from about

June 1852 to May 1865; and it was at Cambridge, walking about in the grounds of the colleges, that I learned it from Stokes" (*Baltimore Lectures*, p. 101). The same inquiry formed the subject of correspondence between the friends in 1854 (*Collected Papers* of Sir G. G. Stokes, vol. iv. p. 367). But Stokes was always loth to accept the credit of this discovery. At the Stokes Memorial meeting at Westminster Abbey in July 1904, Lord Rayleigh commented thus: "Stokes was always very modest upon this subject, and almost repudiated the credit which Lord Kelvin wished to give him. All one could say was that the thing lay between Lord Kelvin and Stokes. The letters which passed between them in 1854 showed plainly enough that the idea was there. Lord Kelvin told them he got his inspirations from Stokes. It might be, and was likely enough, that he developed them; but, at any rate, between the two correspondents they had the whole theory." But by the wholesome rule accepted by men of science, no private communications, oral or written, can establish priority to any scientific discovery; hence it is important to note that, for many years prior to 1860, Thomson was in the regular practice of stating in his public lectures at Glasgow this theory, pointing out that solar and stellar chemistry were to be studied by comparing the dark lines of solar and stellar spectra with the corresponding bright lines in the spectra of artificial flames. A letter of his in 1860 to Helmholtz states this. Thomson's fullest account of his part was given in his Presidential Address (1871) to the British Association, reprinted in *Popular Lectures*, vol. ii. p. 169. In 1902 Lord Kelvin wrote to Professor Königsberger the following additional statement: "I well remember that at that time I was making 'Properties of Matter' the subject of my Friday morning lectures. One Friday morning I had been telling my students that we must expect the definite discovery of other metals in the sun besides sodium, by the comparison of Fraunhofer's solar dark lines with artificial bright lines. The next Friday morning I brought Helmholtz's letter with me into my lecture and read it, by which they were told that the thing had actually been done with splendid success by Kirchhoff." In an appreciation by Professor John Ferguson in the *Glasgow University Magazine* of 1908, is the following passage on the subject of Stokes's explanation of the cause of the dark line D in the solar spectrum, namely, absorption by sodium vapour in the sun's atmosphere, and Thomson's lecture on it in 1859:—

This was explained to us in illustration of a picture of Fraunhofer's lines; and either then or a little later I remember Thomson trying to demonstrate to a few of us the phenomenon by observing the flame of a spirit-lamp made yellow with a little common salt; through sodium vapour got by heating the metal in a platinum capsule. Ten days after this lecture he read to the class a letter he had received the same morning from Helmholtz at Heidelberg. In it was narrated how "Kirchhoff had ascertained that the flame of sodium supplied the line D in the spectrum, of potassium Aa and B in the red, iron Eb in the blue, *und so weiter*," and Thomson was not slow to add that there must be an immense amount of iron in the sun's atmosphere, and throughout space.

At an even earlier period of Thomson's acquaintance with Stokes they had planned together the writing of a series of Notes on Hydrodynamics for the *Cambridge and Dublin Mathematical Journal*, several of which notes duly appeared. In presenting the Copley medal in 1903 to Stokes, Lord Kelvin recalled Stokes's work. "Fifty-two years ago he took up the subject of fluid motion, with mathematical power amply capable to advance on the lines of Lagrange, Fourier, Cauchy, Poisson, in the splendid nineteenth-century 'physical mathematics,' invented and founded by those great men; and with a wholly original genius for discovery in properties of real matter, which enhanced the superlative beauty of the mathematical problems by fresh views deep into the constitution of matter." Thomson could be enthusiastic indeed over the qualities of his friend. "I always consult my great authority Stokes whenever I get a chance" was an aside in the Baltimore Lectures. At the Stokes Jubilee of 1899 Lord Kelvin said:—

When I reflect on my own early progress, I am led to recall the great kindness shown to myself, and the great value which my intercourse with Sir George Stokes has been to me through life. Whenever a mathematical difficulty occurred I used to say to myself, "Ask Stokes what he thinks of it." I got an answer, if an answer was possible; I was told, at all events, whether it was unanswerable. I felt that in my undergraduate days, and I feel it more now.

When Thomson first organized a physical laboratory it was because he found great need for physical data as to the properties of matter. He began with a subject to which his thermodynamic studies had led him—the electric convection of heat, but soon branched off into other work, the determination of Moduluses of Elasticity and the Electrodynamic Properties of Metals.

Early in 1854 he sent to the Royal Society of Edinburgh some account of his experiments on thermo-electric currents, and another on the mechanical theory of thermo-electric currents in crystalline solids. To the French Academy he sent two papers: one on the oscillations of diamagnetic and magnetic needles; the other on the effects of electric currents on unequally heated conductors, producing in them, as he discovered by theory, an actual conveyance of heat by the current from one part of the conductor to the other. A letter of his to M. Élie de Beaumont on the effects of pressures and tensions on the thermo-electric properties of metals was also printed in the *Comptes rendus* of 1854.

Thomson's advice was sought by Clerk Maxwell, who had just taken his degree as Second Wrangler.

TRIN. COLL., *Feb.* 20, 1854.

DEAR THOMSON—Now that I have entered the unholy estate of bachelorhood I have begun to think of reading. This is very pleasant for some time among books of acknowledged merit wh. one has not read but ought to. But we have a strong tendency to return to physical subjects, and several of us here wish to attack Electricity.

Suppose a man to have a popular knowledge of electrical show experiments and a little antipathy to Murphy's *Electricity*, how ought he to proceed in reading and working so as to get a little insight into the subject wh. may be of use in further reading?

If he wished to read Ampère, Faraday, etc., how should they be arranged, and at what stage and in what order might he read your articles in the *Cambridge Journal*?

If you have in your mind any answer to the above questions, three of us here would be content to look upon an embodiment of it in writing as advice. . . .

In conclusion, commend me to the Blackburns and Mrs. Thomson.—Yours truly, J. C. MAXWELL.

Mrs. Thomson's health did not improve, and so in August Thomson wrote to Mrs. King:—

Margaret has not advanced as I had hoped and expected she wd. have done by this time, and it has been very disappointing to feel so little improvement with so much of the summer passed. She is, however, doing somewhat better now, and I think has really advanced during the last three weeks, certainly has suffered less. She *looks* much better and in some respects is much better than when you saw her last, but she has not at all advanced in walking power. I always carry her up and down stairs, and often from one room to another. A walk half round Miss Graham's garden lately knocked her up for several days. But by avoiding all such exertion she keeps tolerably free from pain, and has much the appearance of good health. I take her a drive nearly every day and sometimes twice in a little pony carriage.

I have had hard work on papers for the London and Edinburgh Royal Societies lately, and I now feel much relieved. . . . One of my papers I think will interest David [Dr. King]. It is to prove that the sun's heat is produced by the friction in his atmosphere occasioned by meteoric matter whirling round him (seen, as perhaps you have seen it, as the zodiacal light), and continually being drawn in by his gravitation and incorporated in his mass.

To the British Association at Liverpool he gave also some account of these researches, and a second paper "On the Mechanical Antecedents of Motion, Heat, and Light," being a contribution to the energy-view of the universe. In this he speaks of Joule's discovery of the mechanical equivalence as "the greatest reform that physical science has experienced since the days of Newton," and incidentally denies the nebular hypothesis of Laplace.

But he was far from the end of his labours on the properties of matter. The following letter to his brother deals with matters after the fifth or sixth session of work in the laboratory :—

2 COLLEGE, GLASGOW,
Saturday, Jany. 12, 1855.

MY DEAR JAMES—I have been very dilatory in writing to you since I heard from you last. I would have written from Largs, where I spent the holidays, to wish you a good New Year, but I was occupied all the time I could get for writing with an urgent paper on elasticity. People often ask me about you here, and I say that I believe you are getting a good deal of business, which also seemed to be the impression of those who heard of you otherwise. The water-wheels, too, I hear sometimes spoken of, and I hope they are going to get a name yet of practical value. Have they begun to be profitable? I hope they will sooner or later. I have been keeping experimental work going on with all the hands I can get applied, and have made out one or two new results since the beginning of the session. I find it very difficult, however, to make rapid progress.

A curious thing which has just been coming out yesterday and to-day is, that the whole effect of even a slight tension quite within elastic limits every time it is applied or removed is only gradual in altering the electric conductivity of iron or copper wires. That is to say, for a considerable time the effect of either taking off or putting on weights gradually increases. I am strongly inclined to think that there is a corresponding slowness in the final settlement of the elastic forces in many cases, and possibly that there is not so near an approach to perfect elasticity within even very narrow limits as we have been supposing.

Margaret joins me in best wishes for a happy New Year to yourself and your wife, and I remain, your affectionate brother, WILLIAM THOMSON.

The years 1855 and 1856 were extraordinarily fruitful. To the *Philosophical Magazine* he sent some demonstrations of propositions in the theory of magnetic force, and a communication on the "Magnetic Medium," and on the effects of compression. The former paper was a justification of Faraday's law, that for diamagnetic bodies there is a tendency to move from places of stronger towards places of weaker magnetic force. The later was in the form of a letter to Tyndall, whose views on diamagnetic polarity seemed to Thomson to require correction. Tyndall replied, contesting the necessity of Thomson's inferences, to which Thomson rejoined reasserting his position, but modifying the phraseology. Thomson was also beginning his work on the theory of the telegraph, which is narrated in the next chapter.

In April 1855 Thomson communicated to the

Quarterly Journal of Mathematics an article "On the Thermo-elastic Properties of Matter."

The state of Mrs. Thomson's health, however, interrupted the laboratory work. She was advised to try the curative effect of treatment at a German spa, and accordingly a portion of the summer was spent at Creuznach. Thence he wrote on July 23 to Mrs. King:—

> The doctor here, who I think appears to be a sensible man, is very confident that Margaret will be well, and that the waters here will be most useful in promoting her recovery. . . . As she is not allowed to ride, or to walk more than a few minutes every day, I take her out in a Bath-chair, between which, and a balcony on which she sits a great deal reading or working, she gets a good deal of the open air.
>
> This is a very dull place, and the scenery gives us very poor compensation for our favourite Arran, which we are losing. . . .

Two entries in the mathematical note-book at this time show that he was busying himself with the problem of the origin of solar heat, and the retardations in telegraph cables.

This sojourn at Creuznach brought Thomson within range of Helmholtz, then about to quit Königsberg for the chair of Anatomy and Physiology at Bonn. Thomson had in 1852 read Helmholtz's now famous memoir on the Conservation of Force; and Helmholtz, in a Königsberg memoir, had discussed Thomson's thermodynamic papers of 1852. The two pioneers of physics were now to meet face to face.

Emery Walker Ph. sc.

Margaret Thomson.

From a photograph made in about the year 1858

[W. THOMSON to H. L. F. HELMHOLTZ.]

CREUZNACH, BEI BINGEN,
July 24th, 1855.

SIR—I believe you have sometime since received an official invitation to attend the meeting of the British Association to be held at Glasgow in September. I write now to express personally my anxious wish that you may accept that invitation. I should consider your presence as one of the most distinguished acquisitions the meeting could have, and for this reason, if for no other, it would be satisfactory to me to hear that you would attend, but I would look forward on my own account with the greatest pleasure to such an opportunity of making your acquaintance, which I have been anxious to do ever since I first had the "Erhaltung der Kraft" in my hands. I regretted extremely not having been at the Hull meeting, when I heard you had been there, and I was much disappointed, too, to lose the opportunity of seeing you when you were so good as to call on me afterwards in Glasgow; but I hope that this summer I may be more fortunate. May I ask the favour that you will let me know if you determine to come, and allow me to arrange to have accommodation provided for you during your stay in Glasgow? Will you also let me know if there is any prospect of your being in this part of Germany before September, as I should like, if possible, to make some plan to meet you? I shall remain at Creuznach for three weeks longer, after which my plans are as yet necessarily uncertain.

Allow me to take this opportunity of thanking you for the papers you have been so good as to send me, each of which I need not tell you I value very highly.—I remain, with much esteem, very truly yours,

WILLIAM THOMSON.

Prof. Helmholtz.

P.S.—If you write to me I shall be glad that you do so in your own language, which I shall feel no difficulty in reading, although I do not know it well enough to be able to write it myself.

About a week later Helmholtz journeyed to Bonn, and thence to Creuznach. On August 6th he wrote to Frau Helmholtz that Thomson had made a deep impression on him.

> I expected to find the man, who is one of the first mathematical physicists of Europe, somewhat older than myself, and was not a little astonished when a very juvenile and exceedingly fair youth, who looked quite girlish, came forward. He had taken a room for me close by, and made me fetch my things from the hotel and put up there. He is at Creuznach for his wife's health. She appeared for a short time in the evening, and is a charming and intellectual lady, but is in very bad health. He far exceeds all the great men of science with whom I have made personal acquaintance, in intelligence, and lucidity, and mobility of thought, so that I felt quite wooden beside him sometimes.

Returning to Glasgow for the British Association meeting on September 12th, 1855, Thomson there communicated no fewer than six papers. These were: On the Effects of Mechanical Strain on the Thermo-electric Qualities of Metals; On the Use of Observations of Terrestrial Temperature for the Investigation of Absolute Data in Geology; On new Instruments for measuring Electrical Potentials and Capacities; On the Electric Qualities of Magnetized Iron; On the Thermo-electric Position of Aluminium; and On Peristaltic Induction of Electric Currents in Submarine Telegraph Wires. The adjective "peristaltic" he applied in a picturesque way to denote the action in a submarine cable, whereby any short, sharp electric impulse given to one end of the cable in being conveyed to the other is gradually changed *in transitu* into a less distinct impulse of longer duration, a question to be discussed in the next chapter.

In September 1855, during the B.A. meeting at Glasgow, Maxwell wrote Thomson a long letter concerning the theory of electricity, which Maxwell, judging from the fragments published, conjectured to be lying unpublished in Thomson's desk. It concluded :—

I do not know the Game laws and Patent laws of science. Perhaps the Association may do something to fix them, but I certainly intend to poach among your electrical images ; and as for the hints you have dropped about the "higher electricity," I intend to take them.

Thomson's reply cannot be found, but Maxwell wrote to his father that he got a long letter from Thomson, "and he is very glad that I should poach on his electrical preserves."

All through the autumn of 1855 the laboratory researches were keenly pressed on. We get a notice of them in another letter to James Thomson.

2 College, Glasgow,
December 7th, 1855.

My dear James—I have a good deal going on in the way of experimenting by students under my direction, and some good results obtained already this session ; one that iron wire has its conducting power for electricity diminished by tension. Copper wire shows the same property, but perhaps with a difference. . . .

I am to give the Bakerian Lecture this year in London (R.S.) "On the Electrodynamic Properties of Metals," and at the same time (the end of Febr^{ry}) to give a lecture at the R. Institution (Faraday's place) "On the Origin and

Transformation of Motive Power." If you see Dr. Andrews, will you ask him if he could give me any data regarding the heat of combustion of gunpowder?—Your affectionate brother, WILLIAM THOMSON.

In the preceding November the Rev. J. Barlow, Hon. Secretary of the Royal Institution, had sent Thomson an earnest request to give one of the Friday evening discourses which have made the Royal Institution famous. Already Lyell, Grove, Owen, and Hofmann had promised, besides Faraday and Tyndall. Thomson accepted the invitation, selecting February 29th as date. "A thousand thanks," wrote Barlow; "it will make Faraday jump for joy."

ROYAL INSTITUTION,
January 18th, 1856.

MY DEAR SIR—It so rejoices me to see your name upon our list of Friday evenings that I cannot help but write to congratulate, not you, but myself on the delight I shall have. My head gets weary and dull or else I should often trouble you with a letter, for it seems to me I could often ask and as yet you have always answered. I do not mean in mere form, but to my judgment and understanding I wish I could continually sit under your wing.

I understand Tyndall has undertaken to get all prepared for you that we can do here, or else you know how glad I should be to be useful.—Ever truly yours,

M. FARADAY.

Prof. W. Thomson, etc.

To this Thomson replied:—

2 COLLEGE, GLASGOW,
January 29, 1856.

MY DEAR SIR—Although I hope soon to see you in London, I cannot delay till then thanking you for your letter of the 18th, and for the very kind expressions it contains. Such expressions, from you, would be more than a sufficient reward for anything I could ever contemplate doing in science. I feel strongly how little I have done to deserve them, but they will encourage me with a stronger motive than I have ever had before, to go on endeavouring to see in the direction you have pointed, which I long ago learned to believe is the direction in which we must look for a deeper insight into nature.

I cannot express to you how much I fall short of deserving what you say, but must simply thank you most sincerely for your kindness in writing as you have done.—Believe me, ever yours truly, WILLIAM THOMSON.

Prof. Faraday.

As the day drew near Thomson sent to Tyndall a list of fifteen experiments he proposed to show, and was answered by Tyndall as follows :—

ROYAL INSTITUTION,
Saturday, 23*rd February* 1856.

MY DEAR SIR—Up to the present moment I have been unable to write a word in reply to your last note. On Thursday I have a lecture and shall be very busy ; but on Friday I am quite at your service. We shall have a long day to prepare, and from the nature of the experiments I infer we shall be able to make all arrangements comfortably and efficiently. We have a very beautiful means of showing the heat of congelation, which I will have prepared on Thursday so that it shall be cool on Friday. The subject, so far as I can see, is an extremely promising one ; but remember you have only an hour at your disposal, and if you made all the experiments on your list it would be at the rate of an experiment every four minutes. Would this afford you sufficient time for explanation ? If out of the fifteen you were to choose ten, these ten would afford so many resting-places to the general mass of the audience, and would at the same time allow you a tolerable freedom of explanation. With regard to the experiment on the effect of induced currents, I devised the following means of estimating the comparative conductivities of bismuth and copper some time ago. A cube of each metal was taken and suspended from a twisted string, above the cube and attached to it by a copper wire was a small pyramid, with the base horizontal and its four sides formed by four little triangular pieces of looking-glass. The mirrors rotated with the cube, and the latter was, of course, placed between the poles of an electro-magnet. A beam of light was thrown upon the mirror, and as it went round—slowly at first—the images reflected from its sides followed each other in a circle of about 30 feet in diameter. As the motion quickened the patches of light blended themselves into a continuous line. On closing the circuit with the copper the effect was astonishing—your own term best describes it, the cube was "struck dead." The idea of the top is a beautiful one if you can realise it—if not, we can have the experiment in the above form.—Very sincerely yours,

JOHN TYNDALL.

The discourse was duly pronounced. The notes prepared for it contain various matters not found in the report published in vol. ii. of the *Royal Institution Proceedings*, or in the reprint. The prologue, in the drafting of which Mrs. Thomson helped her husband, was as follows :—

An audience assembled in this place has such frequent opportunities of profiting by the rare faculty of conveying the profoundest views in the clearest manner, that any one who does not possess that faculty must feel that in the position I now occupy he has no easy task before him. I must simply beg for your patience and indulgence, trusting that although accustomed to have the most abstract points of philosophy brought before you, so illuminated that no darkness can be felt, you will nevertheless remember that scientific explanations may be characterized by dulness, and that it is not unusual for a lecturer to fail in that clearness which might consist with the most profound and philosophical treatment of the subject. I fear I must commence by committing the first fault I have mentioned, while I explain some terms and general principles which perhaps require no explanation. If I avoid the second, I shall succeed better than I venture to anticipate.

Joule's researches, with experiments and numerical demonstrations, occupied the chief place in the discourse; and then followed some considerations as to the economy of transformation. In the galvanic engine, where the transformation of energy was not thermal, the equivalent of work done might be 75 per cent or higher. Intensification of heat was impossible, but a portion of the heat might be intensified if the remainder were reduced to a lower intensity by a wider diffusion. What was the mechanical value of cold—for example of ice? Nothing, in winter; but in summer, with thermometer at 92°, the mechanical value of 1 lb. of ice is 15800 foot-pounds, equal to 1 horse-power for half a minute. An engine burning say 4 lbs. coal per horse-power per hour might make 60 lbs. of ice; or 1 ton ice for 60 lbs. coal. Ice might be taken at 1d. per lb., or say £5 a ton. Either ice is too dear, and ought to be made by steam-power; or so cheap that it may be used instead of coals for steam-engines and general heating purposes; or the present system is the best, and ice is cheaper gathered and kept, than made, but not yet so cheaply as to supersede coals. But the time might come when the only way of heating that will be available will be founded on some such process of intensification. The electrical transformation of energy claimed attention, with the query whether all light, and all heat was not electric in its nature. Animal power —the energy of organic life—was probably electric, not thermodynamic. The sources available to man for the production of mechanical effect were examined and traced to the sun's heat and the rotation of the earth round its axis. Published speculations were then referred to, by which it is shown that the motions of the earth and of the heavenly bodies, and the heat of the sun, may all be due to gravitation; or, that the potential energy of gravitation may be in reality the ultimate created antecedent of all motion, heat, and light at present existing in the universe. The peroration was brief.

> The opening of a bud, the growth of a leaf, the astonishing development of beauty in a flower, involve physical operations which completed chemical science would leave as far beyond our comprehension as now the differences between lead and iron, between water and carbonic acid, between gravitation and magnetism, are at present. A tree contains more mystery of creative power than the sun, from which all its mechanical energy is borrowed. An earth without life, a sun, and countless stars, contain less wonder than that grain of mignonette.

No record has been found as to the success of the lecture; of its great originality there can be no question. He had previously published before the Edinburgh Royal Society in April 1854, and the British Association in September 1854, speculations that occupied the conclusion of the lecture, to the effect that the potential energy of gravitation may be the ultimate antecedent of the motions of the earth and of the heavenly bodies, and of the heat of the sun, and therefore of all heat and light and other sources of power available to man.

The Bakerian Lecture is the distinctive title awarded by the Royal Society each year to some original paper of particular merit in the group of physical sciences. Thomson's Bakerian Lecture of February 28, 1856, printed in the *Philosophical Transactions*, vol. cxlvi., is a remarkable memoir extending over 102 pages. By the title "Electrodynamic Qualities of Metals," he wished to connote all properties, such as electric and thermal conductivity, magnetic permeability, retentivity, and thermo-electric rank, together with their variations with temperature change or stress. Starting from the dynamical theory of heat, and applying Carnot's principle to the phenomena of thermo-electricity, he had discovered the unequal thermal effects produced in unequally heated metals by currents passing through them from hot to cold, and from cold to hot. He thus announced the reversible thermal effect of electric currents in which there is a convection of heat, which in the case of the metal iron he found by experiment to be in a direction opposite to that conventionally assigned to the positive flow of current, whereas in copper the convection is in the same sense as the positive flow.[1] This investigation had lasted from 1851 to the end of 1855, and the experimental appliances and operations were described in great detail. The second part of the memoir related to Thermo-electric Inversion, and was directed to elucidate a phenomenon discovered

[1] Another statement of the "Thomson effect" is: An electric current passing in an iron bar or wire from a hot to a cold part produces a cooling, but in copper a heating, effect.

by Cumming that the current generated by heating the junction of certain thermo-electric pairs, such as iron and zinc, the current at first goes from zinc to iron, and, as the temperature is raised, ceases and then passes in the inverse direction from iron to zinc. The neutral temperatures for various pairs of metals were determined, and a diagram of thermo-electric powers plotted. The third part was on the Effects of Mechanical Strain and of Magnetization on the Thermo-electric Qualities. Metal wires subjected to tension showed a change of properties, stretched copper acting as thermo-electrically positive towards unstretched copper; but in iron the reverse was the case. But iron that has been hardened by stretching and then left to itself is thermo-electrically positive with respect to soft iron; also, unmagnetized iron is positive towards magnetized iron if the magnetization is longitudinal, but transversely magnetized iron is positive towards longitudinally magnetized iron. All these effects were investigated and proved with a wealth of patience and ingenuity that are notable. The fourth part on Methods for Comparing and Determining Galvanic Resistances is remarkable, in that, in the attempt to investigate the relative conductivities in a magnetized sheet of iron, Thomson was led to an independent rediscovery of the principle of the device familiar as "Wheatstone's bridge" (discovered by Christie, see Bakerian Lecture for 1843), in which the proportional resistance between the branches of a divided circuit is utilised. The arrangement now known as the

Thomson double-bridge, for testing small resistances, appears to have its origin here. Part five dealt with the Effect of Magnetization on Electric Conductivity, and again a host of ingenious laboratary devices are described. The result deduced was that the conductivity of iron is increased by magnetic force acting across the lines of current. Thus ended for the time this great research. But an addition was made in 1857, extending the results to nickel. And again in 1875, when the resources of a new physical laboratory were available, parts six and seven were added, ending with an account in 1878 of the effects of stress on the magnetization of iron, nickel, and cobalt. In Thomson's reprint (1884) of his Mathematical and Physical Papers this classical research occupied more than half of the second volume.

As if this were not enough, Thomson gave that spring to the Royal Society two more papers. One of these was his scarcely less notable "Elements of a Mathematical Theory of Elasticity," which was awarded a place in the *Philosophical Transactions*, vol. cxlvi., and has become the foundation for all that has since been written on the subject. It lays the theory upon the basis of the proposition, that when any portion of elastic matter is subjected to a *stress*, there results a *strain*; the energy expended in the operation being the product of the stress into the strain.

The other paper proposed a dynamic explanation of Faraday's discovery of the rotation of the plane of polarization of light by the magnet.

When May brought an end to the session of 1855-6, the state of Mrs. Thomson's health prompted a return to the baths of Germany. A letter to his brother contains several points of interest.

CREUZNACH, *July* 2, 1856.

. . . . . .

MY DEAR JAMES—Before we set out for Germany I spent ten days with Joule, and got through some very interesting experiments along with him. Among other things (illustrating the fact, which we had long suspected from other experiments, that a solid against which air is very rapidly flowing always takes a higher temperature than the air), we found that thermometers and thermo-electric junctions, when whirled through the air at from 80-120 feet per second, showed very sensibly higher temperature than when whirled slowly, so as to show simply the temperature of the air. We have found that air forced through small circular apertures follows a very curious law of discharge, according to which the velocity in the aperture, calculated as if the air there were at the same density as on the high-pressure side, has a maximum value of about 550 feet per second, when the high pressure is about 50 inches of mercury above the atmospheric pressures. Either less or greater pressures give less rapid discharge, estimated in bulk on the high pressure side.

I have become acquainted with a Mr. Dellmann here who is "Oberlehrer" in the gymnasium, and has besides charge of meteorological, but especially electrical observations for the Prussian Government. I have seen his mode of observation of atmospheric electricity, which is very simple. . . . There is almost always an effect of one kind (indicating a negative electrification of the earth's surface), but when the sky is much overcast, little or none. Detached clouds often alter the quality of effect quickly, and give a great amount of reverse effect, and the first drops of a shower generally do so. When I was with

Joule we sent up a kite nearly $\frac{1}{4}$ of a mile with a thin iron wire instead of cord from the roof of the house. In an overcast day we got no effect sensible to the knuckles, but on another day with blue sky slightly clouded, we got sparks from $\frac{1}{4}$ to $\frac{1}{2}$ inch long which gave me a shock sensible down to the ankles. I am having an electrometer of Mr. Dellmann's construction, which, I think, is the best yet brought into use, made here to take with me.

It is likely we shall only be a week more here, and then go on to Schwalbach for chalybeate waters, from which a strengthening effect is expected. Margaret is feeling somewhat better and is thought to have made very decided improvement, or else she would not be allowed to try iron waters. . . .

WILLIAM THOMSON.

From Schwalbach Thomson wrote to Helmholtz :—

SCHWALBACH, *July* 30, 1856.

MY DEAR SIR—I have delayed so long to write in reply to your kind letter of the 18th of last month,[1] because I have been hitherto in much uncertainty as to our plans even now. After we are come to this place it is still uncertain how long we shall have to stay, but I see that our homeward journey must be so late that we shall be anxious to make it in the shortest possible time, even if, what it would perhaps be hoping too much from chalybeate waters to expect, my wife would be feeling well enough to undertake any digressions from the way. I should be very sorry, however, to leave Germany without having had any opportunity of seeing you, and I therefore intend to take a run down to Bonn and spend a day there some time either next week or the week after.

My wife did not make any decided improvement during our stay at Creuznach, but in the first few days

[1] This letter announced to Thomson the research on combination-tones, and the discovery of the possibility of the production of objective difference-tones.

after coming to this place she seemed to gain considerably. We have had only a week's experience of the Schwalbach waters, and we can only judge that they are not going to disagree with her. The doctors say that if they do not disagree they are sure to do good, but we have learned to be very sceptical of all "Kurs," and to have very moderate expectations from a few days of improvement. I have only confidence in time.

I shall have many questions to ask you regarding the agency of iron in conveying oxygen to the fire in the animal system, and the quantity and circulation of blood in the human body.

I ought to have thanked you sooner for all the trouble you have taken to procure a wire of measured galvanic resistance for me. Last spring I received a conductor which about four years before I had sent to Germany by a friend, and which was returned with a memorandum of its resistance as determined by Weber, if I remember right, one day of last August. The despatch of that wire for me may have been considered as an answer to your letters, and I should certainly have written to you when I received it if I had guessed that it had any relation to the application you kindly made on my behalf. I have already made use of it to investigate the electromotive force of a cell of Daniell's, with a view to various electrodynamic applications.

Mrs. Thomson sends her regards, and I remain, yours very truly, WILLIAM THOMSON.

SCHWALBACH, *August* 6, 1856.

MY DEAR HELMHOLTZ—I intend to set out to-morrow morning, and go down the Rhine by a steamer leaving Eltville about 10 o'clock, and arriving at Bonn at $4\frac{1}{2}$ P.M. Do not think on giving up any plan you may have made for an excursion on my account, because I should certainly find you on Friday morning when you will have your lecture. I shall remain at Bonn over Friday, and return here on Saturday.—(In haste) yours very truly,

WILLIAM THOMSON.

Thomson went to Bonn as planned; and a fortnight later Helmholtz returned the visit.

EINHORN, SCHWALBACH,
August 11, 1856.

MY DEAR HELMHOLTZ—I find there is no post carriage between this and St. Goarshausen, but only a communication by an omnibus and a post carriage on a cross road which would probably not be convenient.

The omnibus leaves Eltville at 6, or somewhat later, professedly on the arrival of the steamer from Cologne, but when I arrived there about $\frac{3}{4}$ of an hour late I found it had just left, having waited for the Cologne and Dusseldorf steamer, but not for the *Niederländer* which ought to have arrived at the same time. I set out immediately and walked to Schwalbach, taking about three hours to do it. You would probably prefer coming by Bieberich, which is about $\frac{3}{4}$ h. higher up the Rhine, and about a "Stunde" farther from Schwalbach than Eltville. The omnibus leaves Bieberich at 10 o'cl. in the mornings and arrives here about 1 P.M.

If I do not hear otherwise from you I shall expect you by it on Thursday or Friday, and you will come with me and dine at the table d'hôte, for which you will arrive just in time.

Begging to be remembered to Mrs. Helmholtz and to the other ladies.—I remain, yours very truly,

WILLIAM THOMSON.

Helmholtz left for Schwalbach on September 15. He wrote to his father that he was going there "in order to meet Professor Thomson from Glasgow, whom I visited last year in Creuznach, and who has principally concerned himself with the Theory of the Conservation of Energy in England. He is certainly one of the first mathematical physicists of the day, with powers of rapid invention such as I have seen in no other man." He spent with Thomson one day making experiments with the siren, and the next morning renewed the work, with fresh experiments, on combination-tones, that had occurred to Thomson during the night.

Thomson did not return to England in time for the British Association meeting at Cheltenham, but sent a paper on Dellmann's method of observing atmospheric electricity.

His attention now began to be absorbed in the problems of submarine telegraphy, as narrated in the next chapter, and towards the end of the year he sent two papers to the Royal Society on rapid signalling. He spent the new year of 1857 at Belfast, then returned to the laboratory to his investigation of the electric conductivity of copper.

A pleasing touch is to be found in the Letters of Dr. John Brown (author of *Rab and his Friends*). Thackeray in November 1856 was making his second lecturing tour, discoursing on the Four Georges. When in Edinburgh he made his home with Dr. Brown. In Glasgow he met Thomson, whom he visited at his house, No. 2 The College, dining twice with him. After the third lecture Dr. Brown wrote to Miss Jessie Crum (youngest sister of Mrs. Thomson): "I knew Thackeray would go to your heart. . . . He was delighted with your William Thomson; he said he was an angel and better, and must have wings under his flannel waistcoat. I said he had, for I had seen them!"

CHAPTER VIII

THE ATLANTIC TELEGRAPH: FAILURE

HITHERTO Thomson's work had been mainly in pure science, mathematics, the flow and transformations of heat, the mathematical theory of electric equilibrium, the mathematical theory of magnetism, hydrodynamics, and the dynamical problems of bodies in revolution. But in the middle of the 'fifties, while he was still immersed in thermodynamic studies and wrestling in his laboratory with the properties of matter, events were progressing which drew him with irresistible force towards the practical applications of science which made him famous.

Half a century earlier Volta had startled the world with his discovery of the "pile," the primitive battery capable of producing a steady and continuous silent flow of electricity through the conducting wire which constituted a circuit. Oersted had discovered the power of the current to deflect a compass-needle. Ampère and Arago had investigated further the magnetic relations thus revealed. Sturgeon had invented the soft-iron electromagnet —the magnet which is controlled from a distance through the electric wire that conveys the current to it—the magnet which attracts only when the circuit is completed, and, obedient to the hand of the distant operator, ceases to attract from the moment when the circuit is broken. Faraday had laid the foundations for the future development of electrical engineering by his discoveries of the electromagnetic rotations—in the first primitive revolving motors of his design, and of the induction of currents from the motion of magnets—the principle by which the dynamo generates currents mechanically. The first-fruits of all this scientific activity for the purposes of human industry had been the electric telegraph. Men had long thought and speculated on the possibility of transmitting intelligence by signals through an electric wire. The flood of discovery showed how such possibilities might become realities. Electric telegraphy was in the air. The year 1837 saw the telegraph of Cooke and Wheatstone in commercial operation in England, while in America the telegraph of Morse was at work by 1840. The former depended upon the deflexion of a magnetized needle by the influence of the current circulating in a small surrounding coil of wire; the latter was based upon the attraction of an iron keeper by an electromagnet, thereby moving a lever which printed dots and dashes, or gave audible sounds in its movement. Land lines, for the transmission of signals thus spelled out, were soon erected in both continents; and, as experience led to practical improvements, the distances to which telegraphic messages could be sent were extended over hundreds of miles. By the year 1850 overland telegraphy had become a prosperous business, and was rapidly extending. Just at the appropriate moment, too, the discovery of gutta-percha put into the hands of the engineers a material far more fit than any previously known—indiarubber, bitumen, or tarred hemp—to serve as the insulating coating to prevent the electric current from leaking away from the copper wire conductor. Already, in 1849, short lengths of submarine cables had been laid; and in 1851 the successful Dover-Calais line was laid by Crampton, followed by others, connecting England with Ireland, Scotland, Holland, and other countries. In 1856 Newfoundland was joined to Cape Breton, and thence overland to New York. But all these were short lengths compared with the two thousand miles which separated Great Britain from the American Continent, the spanning of which stirred the hopes and ambitions of telegraph engineers, and stimulated the project of an Atlantic cable.

Confronting such a project were great difficulties. The weight and cost of such a cable were enormous. There was much dispute as to whether the cable should be protected by an external armouring of iron wires or not. The manufacture required new machinery and the creation of a new class of operatives. The laying of such a cable across an ocean known to be three miles deep presented an engineering problem of the first magnitude. No single ship existed of sufficient size to hold the cable if it were constructed. But above all these difficulties there supervened an electrical objection of a very discouraging kind. The working speed of signalling through cables of such a length was believed to be very slow. Faraday had predicted the existence of a retardation of the signals in long cables, a retardation arising from the charging of the surface of the gutta-percha coating by the current on its way to the distant end. The Anglo-Dutch cable of 110 miles, the longest then laid, showed this defect slightly. Even the subterranean telegraph wires from London to Manchester were embarrassed by it. What retardation might be expected from a cable 2000 miles long? Would it not so greatly reduce the speed of signalling as to make the undertaking unremunerative? Several electricians had made experiments on lengths of underground cable to investigate the matter, but the results were not decisive.

Thomson's handling of Fourier's mathematics in the problems of the flow of heat through solids had led him from the first to perceive that the diffusion of an electric current through a conducting wire, though immensely quicker, obeyed the same laws and was amenable to similar calculations. In 1854, when his attention had been directed to the problems of submarine telegraphy by the experiments of Mr. Latimer Clark on retardation in the Anglo-Dutch cables, and by Faraday's investigation of the same, he began new calculations. These he communicated to Stokes in two letters, the substance of which was subsequently given[1] to the Royal Society in a paper "On the Theory of the Electric Telegraph."

In this very important paper he showed that in cable signalling there is no regular or definite velocity of transmission, but that a signal which is sent off as a short, sharp, sudden impulse, in being transmitted to greater and greater distances is changed in character, smoothed out into a longer-lasting impulse, which rises gradually to a maximum and then gradually dies away. Even though at the distant station the commencement of the signal may be practically instantaneous, an appreciable time may elapse for the retarded impulse to reach its maximum; and so the signal is for effective purposes retarded. Thomson showed now for the first time the law governing such retardation: that it varied in direct proportion with the "capacity" and with the "resistance" of the cable. As each of these quantities is, in a cable of given type of construction, proportional to its length, it followed that the retardation would be proportional to the *square* of the distance. If a cable 200 miles long showed a retardation of $\frac{1}{10}$th second, one 2000 miles long, if of similar thickness, would have a retardation 100 times as great, or 10 seconds. It became immensely important, therefore,[2] to know how to combat this

[1] *Proc. Roy. Soc.* vii. p. 382, May 1855, and *Math. and Phys. Papers*, vol. ii. p. 61.

[2] Thomson emphasized this in his *B.A. Paper* of 1855 (see p. 310) on peristaltic induction, by reference to the shorter cable between Varna and Balaclava, urging experiments to be used hereafter for estimating the proper dimensions for future long cables. "Immense economy," he concluded,

effect. By increasing the thickness of the copper the "resistance" could be reduced. By increasing the thickness of the gutta-percha coating the "capacity" could be reduced. Or, by increasing the diameter of the wire and of the gutta-percha coating *in proportion to the length* of the cable, the retardation of the long cable could be kept the same as that of the short one. This "law of squares" discovered by Thomson did not pass unchallenged; neither were his practical deductions accepted at once. Mr. O. E. Wildman Whitehouse, a retired medical man, who had taken up electrical studies, and was interesting himself in cable projects, read to the British Association meetings in 1855 at Glasgow, and in 1856 at Cheltenham, papers in which he professed to disprove by experiments[1] the doctrine of squares, declaring that, if it was true, signalling through long cables would be quite impracticable. Thomson was not present, being in Germany with his wife; but the matter being reported in *The Athenæum* of August 30th, 1856, he addressed to that journal, on September 24th, from the Isle of Arran, a letter which appeared on October 4th. Whitehouse had objected to Thomson's warning, based on the theory which he had worked out, that in a cable of this

"may be practised in attending to these indications of theory in all submarine cables constructed in future for short distances; and the non-failure of great undertakings can alone be *ensured* by using them in a preliminary estimate."

[1] Whitehouse's paper was issued as a pamphlet in 1855. Articles by Whitehouse on his experiments and instruments were also published in the *Engineer* of Sept. 26, 1856, and Jan. 23, 1857. In Lord Kelvin's copy of the last named he has written:—"The best account of what is good in Whitehouse's experiments and instruments. The conclusions, however, are fallacious in almost every point."

length it might be necessary to provide more than ordinary lateral dimensions of copper wire or of insulating coating, if sufficient rapidity of signalling were to be attained. Thomson now wrote to say that Whitehouse's experimental results, if rightly interpreted, were consistent with the true theory—adding that he came to this conclusion from "a knowledge of the theory itself, which, like every *Theory*, is merely a combination of established truths." He pointed out that Whitehouse's observations, as reported, did not show at what speed such a succession of signals as is required for the letters of a word can be sent through the greatest length of wire which he used; and that experiments of a more practical kind were required to show at what rate the irregular non-periodic alternations of currents required to spell out words may be reproduced at the distant end of a long cable. Whitehouse replied somewhat testily, reasserting his statements as to the retardations he had observed, though these were in reality due in part to his particular mode of using the current, and in part to the sluggishness of his own heavy instruments. Thomson returned to the matter in a second letter in *The Athenæum* of November 1st. Courteously admitting the accuracy of the observations of Whitehouse, he pointed out that if instead of the operations therein used, certain definitely specified electrical operations in signalling (such as the application of the electromotive force of a powerful battery for one-twentieth of a second) had been employed, and the retardations observed in cables of various lengths from 150 to 2400 miles, the law of squares would have been almost exactly fulfilled. He therefore concluded that the receiving instruments used by Mr. Whitehouse had, by reason of the sluggishness of their own electromagnetic action and inertia, masked the result. He also noted that since in some of Whitehouse's tests the battery current was applied for a whole second at a time, the inconstancy of the battery itself would influence the phenomena. He insisted on the need, in a strict test, of a constant battery (such as Daniell's), and of short sharp sending at the battery end of the cable. Then turning to the question of the dimensions of conductor and insulation, he reiterated his preference for a thick copper wire, or one made thick by stranding together a number of small wires; pointing out that Whitehouse's test, in which three thin separately insulated wires were used, furnished no disproof of his calculations.

In November 1856 Thomson communicated to the Royal Society a paper "On Practical Methods for Rapid Signalling by the Electric Telegraph," followed by another on December 11th. Here he explained a proposed system likely to give nearly the same rapidity of utterance by a submarine one-wire cable of ordinary lateral dimensions, between Ireland and Newfoundland, as is attained on short submarine or land lines. The plan of signalling was to employ a regulated galvanic battery to impart during a limited time a definite transient rise of potential; the end of the cable being immediately put to "earth." Guided by Fourier's theorems of the conduction of heat, he proposed to regulate the time of contact with the battery so that the coefficient of the simple-harmonic term, in the Fourier-series which expresses the distribution of electric potential, shall vanish—resulting in the retarding electrification of the coating of the cable being discharged four times as fast as would be the case for an ordinary electrification. For receiving the message he proposed a form of Helmholtz's galvanometer, in which the suspended magnet is provided with a copper damper, adjusted so that during the reception of an electric pulse the magnet would turn to its position of maximum deflexion, and fall back to rest. Adopting the "subjective" method of observation, the observer will watch through a telescope the image of a scale reflected from the polished side of the magnet, or from a small mirror carried by it, "and will note the letter or number which each maximum deflexion brings into the middle of his field of view." This method was soon abandoned.

The plan of letter signals first suggested was to inscribe on the scale of the instrument the 26 letters of the alphabet, and to arrange 13 positive and 13 negative strengths of current, such as to give deflexions that would bring one or other of the letters into the observer's field of view. But he foresaw that this might be impracticable, and assuming that only 3 or 4 different strengths of current might be available in practice, he suggested combinations of two or three signals for each letter, assigning the simplest combinations to the most frequently used letters.

He also suggested the possibility of a plan for rapid self-recording of signals by an apparatus, the description of which he reserved for a future time.

In the second communication he suggested a modification of the plan of sending, attaining greater rapidity by use of the third harmonic term; being a foreshadowing of the plan of curb-signalling by which, after the application of the battery, the cable is momentarily connected to a reversed electromotive force before being put to earth. He added an ingenious device suitable for rapid signalling on short lines, by introduction of electro-chemical relays.

The galvanometer was strongly on his mind, for he saw that the heavy electromagnetic relays proposed by Whitehouse introduced retardations of their own. In an enterprise involving so great a capital expenditure as an Atlantic cable, the earning power depended mainly on the attainable rapidity of signalling. A light, quick-moving instrument that would give instantaneous response at the distant end was a prime desideratum; and, if it would work with minute currents, the delays in the cable itself (due to the accumulating electric charges in transit) would be reduced, thus further augmenting the speed. Accordingly he looked around for suitable instrumental aids, as the following letter to Helmholtz shows:—

FORTBREDA, BELFAST, *Dec.* 30*th*, 1856.

MY DEAR HELMHOLTZ—I have been long wishing to write to you regarding your galvanometer, but have been prevented by pressure of business that would not bear delay.

Could you give me any idea of the dimensions of the two sizes of wire referred to in the enclosed letter from Siemens and Halske, and of the sensibility of the galvanometer with one or the other? For instance, what are the lengths of wire in the coils of the galvanometer in the two cases? What is the diameter of each copper wire, or the weight of copper per metre of length? How many turns of wire are there on each coil of the galvanometer with one wire and with the other? What are the dimensions of the coils themselves, and what their distances from the centre of the suspended magnet? Do you know the amount of deflection produced by any particular electromotive force—that of a single cell of Daniell's, for instance, or any submultiple of that of a single cell of Daniell's, or a copper and bismuth thermo-electric element with stated temperatures, or two plates of copper, one dipped in a solution of sulphate of copper and the other in a porous cell immersed in the same, and containing sulphuric acid? . . .

Can the instrument be made so that different coils can be substituted for one another upon it? If making the coils removable would introduce any inconvenience or defect I would rather have them fixed, as I believe they are in your instrument. But in this case I should probably want two instruments, as I shall certainly wish to have coils of small resistance for thermo-electric measurements, and I shall probably wish to have an instrument giving indications with very small absolute strengths of current to test a plan for telegraphing through great lengths of submarine wire which I have proposed.

The Atlantic Telegraph is now in the process of manufacture, 2500 miles of cable are to be finished and ready to go to sea by the end of May (the distance between Valencia in Ireland and Trinity Bay near St. John's being only 1900 miles), and if no accident happens electric messages will be passing between Ireland and Newfoundland before July. I have been appointed one of the directors, and what I feel most anxious about now is the laying of the cable. The plans must be better arranged than they have been in all such operations hitherto, in which there have been almost as many failures as successes. However, the circumstances are in some respects more favourable than they have been in former cases. We have a soft level bottom (consisting of fine sand and microscopic shells) the whole way across, nowhere more than $3\frac{1}{3}$ miles deep, which will be much better than the Alpine precipices and valleys below the waters of the Mediterranean. The cable is much lighter than any hitherto laid, weighing only 18×112 lbs. per mile, or in water only 10×112. The practical men engaged have all the experience of previous failures, and it is to be hoped have learned some of the causes and will know to avoid them. Altogether, I think there is a good chance of success.

I have been very much occupied since our return from Germany, chiefly bringing out a paper[1] on "Mathematical Theory of Elasticity," and a long paper describing those electrical experiments I have spoken to you about, through the press.

I have worked a good deal, too, at the solution of problems (exactly like those of Fourier) regarding the propagation of electricity through submarine wires. It is the most beautiful subject possible for mathematical analysis. No unsatisfactory approximations are required; and every practical detail, such as imperfect insulation, resistance in the exciting and receiving instruments, differences between the insulating power of gutta-percha and the coating of tow and pitch round it, mutual influence of the different conductors (when, as is not the case of the Atlantic cable, more than one distinct conductor is used),

[1] See p. 319, above.

attempts to send messages in both directions at the same time, gives a new problem with some interesting mathematical peculiarity.

I am here for a few days of "Christmas holidays," and I return to Largs (in Scotland) to-morrow, where I left my wife. She is, on the whole, much better than last winter, but is still much of an invalid. I hope, however, that she is really advancing to a complete recovery of health.

I must be at my post in Glasgow on Monday next, so if you write address me No. 2, College, Glasgow. Will you at the same time enclose Siemens and Halske's letter? The price they mention is much more, I think, than you told me (45 Th., I believe) they had charged for the instrument. I think they ought not to charge more, or not so much more, as they have drawings and other facilities in making a fourth and fifth specimen of the instrument, which should make it less expensive to them than the first.

Give my respects to Mrs. Helmholtz, and believe me, —Yours very truly, WILLIAM THOMSON.

P.S.—When will your book on the eye be completed, or is it so already? I find people greatly interested in it, especially regarding the adjustments.

I was out with a shooting party a few days ago at Largs, and looked into the eyes of various birds immediately after death. I saw the three images of the sun well in a woodcock's eye, but was puzzled by the position of the image by reflection at the posterior surface of the lens. I had a very curious view of the interior by simply pressing my eyeglass on the front of the cornea so as to nearly flatten it. Have you seen an owl's eye? It is a splendid thing. I cut one open, but learned nothing more than that the cornea is very tough.

This letter shows that events had been moving rapidly towards the great enterprise of the Atlantic Cable. Before the end of the year the project took definite shape.

Certain concessions having been obtained in Newfoundland, and arrangements having been made with the British and United States Governments for aid, in the form of ships, to assist in the laying, and of guarantees of subsidies for the working of the proposed cable, an association was formed on October 20, 1856, called The Atlantic Telegraph Company. The active promoters of the enterprise were Mr. Jacob Brett, Mr. (afterwards Sir) Charles Bright, Mr. Cyrus Field of New York, and with them Mr. O. E. W Whitehouse. The flotation of this company was remarkable. No prospectus was issued, no promotion money paid; there was no advertising, nor any commissions to brokers; the directors were elected by the shareholders after allotment; and the promoters were to receive no remuneration until the shareholders' profits should reach 10 per cent per annum. The capital was £350,000, in shares of £1000 each, of which less than one-twelfth was subscribed in America. Many of the subscribers were shareholders in the earlier telegraph companies; but many outsiders, parliamentarians, lawyers, and literary men, including Thackeray, subscribed for shares. Of the 18 directors elected in December 1856, 7 were from London, 6 from Liverpool, 2 from Manchester, and 2 from Glasgow, including Professor William Thomson.

Often as the story of the Atlantic Cable has been told, the precise part which Thomson played in the enterprise has never been fully stated. The work which he undertook for it was enormous; the sacrifices he made for it were great. The pecuniary reward was ridiculously small. The actual position which he held was relatively subordinate, and must have been at times galling. Yet he bore himself throughout with the most unswerving courtesy and delicacy of feeling. Never was more conspicuous that "laborious humility" which fifty years afterwards was noted by Lord Rosebery as the keynote of his career. Thomson joined the enterprise simply as a director, selected by the suffrages of the Scottish shareholders. He held no technical position.

The board appointed Bright as engineer-in-chief, Whitehouse as electrician, and Cyrus Field general manager. This staff had no sooner entered on its work than it discovered that the provisional committee which registered the company had, in its anxiety to save time, already entered into contracts for the manufacture of the cable. They had had before them some sixty-two samples of proposed types of construction, and had adopted a very light core with only 107 lbs. of copper per nautical mile for the conductor, and only 261 lbs. of gutta-percha per nautical mile as insulation. Bright advocated a copper conductor of 392 pounds, and an equal weight of gutta-percha. Whitehouse supported the adoption of a small core; while Varley and Thomson urged a larger one. It was, however, too late to change the contracts. With its sheathing of stranded iron wires the weight was about one ton per nautical mile. The contract for the core (the copper conductor and its coating) was assigned to the Gutta-Percha Company, while that for the iron sheathing was divided between two firms, Messrs. Glass, Elliot, and Co., of Greenwich, and Messrs. Newall and Co., of Birkenhead. So loosely was the specification drawn up, that not until the manufacture was completed was it discovered that the two halves had been made with opposite directions of twist in the armouring! The cable was manufactured in 1200 pieces of two miles length each. It was then joined into eight pieces each 300 miles long. Although various deep-sea cables had already been made, the processes of manufacture were still very crude. There was no regular test for the conductivity of the copper; the gutta-percha was laid on in a manner that would not be tolerated to-day, and which did not ensure that the copper conductor should remain central within its coating.[1] The manufacturers were bound to complete their work within six months, and, in fact, the delivery was completed by July 6, 1857.

All through the preceding months Thomson, with his laboratory corps, had been engaged upon a research on the conductivity of copper. He had realized the prime importance, in speed of signalling, of reducing the resistance that is offered by the

[1] A curious commentary on the state of knowledge is afforded by the fact that the Prince Consort urged on one of the directors of the Atlantic Telegraph Company that the proper plan would be to enclose the copper wire in a flexible tube of glass throughout its entire length. On being told that this was impracticable, he took down a volume of the writings of *Petronius Arbiter* to prove that flexible glass was a known substance!

conducting wire. To raise the speed you must have a better conductor; and the conductor could be bettered either by increasing its cross-section (and weight per mile) or by improving the conductivity of the metal itself. On procuring samples from various manufacturers of copper, and measuring the wires, he "was surprised" to find great differences in their conductive quality. Even a small percentage of impurity might reduce the conductivity by as much as 30 or 40 per cent. In a paper "On the Electric Conductivity of Commercial Copper," read to the Royal Society in June 1857, he detailed his investigations, giving, in terms of an "absolute" system of measurement, a table of the results of his own experiments, and comparisons with the standard wires used by Weber, Kirchhoff, and Jacobi.

In July 1857 there was issued by order of the Board a pamphlet of 70 pages, signed "R. J. M.," describing the project. It eulogized Mr. Whitehouse's particular inventions. To test the current he proposed to supersede the galvanometer with an instrument named the "magneto-electrometer," a sort of steelyard with sliding weights to counterpoise the pull on a suspended iron armature of a soft iron electro-magnet, the coils of which were traversed by the current. It was this very instrument which had in 1856 led him to dispute the law of squares, which law was still bluntly dismissed with the words: "Nature recognizes the existence of no such law"! He also designed some induction-coils

36 inches long, with fine-wire secondaries capable of yielding discharges at a high electromotive force, with which he purported to have proved that magneto-electric currents travel more quickly than voltaic currents along gutta-percha-covered conductors. For testing the insulation and continuity of the cable he proposed a species of relay, working an alarm bell. To operate his induction coils the source of current was to be a voltaic battery of ten giant cells, called the "Whitehouse laminated or perpetual maintenance battery," the cells being large Smee cells with multiple replaceable plates. As receiving instrument he devised a relay having a small permanent magnet of horse-shoe form suspended between the poles of a large soft-iron electro-magnet, the coils of which were to be connected at the receiving end of the cable. This relay was to work a Morse embossing recorder.

By a sort of irony, the compiler of this pamphlet inserted the statement that "the scientific world is particularly indebted to Professor W. Thompson (*sic*), of Glasgow, for the attention he has given to the theoretical investigation of the conditions under which electrical currents move in long insulated wires, and Mr. Whitehouse has had the advantage of this gentleman's presence at his experiments, and counsel, upon several occasions, as well as the gratification resulting from his countenance and co-operation as one of the Directors of the Company." There was another side to that story.

Meantime arrangements had been made for the

laying of the cable. The British Government furnished H.M.S. *Agamemnon*, a screw-propeller battleship (which had been the admiral's flag-ship in the Crimean war), and the United States Government lent the U.S. frigate *Niagara*, a screw-corvette of 5200 tons. The paying-out gear had to be rather hurriedly constructed. Its brake-wheel was governed by a clutch controlled by hand; and it was so heavy, that its grip of the cable was difficult to relax, nor could it accommodate itself readily to sudden strains on the cable due to pitching of the ship. No opportunity was afforded Thomson for testing the cable before laying.

The ships met at Queenstown on July 30th and proceeded to Valencia Bay, where on August 5th the shore-end was landed. The *Niagara* began to pay out her part; it having been arranged that, when half-way over, the *Agamemnon* should splice her half of the cable to that already laid, and complete the laying. The electrician of the company, Mr. Whitehouse, did not accompany the expedition, and excused himself at the last moment on the score of ill-health, the chief of the electrical staff on board being Mr. De Sauty. When it was found that Mr. Whitehouse could not undertake the voyage, at the request of the directors, and without any salary or position other than his membership of the Board of Directors, Thomson agreed to join the expedition, and was quartered on the *Agamemnon*. On August 5th the shore-end was landed from the *Niagara*, while great cheers went up from the other ships, gaily dressed from stem to stern in bunting, and from the party who, headed by the Lord-Lieutenant, stood on the shore to watch the American sailors haul up the end. After the *Niagara* had laid 330 nautical miles, owing to mismanagement on the part of a mechanic attending the paying-out brake, the cable parted in water of 2000 fathoms, and the expedition returned to Plymouth. The cable was unloaded at Keyham Harbour, and re-coiled in tanks in a shed, to be stored for the winter. An additional length of 700 miles was manufactured, and new paying-out machinery was designed in readiness for the renewal of the enterprise the next year.

Thomson went to the British Association meeting at Dublin, and there on August 28, 1857, he gave an evening lecture on the Atlantic Telegraph. In the sectional meetings he read three papers. One was "On the Effects of Induction on Long Submarine Cables"; a second "On Mr. Whitehouse's Relay and Induction Coils in action on Short Circuit." That Thomson should sink his *amour-propre* in the interests of the Atlantic Telegraph Company by thus associating himself with Whitehouse's apparatus, is a surprising but characteristic action. The third, "On Machinery for Laying Submarine Telegraph Cables," was printed in the *Engineer* of September 11th. In it he discussed the curvature of the portion hanging from the stern of the ship during laying, and the forces acting upon it, and declared that one necessity was a mechanism that would, like the action of the fly-fisher in yielding to any sudden tug on the part of the fish, afford play in the event of any sudden strain that otherwise might snap the cable. After leaving Dublin he wrote on the question to his brother James, to whose experience as a practical engineer he often resorted :—

ELK STR. OFF CARRICKFERGUS,
Sep. 3, 1857.

MY DEAR JAMES—I have been thinking of the cable settling curve, and I find that after all it was right that it will be really a curve even when the tension is equal to the weight of a length going perpendicularly to the bottom from any point. Mr. Hart was wrong in saying that the motion is rigorously perpendicular to the length when the paying out is uniform and there is no slack. Universally, if the paying out is uniform with no slack, the direction of the absolute motion of any part of the cable must bisect the angle between the line running horizontally forwards and the direction of the cable obliquely downwards, as may be inferred from the consideration of the preceding case, or as we see by considering that the velocity of the cable relatively to the ship (that is, the relative velocity with which it passes away from the stern pulley), is equal to the velocity of the ship, and that the absolute velocity of the cable is the resultant of these two. Hence there is always in uniform paying out with no slack a motion which will give rise to tangential component of resistance of the water helping to bear the weight. This conclusion is *very important*, as it shows that less resistance than the weight of a portion hanging vertically to the bottom will always suffice to stretch the cable firmly on the bottom. It is clear still that if the resistance be such as to give, first, no tension at the bottom, but no slack, the cable will go down in an inclined straight line ; but if it be resisted in leaving the ship with any greater resistance than that, it will go out in a curve which is convex upwards in its upper part and concave upwards in its lower part. The inclination of this curve to the vertical at its point of inflection will be exactly the same as that of the straight line in which the cable goes down when the speed is the same, and the resistance only just enough to prevent slack.

If my paper is not gone to the *Engineer*, will you alter the part in which I say that the motion is rigorously perpendicular to the length? If not, will you send an extract of this after it as an antidote? Of course in this case you would cut out the sentence in which I speak of Hart being wrong. I write this to give it to the post at Greenock in the morning for you, so it may reach you on Saturday.—Your affectionate brother,

WILLIAM THOMSON.

The *Engineer* of October 16 contained the amending addition. To his sister Mrs. King, then in Vevey, he wrote:—

INVERCLOY, ARRAN, BY ARDROSSAN,
Sep. 22, 1857.

MY DEAR ELIZABETH—I am afraid I have done little to improve my character as a correspondent since you left England, but my old excuse has been more valid than ever this summer. Ever since I last saw you in London I have had a very disturbed time with a few short intervals. Latterly the Atlantic Telegraph business has been very urgent, and I have been repeatedly called by it to London. I had only returned here three days from the meeting of the British Association, which kept me a week at Dublin, and was just beginning to feel settled with the prospect of an unbroken time of quiet till the beginning of the session, when I was called away to London to attend a meeting of the Board of Directors, and I had to make a journey to Devonport to look after the cable on board the ships there, and arrangements for the electrical department before I got home again. I trust I shall now have no more such disturbances, and get a little rest before the beginning of my winter's work. I quite feel the necessity of getting myself freed from the various engrossing occupations which have for a long time prevented me from ever having my mind off work, and I intend for some time to come to limit myself very strictly to the duties of my professorship. It will be a novelty to me to have no paper in progress and no proof sheets coming in at all unreasonable times, and I quite enjoy looking forward to it. . . .—Your affectionate brother, W. THOMSON.

Whatever Thomson's expressed intention as to devoting himself more exclusively to the duties of his Chair, he could not keep his mind from the cable problems. He had begun with a Helmholtz galvanometer. Now he was to step forward with his own. He had been impressed with the necessity, if signalling was to be rapid, of working with the smallest possible currents, so that time might not be lost while the pulse received from the sending end rose to its full value. He wanted an instrument that would work with a smaller fraction of current. So he determined to lighten the moving part—the suspended magnet—substituting for the heavy needle a minute bit of steel watch-spring (or two or three such bits), which he cemented to the back of a light silvered glass mirror suspended within the wire coil by a single fibre of cocoon silk. Then he got rid of the observing telescope used with the German galvanometers, by the device of directing upon the mirror a beam of light from a lamp, which beam, reflected on the mirror, fell upon a long white card, marked with the divisions of a scale, which was shaded from daylight, or set up in a dark corner. When on the arrival of an electric current the suspended magnet turned to right or left it deflected the spot of light to right or left upon the scale, and so showed the signal; the beam of light serving as a weightless index of exquisite sensitiveness, magnifying the most minute movements of the "needle." It is said—and the story is believed to be true—that this happy idea of using the mirror on the "objective" plan arose from noticing casually the reflexion of light from the monocle which, being short-sighted, he habitually wore hung around his neck with a ribbon. This *mirror-galvanometer*, which with sundry modifications[1] was embodied in his patent taken

[1] The patent specification, No. 329 of 1858, describes two early forms of electrometer for testing, and several varieties of the mirror galvanometer, with single, bifilar, and anchored suspensions, and with arrangements of controlling magnets or controlling currents in auxiliary coils. It also contains several suggestions for duplex working; and describes an early plan of curbing the signals by applying first a positive electromotive force for a short time, then a slightly greater negative electromotive force for an equal duration, followed by a much smaller positive one for a third equal period. It also suggested a device for utilizing a cable that might be able to transmit messages twice as fast as any one clerk could send by providing at each end synchronous circuit charges, so that the letters of two messages should be alternated in transmission, and gives the following example:—"Suppose that at one station there are two instruments and two clerks, and that at the other station the following series of signals arrive, PFRROETEETCR TAIDOEN, a clerk at the second station taking these letters alternately, finds that the first clerk transmitted the word PROTECTION, and the second the words FREE TRADE."

A surprising addendum to the specification is a telewriter for transmitting lines, figures, letters, or symbols by means of two separate circuits, acting at the distant end on a beam of light by the movement of two mirrors to combine two component movements at right angles to one another, with a third wire to cut off the beam when the sender lifts his tracing point from the writing surface.

In the *Encyclopædia Britannica* (8th edit., of 1860), Thomson says:—"The 'marine galvanometer' constructed by the writer for use on board the *Agamemnon* and *Niagara*, differs from the land mirror galvanometer only in the use of still higher directive force on the needle, and in the mode of suspension of the mirror and needle, which was by means of a fine platinum wire, or, as has since been found better, a stout bundle of twenty or thirty silk fibres, firmly stretched between two fixed points of support." The substitution of a silvered bit of microscope glass for a metal mirror was suggested by James White the optician. The mirror weighed but one-third grain, and the magnet

out February 20, 1858, proved to be of enormous importance in the subsequent history and development of submarine telegraphy. It served not only as a "speaking" instrument for receiving signals, but as an absolutely invaluable appliance both at sea and in the laboratory[1] for the most delicate operations of electric testing. He also patented an improved brake for paying out cables from ships.

To Glasgow he took with him some sample pieces from the unlaid portion of the cable, and from other cables, to investigate the conductivity of the copper. Finding them to differ greatly amongst themselves, he selected five specimens,

and counterpoise attached at the back about as much, the whole being less than one grain.

The above patent was not actually Thomson's first. He had in 1854, along with Rankine, taken out a patent, No. 2547, for improvements in electrical conductors for telegraphic communication, in which they proposed a stranded multiple conductor. For some reason this patent was abandoned. The 1858 patent was, in 1871, extended for another eight years by decision of the Privy Council.

[1] Thomson's galvanometer stimulated Clerk Maxwell to pen the following characteristic parody, which appeared in *Nature* of May 16, 1872:—

A LECTURE ON THOMSON'S GALVANOMETER.

[*Delivered to a Single Pupil in an Alcove with Drawn Curtains.*]

The lamp-light falls on blackened walls,
And streams through narrow perforations;
The long beam trails o'er pasteboard scales,
With slow decaying oscillations.
Flow, current! flow! set the quick light-spot flying!
Flow, current! answer, light spot! flashing, quivering, dying.

O look! how queer! how thin and clear,
And thinner, clearer, sharper growing.
This gliding fire, with central wire
The fine degrees distinctly showing.
Swing, magnet! swing! advancing and receding;
Swing, magnet! answer, dearest, what's your final reading?

O love! you fail to read the scale
Correct to tenths of a division;
To mirror heaven those eyes were given,
And not for methods of precision.
Break, contact! break! set the free light-spot flying!
Break, contact! rest thee, magnet! swinging, creeping, dying.

the conductivities of which, in terms of a certain standard, were respectively 42, 71·3, 84·7, 86·4, and 102 per cent; he sent them for analysis to Hofmann, who pronounced the percentages of copper in them to be 98·76, 99·20, 99·53, 99·57, and 99·20 respectively. Observing that the conductivity was in the order of the purity of the copper, he procured over forty other specimens that had been specially alloyed with various minute proportions of other metals. These he later described to the Royal Society (Feb. 1860). A sample of Rio Tinto copper examined by Matthiessen had a conductivity no better than that of iron! Calling the attention of the Directors of the Atlantic Cable Company to these deficiencies of conductivity, he was vexed to find them apathetic. They had been assured by their official electrician that the speed of signalling was not affected by the resistance of the conductor. Not until he took up an obstructionist attitude, and persevered at successive meetings in opposing all other business, did they listen to his insistence on proper testing of material; and he ultimately secured the insertion of a clause requiring "high conductivity" in the contract for 700 miles of new cable ordered in the late autumn of 1857 to supply the place of the portion lost in the first expedition. When this clause was first submitted to the contractors they declared[1] it impossible, but eventually consented

[1] "I fear," wrote the secretary, Mr. Saward, to Thomson on October 6, "that difficulty in carrying out your views about copper will arise from the Gutta-Percha Company, who state that they cannot contract to deliver within a specified time if the close examination you contemplate is to be undergone."

at an increased price. To ensure compliance with this provision, Thomson set up at the factory a set of testing appliances, thus establishing for the first time a factory testing laboratory. After this practical engineers came to believe in the reality of the existence of these differences in quality. "From that time to the present," wrote Thomson in 1883, "there has never been a question, on the part of either companies or contractor, as to the necessity for the stipulation of 'high conductivity'; and a branch of copper manufacture has grown up in the course of these twenty-six years for producing what is called in the trade 'conductivity copper.'"

Still occupied with the questions of cable-laying, Thomson wrote, in March 1858, to his brother James:—

BIRKENHEAD, *March* 19, 1858.

MY DEAR JAMES—I was so busy the whole week after you left that I could not manage to write a line in reply. I was incessantly occupied getting my model ready, and I had a new one made with various appliances to illustrate how I would work in reality—shifting the chain when worn, regulating the pressure, etc. The plan you propose does not fulfil the condition which I think of greatest importance in connection with my plan—that is, having the part of the chain which is under small tension *simply fixed*, while the regulation of stress is to be effected at the end under heavy tension. My reason for making this of importance is that the whole resistance cannot possibly exceed the tension at the heavily stretched part,

A fortnight later he wrote: "The conductivity question is occupying the best attention of all of us—every hank of wire is being carefully tested to a given standard. The attention of the most eminent smelters is being directed to the matter, and they have been invited to the gutta-percha works to see for themselves the variations in conducting power of the several parcels sent in."

and will always fall short of it by an amount (the tension at the slightly stretched part) which is necessarily only a small fraction of the whole, and which may therefore be safely left to vary with friction; though it should double itself the diminution of resistance on the whole need not be more than 3 or 4 per cent.

The plan of a self-acting regulator by a movable pulley has been much considered, and may possibly be used. I rather think it will be better, however, to be content in reply with an *indicator* of that kind, to show the tension of the cable leaving the machinery, and to regulate when necessary by hand. With my brake nothing will be required but to put on a safe load on the heavily stretched end and leave it to itself. I propose, however, to have a plan by a lever at the lightly stretched end by which in an instant this weight may be let down and the brake entirely released—a plan, in fact, like yours, but worked by the hand and (*what is the essential feature of my plan*) incapable of doing more than allow the full load at the other end to act. I have altered the arrangement of chain so as to give it only one straight lead from drum to drum; this makes the pressure on the bearings as little as need be, or as can be, with any form of brake. I put a guard and support to keep this horizontal part in its place and prevent it from rocking when the tension is slack, and the effect is *excellent*. I have not a doubt of the whole plan being right.—Your affectionate brother,

WILLIAM THOMSON.

2 COLLEGE, GLASGOW, *April* 19, 1858.

MY DEAR JAMES— . . . I am having instruments made for both Valencia and Newfoundland to work the telegraph, and must have them finished and tested before the 14th of May. I only commenced last Wednesday, having just returned from Devonport, where I verified the mathematical theory (with marvellous closeness, partly by chance), and got, with care, a letter every three seconds through 2700 miles.—In haste, your affectionate brother,

W. THOMSON.

The early summer of 1858 found the "wire squadron" preparing for the second attempt to lay the cable. The *Agamemnon* and the *Niagara* were again commissioned; but the plan of campaign was altered. This time they were to proceed to mid-Atlantic, there effect a splice, and begin to pay out in both directions at once. Also it was determined to have a preliminary trial trip to the Bay of Biscay to gain experience. Immense activity was manifest. The cables were once more being coiled in the ships. The Directors throughout April and May were meeting daily, as the minutes of the Board show. On April 7 the Directors were asked to say which of them would accompany the expedition. Disquieting reports were received on April 10 and 12 as to defects found in the cable in the process of coiling it in the tanks on board at Devonport. Thomson, still at his duties at Glasgow until May 1, was writing about improvement for quicker signalling, and about his improved paying-out brake. On April 13 he asked the Directors to grant £2000 to cover the cost of constructing new signalling instruments. On April 21 he gets his reply:—"The Directors, having regard to the reports and observations of Mr. Whitehouse, and particularly to the financial state of the Co., are of opinion that it would not be expedient to advance so large a sum under present circumstances . . . and, therefore, desire to postpone their decision upon the form of instrument to be ultimately used for working through the cable until the task of submerging it shall have been successfully accomplished." On 23rd he replies, asking for £500 for instruments. The Board gives him—by a bare majority—authority to complete the instruments he has on hand, but entreating him to keep down the cost. On 27th the Board receives from him an acknowledgment, when he adds that he will not expect the Company to take his instruments unless they realize a speed of three words a minute after the cable is laid; he also requests that he may be allowed the use of the cable (as coiled on ship) for two or three hours on the 3rd or 4th of May to make some tests. The Board foreseeing difficulties makes a formal minute:—"Ordered that Mr. Whitehouse be requested to desire his assistants at Devonport to render Professor Thomson all possible attention and facilities, and to follow his directions with respect to the experiments he desires to make on the cable." The same day the secretary is directed to address to Mr. Whitehouse a letter, requesting him to inform the Board on which of the ships he intends to embark. On May 14, a letter is read from Mr. Whitehouse, advising the Board of his intention to embark on the *Agamemnon*, but expressing doubts as to his accompanying the experimental trip. "Ordered that Mr. Whitehouse be informed that the Board desires his presence on board H.M.S. *Agamemnon* during the experimental trip, and that he be requested to make his arrangements accordingly." But on May 29, when the fleet set out for the Bay of Biscay, Whitehouse excused himself. Thomson, waiting anxiously for the arrival of the latest instruments from Glasgow, was almost the last man to go aboard, a precious package being handed up to him at the last moment by his

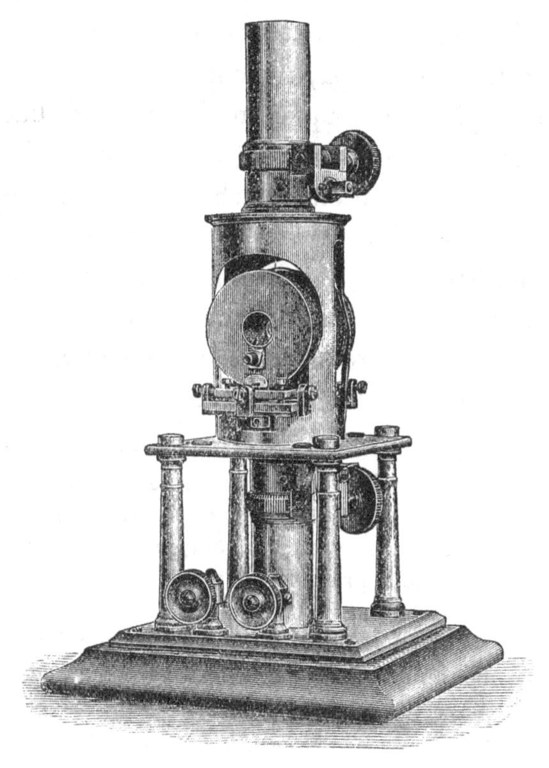

THOMSON'S MARINE MIRROR GALVANOMETER.
Used on board the *Niagara*, 1858.

assistant Donald MacFarlane, who had brought it by express. It contained an object like a small brass pot standing on four legs—the "marine mirror galvanometer," constructed during the preceding fortnight. One of these instruments is preserved in the Natural Philosophy Laboratory of the University of Glasgow. The cut is from a photograph of the instrument by kind permission of Professor Andrew Gray.

All went well on the trial trip. In water nearly three miles in depth they practised making splices and lowering them, paying out and hauling in the cable, changing from upper to lower hold, buoying the cable and picking it up, and passing the cable from stern to bow—operations which might be necessary if anything should go wrong in the laying of the line across the Atlantic. This trial trip being successfully over, the ships returned to Plymouth on June 3.

Saward, the Secretary of the Company, has put on record the following comment:—

> On arriving at Plymouth the condition of the electrical department was found to be such as to cause great anxiety to the Directors. They had given instructions to the electrician during the winter to employ the Company's operators in constant practice upon the instruments which were supposed to be in preparation for final use in working through the cable; and as the whole of the latter was coiled in one building at Keyham, it was supposed that they would have the opportunity of sending and receiving messages through its whole length during several months, and be thus prepared for all the peculiarities of a conductor so long and special in its character. The Directors were, therefore, greatly disappointed to find that not only had this not been done, but they found on their assembling at Plymouth that the instruments were not in a state, nor of a nature calculated to work the cable to a commercial profit.

The expedition sailed westwards on June 10. Charles Bright was again chief engineer, on board the *Agamemnon*, with Canning and Clifford as assistants; and was represented on the *Niagara* by Everett and Woodhouse. Cyrus Field was on board the *Niagara*; De Sauty was electrician on board the *Agamemnon*. Whitehouse, being again unable to go to sea, repaired to Valencia to await the arrival of the Irish end of the cable. At the request of the Directors Thomson consented to supervise the arrangements of the testing-room on the *Agamemnon* in an honorary capacity. It was a difficult position, for he had had little opportunity of making any tests on the state of the cable since it left the factory. His urgent representations as to the necessity of providing proper resistance coils and other testing apparatus on board had been ignored. Even his instruments had only been allowed on sufferance to be inserted in the circuit on his arrival in Ireland. He was in no way responsible for the instruments provided by the official electrician for the tests on board, upon which reliance was supposed to be placed. But before consenting to go to sea he made it a condition that permission should be given to have his marine galvanometer in circuit, because by it he expected to have more definite information than he could have by watching the tests that were prepared.

When less than two days out the *Agamemnon* encountered an Atlantic storm[1] of most exceptional

[1] A wonderfully graphic account of this really terrible storm was written by Mr. Nicholas Woods, the *Times* correspondent on board the *Agamemnon*, in the *Times* of August 11, 1858. It is reprinted in Mr. Charles Bright's *Life Story of Sir Charles T. Bright*.

violence, which lasted for eight days. Laden as the ship was in her hold, and with a dead weight of some 250 tons coiled on her upper deck, the *Agamemnon* was ill fitted to encounter such a trying occurrence. Her deck planks already gaped an inch apart, and when rolling in the heavy seas they opened wide and closed again at every oscillation. The electrical testing cabin was flooded. A portion of the coiled cable was seriously displaced and damaged; the cargo of coal broke loose; and ten of the crew were injured. On the return of fine weather the squadron met at their rendezvous in mid-ocean fifteen days after leaving Plymouth. It took two days to uncoil and re-coil the tangled section, nearly 100 miles long, of cable. On June 26 a splice was made and paying-out began; but the cable broke on board the *Niagara* when about 6 miles had been paid out. A second splice was made, but again the cable failed when some 80 miles had been laid. On the 28th a third splice was effected, but the day after, when each ship had paid out over a hundred miles, the cable parted at the stern of the *Agamemnon*, giving way at a point where it had been damaged by the disturbance of the flooring of the tank during the storm. During foggy days which succeeded, the *Agamemnon* failed to find her sister ship at the rendezvous, and therefore returned to Queenstown to meet her. Queenstown was reached on July 12, after being thirty-three days out of sight of land. No sooner did the ships meet here than Thomson had the two halves of the cable temporarily spliced, and transmitted signals through the entire length, thus showing once more that the difficulties were mechanical, not electrical.

Faced with this failure the Directors were divided; some of those most concerned commercially, being convinced that the project was impracticable, now advocated a realization of assets and winding up of the Company. Against these despairing counsels the men of action stood for a final effort. Bright, Brett, Field, Whitehouse, and Lampson were all eager to go on. Thomson, who had never once despaired of ultimate success, was emphatic for continuing the attempt. After keen discussion, followed by the resignation of the Vice-chairman, the Board ordered the immediate sailing of the expedition on what to all but themselves seemed truly a forlorn hope.

After taking in coal the ships steamed away from Queenstown, the *Niagara* and the two consorts, the *Gorgon* and the *Valorous*, leaving on July 17, and the *Agamemnon*, with Thomson on board, on the day following. They met at the rendezvous on July 29, and the same day spliced ends and began paying out. The weather was mostly bad, and there were several temporary mishaps which only the vigilance and energy of the staff prevented from becoming disasters. Owing to the inertia of the paying-out wheels and machinery, the stress on the cable fluctuated with every rise and fall of the ship's stern, so that to save the cable from snapping the run of the wheels had to be regulated by the sailors' hands, under the immediate supervision of officers, without intermission day and night. Both ships reached land on the same day, August 5, and landed their respective ends of the cable.

Of the resourceful energy of the chief engineer, and the devotion and ability of the captain, it would be impossible to speak too highly, and they are rightly praised in the contemporary narratives where for the most part the unassuming services of Thomson are but slightly recorded; and little is said of the personal part he took. Happily a contemporary account in the *Sydney Morning Herald*, in a letter written by a junior member of the electrical staff, supplies a few notes that are of more than personal interest:—

The electrical room is on the starboard side of the main deck forward. The arrangements have been altered several times in order to avoid the water which showers down from the upper deck. At one end of the little place the batteries are ranged on shelves and railed in. At the other stands a table with the various instruments arranged in electric series. On one side stand the "detectors" of the old system, so called from being chiefly used in testing for faults; and Whitehouse's beautiful "Magnetometer," called by some one his "pet child." These are under the eye of one of the clerks on duty. On the opposite side of the table is Thomson's marine galvanometer, so called because it combines delicacy with perfect stability at sea. It is closed up in a plain deal box, which is placed on a frame, equally primitive, attached to springs. Yet this little "Jack-in-the-box," as we often call it, does the work of every instrument on the table in its own peculiar way, and a deal more accurately. The indication is given by a little mirror which reflects the light of a paraffin lamp, through a lens, on a scale. This little mirror is fixed to a very small magnet, which, being influenced by the current, moves it, and therefore the spot, over so many degrees. We send and receive during alternate ten minutes. The current is sent through Dr. Thomson's galvanometer to the lower end of the "orlop" coil, which will be brought to shore, through all the cable on board, over the stern, under the sea, to the *Niagara*, where it traverses all her cable before reaching instruments exactly similar. The most valuable observation is taken in sending on the marine galvanometer. Three seconds before it is taken, the clerk at the opposite side of the table who times all the observations by a watch regulated by a chronometer too valuable to bring into so wet a place, says "Look out." The other clerk at once fixes his eye on the spot of light, and immediately the word is given "Now," records the indication. This testing is made from minute to minute, so that a flow is detected the moment it occurs. Indeed, on one occasion, a break which happened between the taffrail and the water was observed before it reached the sea, which of course made it at once evident enough.

July 29.—It is rather an exciting occupation to watch the tell-tale signals as we pay out. Even the most indifferent "holds his breath for a time" when their story is of dubious or ominous import. We are regarded by the engineers about the paying-out machinery as birds of evil omen. If one of our number rushes upon deck or approaches with a hurried step, they look as a Roman husbandman might have done at a crow on a blasted tree. Indeed it is almost impossible to realise the anxiety and *heart-interest* everybody manifests in the undertaking. No one seems to breathe freely. Few, but the men, even sleep soundly. Professor Thomson frequently does not put off his clothes at night.

To-night, but a few hours after starting, we had an alarming crisis. We had signalled to the *Niagara*, "Forty miles submerged," and she was just beginning her acknowledgment, when suddenly, at 10 P.M., communication ceased. According to orders those on duty sent at once for Dr. Thomson. He came in a fearful state of excitement. The very thought of disaster seemed to overpower him. His hand shook so much that he could scarcely adjust his eyeglass. The veins on his forehead were swollen. His face was deadly pale. After consulting his marine galvanometer, he said the conducting wire was broken, but still insulated from the water. Mr. Bright noticed the Professor hurrying to the electrical room, and followed close on his heels. He supposed the fault might lie in a suspicious portion which had been observed in the main coil. The cable was tested on both sides of this place, but it was all right there. The fault was not on board. but between the ships. There did not seem to be any room to hope; but still it was determined to keep the cable slowly going out, that all opportunity might be given for a resuscitation. The scene in and about the electrical room was such as I shall never forget. The two clerks on duty, watching, with the common anxiety depicted on their faces, for a propitious signal; Dr. Thomson, in a perfect fever of nervous excitement, shaking like an aspen leaf, yet in mind clear and collected, testing and waiting, with half-despairing look for the result; Mr. Bright, standing like a boy caught in a fault, his lips and cheek smeared with tar, biting his nails as if in a puzzle, and looking to the Professor for advice; Mr. Canning, grave, but cool and self-possessed, like a man fully equal to such an emergency; the captain, viewing with anxious look the bad symptoms of the testing as indicated on the galvanometer and pointed out by Dr. Thomson. Behind, in the darker part of the room, stood various officers of the ship. Round the door crowded the sailors of the watch, peeping over each other's shoulders at the mysteries, and shouting "gangway!" when any one of importance wished to enter. The eyes of all were directed to the instruments, watching for the slightest quiver indicative of life. Such a scene was never witnessed save by the bedside of the dying. Things continued thus. Dr. Thomson and the others left the room, convinced they were once more doomed to disappointment. Still the cable went slowly out, while in the hold they were re-splicing the suspected portion. The clerks continued sending regular currents. All at once the galvanometer indicated a complete breaking in the water. We all made the dread interpretation, and looked at each other in silence. Suddenly one sang out, "Halloa! the spot has gone up to 40 degrees." The clerk at the ordinary instrument bolted right out of the room, scarcely knowing where he went for joy; ran to the poop, and cried out, "Mr. Thomson! the cable's all right; we got a signal from the *Niagara*." In less than no time he was down, tested, found the old dismal result, and left immediately. He had not disappeared in the crowd when a signal came which undoubtedly originated in the *Niagara*. Our joy was so deep and earnest that it did not suffer us to speak for some seconds. But when the first stun of surprise and pleasure passed, each one began trying to express his feelings in some way more or less energetic. Dr. Thomson laughed right loud and heartily. Never was more anxiety compressed into such a space. It lasted exactly one hour and a half, but it did not seem to us a third of that time.

. . . .

As we drew nearer Ireland the storm began to abate, and things became altogether so cheerful in aspect that we dared to hope. Still, to the last, we never were entirely free from anxiety from one cause or another. The signals failed once altogether. The only instrument which kept us from despair was Dr. Thomson's. I could compare him, watching the slight quivering indication, only to one holding a mirror to the lips of a dying relative to see whether so much of breath remained as would dim its surface. Twice, also, we narrowly missed having a collision with vessels, in each case American. On one occasion the unlucky craft was a small anomalous-looking schooner. But the crew made all possible amends, as soon as they saw the cable at our stern, dipping their ensign some half-dozen times, and cheering lustily.

On the night of the 3rd August we got into shallow water. About ten Dr. Thomson came into the electrical cabin, evidently in a state of enjoyment so intense as almost to absorb the whole soul and create absence of mind. His countenance beamed with placid satisfaction. He did not speak for a little, but employed himself stretching scraps of sheet gutta-percha over the hot globe of our lamp; watching them with an absent eye as they curled and shrank. At last he said: "At half-past eleven you may send the 200 fathoms' soundings signal." He then proceeded to congratulate those present on being connected with such an expedition, regarding its object as already *un fait accompli*.

. . . .

When we got close inshore we threw off the cable-boat. Before our prow grated on the strand her impetus had taken her ashore. The *Valorous*, in the distance, fired her guns. The end was seized by the jolly tars and run off with; a good-humoured scuffle ensued betwixt them and the gentlemen of the island for the honour of the pulling the cable up to the office. The Knight of Kerry was upset in the water. As soon as it got fairly on *terra firma* a bevy of ladies gave it a make-believe haul—just so much as to tar their gloves or white hands, and give occasion for a nice business-like little fuss in getting butter or other oleaginous matter to remove the stain! Meanwhile we were thrown rather behind. Seeing that those who most deserved the honour were likely to lose it, Dr. Thomson, followed by Anderson a clerk, and myself, jumped out of the boat and waded ashore, but in time only to tar our hands ineffectually, like the ladies. The end was taken to the slate works, where the Company's offices are temporarily fixed. About five minutes to four Dr. Thomson sent the first current from shore to shore, to test the state of the cable. All was right. At four we received the first current.

Thus was the grandest undertaking of the century terminated with success, and just a year after the commencement of last expedition. The ships started first 5th August 1857; we brought in the cable 5th August 1858.

Unbounded were the rejoicings on both sides of the Atlantic. In every American city there were wild celebrations and gay displays and processions. In New York the church bells were rung, and a salute of a hundred guns fired. The success of the cable became "the theme of innumerable sermons and a prodigious quantity of doggerel." In England the rejoicings were more restrained, but the Irish cities were scenes of wild delight. Bright received knighthood at the hands of the Lord-Lieutenant of Ireland, and the Directors congratulated themselves on the success which crowned their efforts.

Thomson has recorded, in his Presidential Address of 1889 to the Institution of Electrical Engineers, the following generous note :—

> The first Atlantic cable gave me the happiness and privilege of meeting and working with the late Sir Charles Bright. He was the engineer of this great undertaking, full of vigour, full of enthusiasm. We were shipmates on the *Agamemnon* on the ever-memorable expedition of 1858, during which we were out of sight of land for thirty-three days. To Sir C. Bright's vigour, earnestness, and enthusiasm was due the laying of the cable.

And Bright has given us the following little silhouette of Thomson :—

> As for the Professor . . . he was a thorough good comrade, good all round, and would have taken his "turn at the wheel" of the paying-out brake if others had broken down. He was also a good partner at whist when work wasn't on ; though sometimes, when momentarily immersed in cogibundity of cogitation, by scientific abstraction, he would look up from his cards and ask, "Wha played what?"

Throughout the voyage Thomson's mirror galvanometer had been used for the continuity tests and for signalling to shore with a battery of seventy-five Daniell's cells. The continuity was reported perfect, and the insulation had improved on submersion. On August 5 the cable was handed over to Mr. Whitehouse and reported to be in perfect condition.

On August 6 the Secretary wrote to Thomson:—

"The Directors desire very heartily to congratulate you upon the successful accomplishment of the great work in which we have been engaged, which has created the greatest and most lively wonder and satisfaction among all classes of society here." He further enclosed an extract from the Board minutes, giving strict orders that no person was to send messages of any description except Professor Thomson and Mr. Whitehouse in Ireland, and Mr. De Sauty and Mr. Laws in Newfoundland, followed by other regulations pending the opening of the regular service. On the 10th the Directors passed a formal vote of thanks, which was later presented to Thomson, framed in decorative form. It ran:—

Resolved: That the sincere thanks of this Board be given to their distinguished colleague Professor Thomson of Glasgow for his exertions on behalf of the Company, more especially for the great sacrifices he has made in twice accompanying the Atlantic squadron during its perilous course, and for the important aid he has rendered towards bringing this great undertaking to a successful issue.

Whitehouse at once abandoned the Thomson

mirror instruments, and began working with his own patented apparatus, using heavy relays and his special transmitter with induction coils. He sent in no report to the Directors for a week, while he made ineffectual attempts with bigger induction coils to get his apparatus to work. To the Directors he sent the excuse that he was still "adjusting" the instruments. Days passed, and still no messages were transmitted. The public began to murmur. The Directors became uneasy. Thomson, who had remained at Valencia to assist Whitehouse—when permitted—had left early on the 10th August for Glasgow—from which he had been absent since 2nd May. Board-meetings held in London on the 9th, 10th, and 11th of August awaited impatiently for reports from Whitehouse as to the state of things. After more than a week the reflecting galvanometer and ordinary Daniell cells were resumed at Valencia, and only on August 13 the first clear messages were interchanged and international congratulations passed. But at Newfoundland they were still using the older instruments, and the signalling was very slow, and was often interrupted. A congratulatory message from the Queen to the President of the United States, containing but 99 words, took 16½ hours to transmit! Yet the same message was repeated back from Newfoundland without a mistake in 67 minutes. Clearly there was something seriously wrong. Whitehouse had told Thomson before he left that he suspected a fault about two miles away in the shallow water of Valencia harbour mouth. Thomson assured him that there was no evidence in the tests for this, but that probably the fault lay about 300 miles to the west, where the ocean deepens. Whitehouse proposed to raise the cable in the harbour to look for the supposed fault. The Board issued strict orders that he must not attempt this. They also sent Mr. France, a skilled cable operator, to aid in unravelling the difficulties; but on reaching Valencia he was refused admission by Whitehouse. Receiving only evasive replies, and learning a few days later that, in contravention of their orders, Whitehouse had underrun the cable to the harbour mouth, and had cut and buoyed it at Doulas Head, the Board on the 17th declared his appointment terminated, and summoned him to London.

The Board then addressed to Thomson the following authority to take sole charge:—

ATLANTIC TELEGRAPH COMPANY,
22 OLD BROAD STREET,
LONDON, *August* 18, 1858.

DEAR SIR—You are hereby authorised and empowered to take charge and possession (until further arrangements can be made) of this Company's office and Electrical apparatus at Valencia, and to issue in respect to the adjustment and working of the instruments such instructions as you may deem best. It is also hereby ordered and authorised that no person whatever is to be allowed on any pretence to enter the Company's electrical department without your special order and permission.—We are, dear sir, yours truly,

C. M. LAMPSON,
Vice-Chairman.
WILLIAM LOGIE,
HENRY HARRISON,
GEORGE SAWARD,
Secretary.

Professor W. Thomson, LL.D.,
etc. etc.

Thomson, on arriving at Valencia on the 21st, was at first disposed to defend Whitehouse, telegraphing twice, and then dispatching a long report. To his surprise he had found his own instrument installed instead of Whitehouse's!

In the subsequent enquiry it came out that from the first Whitehouse had been unable to receive messages satisfactorily with his instruments. Part of his method of working was to relay the signals, as received, upon an automatic Morse recorder, which printed dots and dashes of the message upon a tape. When his own relays proved too clumsy for the delicate impulses that came from Newfoundland, he had them read on Thomson's galvanometer by a clerk who watched the movements of the spot of light, and who, working by hand a signalling key on the table before him, joined by wires to the Morse recorder on the other side of the table, printed the dots and dashes on tape as though they had been received by relay. The receiving clerk, in fact, acted the part of relay. These slips were sent by Whitehouse to the Directors in London, who were thus misled into supposing that the messages had been received by means of his own apparatus. But it was otherwise at Newfoundland; for there the operators, obedient to instructions, as soon as the Thomson instrument at their end showed that signals were being transmitted, at once threw it out and substituted a common telegraph detector and the Whitehouse relay; and finding the incoming currents too weak to read, signalled back to Valencia for slower currents to be sent. Whitehouse at Valencia put on bigger induction-coils, emitting more powerful discharges.

Thomson now changed all this. After trying fruitlessly for a whole day to get the Newfoundland operators to understand, he succeeded in directing them to substitute his galvanometer for reading the signals, and at once communication was restored. Thomson had previously been led to pronounce the opinion that Whitehouse was to blame for the failure of Newfoundland to receive signals. He now with an excess of generosity asked the Board to condone Whitehouse's errors of judgment, and to reconsider the resolution of the 17th. The Directors, on the 25th, sent a reply to Thomson that his own benevolent and kindly feelings must have obscured his more reflective judgment. Their criticism of Whitehouse was unsparing. The following is taken from the official letter :—

> Mr. Whitehouse has been engaged some 18 months in investigations, which have cost some £12,000 to this Company, so that he has been in a position to avail himself of every resource that would tend to accomplish the objects on which he was at work, and now! when we have laid our cable, and the whole world is looking on with impatience to realise some results from our success, we are, after all, only saved from being a laughing stock because the Directors are fortunate enough to have an illustrious colleague who has devoted his mind to this subject, and whose inventions produced *in his own* study —*at small expense*—and from his *own resources*, are available to supersede the useless portions of apparatus prepared at great labour and enormous cost for this special occasion.

It cannot, therefore, on your own evidence, be admitted, that Mr. Whitehouse "*has been conducting his own proper business in a thoroughly efficient and successful manner,*" but on the contrary the evidence, as given by yourself, demonstrates that there has been neither "efficiency" nor "success," nor can the Directors agree with you in believing him to be "one of their most devoted officers." On the first sailing of the Atlantic squadron he abandoned his post—without notice—without any indication to any one up to the last moment that he intended to do so On the last occasion he again refused to go out, and had it not been for yourself, we should consequently have been placed in serious difficulty. He has run counter to the wishes of the Directors on a great many occasions—disobeyed time after time their positive instructions, in respect to incurring liabilities on behalf of the Company—thrown obstacles in the way of every one, except yourself, whom the Directors desired to consult, and acted in every way as if his own fame and self-importance were the only points of consequence to be considered in dealing with his department. These matters, if you will give them due reflection, will, it is trusted, show you that the quality of devotion to this Company which you claim for our late electrician does not exist.

But your letter leaves undefended the chief and most important point in Mr. Whitehouse's conduct, and which rendered his dismissal inevitable and imperative, viz., his recent acts in reference to the under-running of the cable and his treatment of Mr. France. These acts are only the climax of many previous and similar ones, and it has now become a question whether Mr. Whitehouse or the Directors are to govern this undertaking.

What are the facts? From the 5th to the 13th instant Mr. Whitehouse has undisturbed possession of the cable, he finds difficulties and perplexities which he cannot combat, and towards the solution of which he calls in no extraneous aid. On the 13th he sends to the Directors a message which conveys gratifying and satisfactory intelligence respecting the working of the cable.

They publish it, little thinking that at the moment this very message was on its way to them another and secret message was being sent. To whom? Not to the recognised engineer of the Company! Not to the Directors! but to Mr. Canning! ordering him to come to Valencia to under-run the cable, thus jumping at once to the conclusion—which no subsequent circumstances have proved—that a fault existed in the work of another department of the Company, but denying to the head of that department the opportunity of showing that his work was done properly, and taking upon himself the ungenerous task of overhauling another's labours, without seeking or obtaining the least approval or assent from the Directors in doing so, an assent which he could have obtained in a couple of hours.

On Mr. Whitehouse's arrival in town his demeanour is precisely on a par with his previous behaviour. He tells the Vice-chairman that the Directors have grossly neglected their duties, that they have insulted him, and that he would on similar occasions act exactly as on the last. It is plain either Mr. Whitehouse or the Directors must resign, and we do most respectfully beg of you to reconsider your whole judgment of this matter, and to support, as it is thought you are in justice bound to do, the dignity and authority of your colleagues.

If anything could prove Thomson's entire disinterestedness,—for even this letter did not open his eyes,—it is the circumstance that his reply pleaded that Mr. Whitehouse's name should still remain connected with the Company!

Meantime it became evident that the cable was showing distinct enfeeblement under the treatment it had received. The tests showed its insulation to be far worse than it was before submergence.

More feeble grew the signals. Official messages

could be sent to Newfoundland only with the greatest difficulty and many repetitions of words; yet the messages back were always clear and distinct, even if weak. The President's reply to Queen Victoria's message was read with comparative ease. News of peace with China and of the end of the Indian Mutiny was transmitted with difficulty. On August 31 a Government message was sent to the Canadian government countermanding the sailing of troops from Canada; and a second, of great importance, on September 1st. Then a great change for the worse came, and it was almost impossible to make the signals understood at Newfoundland. On September 3rd a dinner was given at Killarney to the Chairman and Directors of the Company at which Thomson spoke. He was as ever modest as to his own part, optimistic, and generous. Dealing with the special difficulty that beset cable signalling—the retardation due to capacity—he said:—

The genius of Faraday anticipated the difficulty; mathematicians calculated it. They gave a necessary warning, and, as events have proved, a perfectly correct estimate of the amount of embarrassment to be overcome. The elaborate and preliminary experiments of Mr. Whitehouse had a large influence in removing from the public mind doubts felt as to the practicability of telegraphing through 2000 miles of submarine cable at a sufficient rate for practical purposes; and to him in a great measure is due the present existence of the Atlantic telegraph. . . . Sir Charles Bright was the man who had undertaken and executed the impossible task. . . . The possibility of sending currents through 2500 miles of submarine cable was not reasonably doubted.

Slowly matters drifted from bad to worse. On the twenty-third day from landing, the last consistent message was received at Valencia from Newfoundland ending with: "Forty-eight words. Right. Right." The last message received at Newfoundland from Valencia ended with the words "now in position to do best to forward." Altogether 732 messages had been conveyed. After that the cable spoke no more, though for several weeks Varley and other electricians who were called in attempted with partial success to restore operations. The very last intelligible phrase came through on October 20: "Two hundred and forty tk . . . two Daniells now in circuit." Almost as dramatic as its success was its end.

Unhappily there now broke out angry recriminations between Whitehouse and the Directors. In *The Times* of September 7, Whitehouse attacked the Board in a number of wild statements. Thomson telegraphed from Valencia to the Directors :—

It is not correct, as stated in Whitehouse's letter to *Times*, that the President's reply was received and recorded here under Whitehouse's patent. It was received on my land reflection galvanometer and recorded by hand. Signal book contains full evidence. For instance, take following extract. . . .

I trust you will send an authorized correction on this point to *The Times* if you have not already done so. Please reply.

The Directors did reply, and very emphatically, in *The Times* and other journals, showing how their late electrician had misled and disobeyed them;

and Whitehouse rejoined unsparingly with bitter personalities.

Thomson himself wrote a private letter to Whitehouse, of which the draft has been preserved :—

Your letter to the *Times* has given me more pain than any one of the many painful incidents of the last three weeks. How could you possibly have allowed yourself so far to forget the facts as to say that the President's reply had been received and recorded here under your patents? Should the Directors not give a decided reply on this point, it would become necessary for me to reply in my own name. I regret extremely, too, that you should have committed yourself so, as regards a fault in the harbour, to an opinion which is certainly mistaken, and which cannot but have a most hurtful effect on the public as regards the value of the Co.'s property. I was most anxious to pass over what I knew to be a mistake all along, and to have as little said about it as possible, but now it becomes quite necessary that the public should be told what the amount of the fault in the cable really was, which was supposed to have caused the break of signals complained of at the other end. I wrote to you from London a warning, in the anxious hope that you might not commit yourself further, and felt greatly disappointed that you did not meet me in the spirit of candid inquiry into the truth, rather than go on defending a hasty step, for which the extreme perplexity of the circumstances in which you acted afforded the only justification. You will, I hope, believe me, that I say all this in no unfriendly spirit, but in self-justification. There would be a want of sincerity now in concealing from you the views which I cannot longer conceal from others without a violation of truth. I cannot forget the regard and esteem I have felt for you, nor the pleasure which intercourse with you has given me; and although I cannot understand the course you are now taking, I hope that we shall meet again without allowing anything that may come before the public to alter the kindly relations between us.

Whitehouse's reply to the foregoing is unknown, but its contents and tone may be judged of by the strong rejoinder it drew from Thomson, here given from the carefully corrected draft in his own handwriting. The passage here printed in brackets is scored out in this draft:—

VALENCIA, *September* 23, 1858.

MY DEAR WHITEHOUSE—Your reply to my last letter has made me feel how widely we differ in much that is essential for true friendship. You had surely learned long before that "no difference of opinion need separate friends," in the example of our own intercourse. It is not now a question of difference of opinion, but of truth and honour that has risen.

The impression conveyed by your second letter to the *Times* is that your first statement regarding the President's reply was a mere inadvertence, attributable to your absence from Valencia, and that what you had said of it was true of other messages. I can find no evidence, after the most minute investigation, that it was true of any one complete message, or that any other system than that by which the President's reply was received, was ever depended on for doing business at this station from the time when I showed you the dots and dashes on my galvanometer, on the night when telegraphic signals first came from Newfoundland.

I want no one to explain my galvanometer to the world, or to boast of its capabilities. I look for no acknowledgment in the newspapers of what I have done [either for you or for the Company]; but I want truth. If this had been maintained not only with reference to what more immediately concerns myself, but by the correction of such falsehoods as appeared in a letter signed T. S., in a Liverpool paper in which telegrams were dealt with, in a manner implying that the writer had had access to them from you, I should gladly have left to time the settlement of all questions between us, either as to science or as to the art of telegraphing. Three weeks ago I could not have believed it possible that under any circumstances you could act as you have done, and a most painful feeling of disappointment would be removed, if it could still be proved that I was not thus mistaken.—I remain, yours faithfully, W. T.

E. O. Wildman Whitehouse, Esq.,
British Association, Leeds.

The next day Thomson wrote to the Secretary :—

24th Sept. 1858.

MY DEAR SIR—The results of experiments on the cable, quoted by Mr. Varley as having been communicated to him here, were not, as he naturally conjectures, made at Queenstown, but chiefly on board the *Agamemnon* during her last cruise. By a comparison which I made of them after landing, with a few hasty memoranda of observations I had made at Keyham (the last set quoted by Mr. Varley), I was led to infer that the part of the cable on board the *Agamemnon* had all along been in a much worse state as to insulation than the average of the whole cable. Even the observations at Keyham alone showed that the *A.'s* portion had some much worse insulation in some part of it than the *Niagara's*. I mentioned this to Mr. Whitehouse, and he said he was aware of it, and that it was owing to the improved manufacture of the cable of this year, which was then being put on board the *Niagara*, and chiefly influenced the tests I applied. A thorough system of testing applied to the cable from the beginning, and regularly continued during the various changes to which it has been subjected, would probably have obviated the disaster by which the whole has been ruined. Even the doubt which is now felt as to whether the fault is 240 or 300 miles off, could not have existed if each part of the cable had been tested for inductivity. I mention this not to imply censure on the past management, but rather to call attention to what we should aim at if we are to continue the great undertaking. The thorough testing to which I refer has not yet been fully carried out on any cable or other telegraphic conductor, but I believe that I shall be able to show that it can be done with ease, and without any increased expense, by a little arrangement beforehand.

There is one part of Varley's opinion in which I cannot quite agree with him—that the resistance of the fault is equal to at least 10 miles of the cable. From observations I have made, both before and after his visit to Valencia, I feel convinced that it must be much less (or the effect of the fault much greater) than he supposes. I do not think the resistance can possibly be greater than from 2 to 5 miles, but I shall repeat my observations with care, and let you know the result. It may be still that there is ground for Varley's conclusion that another fault is necessary to account for the extreme weakness of the signals, but I believe the removal of the fault about 300 miles off, if successfully accomplished, will be sufficient to restore the cable to working order. It is chiefly with reference to anticipations on this score that a more accurate determination of the resistance of the known fault is desirable.—I remain, etc., W. T.

P.S.—Varley's report is, in my opinion, evidence of high scientific and practical talent.

Thomson worked on at Valencia, still nursing the unconquerable hope. On September 25th he wrote to Joule:—

Instead of telegraphic work, which, when it has to be done through 2400 miles of submarine wire, and when its effects are instantaneous exchange of ideas between the old and new worlds, possesses a combination of physical and (in the original sense of the word) *metaphysical* interest, which I have never found in any other scientific pursuit—instead of this, to which I looked forward with so much pleasure, I have had, almost ever since I accepted a temporary charge of this station, only the dull and heartless business of investigating the pathology of faults in submerged conductors. A good deal that I have learned in this time has, I believe, a close analogy with some curious phenomena you have described, and which you partially showed me last winter, regarding intermittent effects of resistance to the passage of an electric current between two metal plates in a liquid. . . . On the fourth day after the cable was landed here, I found that a positive current entering from ten cells of constant battery fell in the course of a few minutes to half strength. When the battery was next suddenly reversed the negative current rose, and remained after that nearly constant, at about the same degree of strength as that at which the positive current had commenced.

Incidentally, the phenomenon mentioned in the closing sentences of this letter furnishes the first-recorded instance of an electric valve which gives more ready passage to an electric current flowing in one direction than to a current flowing in the reverse.

Thomson also wrote to the Board respecting the sum granted in May for constructing instruments, telling that he had personally expended upwards of £1000, and that his three months' absence from England had entailed on him additional expenses which he could ill afford.

By thus remaining at Valencia he missed the British Association meeting at Leeds, where the Abbé Moigno addressed to him a letter asking for the latest news. The reply gives additional facts at first hand:—

INVERCLOY, *Oct.* 23, 1858.

MY DEAR SIR—I have only this day received the copy of *Cosmos* for the 17th September, which you were so good as to address to me. Your letter of the 18th preceded it in reaching me, but both were much delayed, being forwarded to Glasgow at the conclusion of the Leeds meeting, thence to Valencia, and from Valencia by very slow posts to this other island, where I try to escape from "the cable," and get some rest before recommencing my professional duties at Glasgow at the beginning of November. I have endeavoured to lose no time on collecting facts to reply; and I can now assure you that you were perfectly correct in stating that the 99 words of the Queen's message had come through the cable in 67 minutes. It is true that the transmission of her Majesty's message from Valencia to Newfoundland was very tedious. It occupied 16½ hours in consequence of the great difficulties experienced at that time by Newfoundland in reading. As soon as they had received it complete, they repeated it back, every word spelled in full, within a period of 68 minutes, including the initial "attack" signal and the "finis" signal, without a single mistake. A word and a half per minute was the common rate at which messages came from Newfoundland to Valencia, and it was sometimes considerably exceeded. A message of as many as 76 words—a message from the Directors of the New York, Newfoundland, and London Telegraph Company to the Directors of the Atlantic Telegraph Company—occupied only 36 minutes in its transmission from Newfoundland, and was received correctly at Valencia. Again (August 20), I find a short message of 13 words, containing 60 letters, completed and read with perfect ease at Valencia within 7 minutes. While messages were received at so good a speed, and read so perfectly at Valencia, the greatest difficulty was experienced in conveying any intelligence whatever to Newfoundland through the cable, and after the 20th August the rate of half a word a minute had been rarely, if ever, attained. On the night preceding that day they replied in the affirmative to a message from Valencia directing them to read on "Thomson's galvanometer." On Saturday the 21st we learned from them that they had introduced my instrument. Their message to us was, "Land galvanometer in circuit. Signals beautiful." Immediately afterwards we sent them a message of 35 words in 34 minutes, which they understood at once; and thenceforward they received with nearly the same ease as we, never requiring a slower rate from us than one word per minute. At Valencia, on the night of the 9th of August, when letters and words first began to come through the cable, I put one of my land galvanometers into circuit, and called Whitehouse's attention to its availability as a receiving instrument for Morse signals. I left next morning for Glasgow and London, and returned to Valencia on the 21st, when I found my instrument regularly installed, and learned from the clerks that every message from the beginning [had] been read on it. During the first two days (the 10th and 11th) Mr. Whitehouse's relay was also in circuit, and several letters were recorded by it, but Mr. W. himself was led to abandon it in consequence of the comparison, and during the night following the 11th sent the message, "Use T.'s galvanometer," which was repeated over and over again to Newfoundland from day to day, but was never replied to till the 19th, after which, as I have said, my instrument came to be used at both ends.

After repeated trials during the 10th and 11th, the use of the induction coils at Valencia was abandoned, and my battery (the sawdust Daniell's) was brought into use as the only available means by which intelligible signals could be conveyed to the other end, and after the 11th it has been exclusively used, with the exception of several further trials of the coils, which confirmed the previous result. Since the failure of the line a large magneto-electric machine of Mr. Henley's has been tried, but as yet it is not known whether with any effect. From the beginning I have always advocated the use of

D.'s battery, applied direct to the line, as preferable for every kind of submarine signalling through a long submarine line to induction coils, but Mr. W.'s system was part of the basis on which the Atlantic Telegraph Coy. was founded, and he never would be persuaded that any other system could reach the advantages he supposed to be derivable from properly constructed coils. It appears now that my battery at Newfoundland has succeeded in conveying a perfectly distinct communication ("Daniell's in circuit") thro' the cable after a period of seven weeks, during which the coils had not sent a single word intelligibly. As regards the instruments actually used at the two stations, the state of the case is shortly this. At Valencia the whole work has been done by my battery and receiving instruments. Not a single message has been sent otherwise. Every letter, word, and message that came was read on my galvanometer. No complete message was received by the relay, but only trial letters, and a few words or parts of words. Newfoundland, until the last few days, has always sent by Whitehouse's coils, and until the beginning of September, when the thing failed, we had no difficulty in reading their signals. There they used only their own instruments for reading also, at first, and scarcely succeeded in reading at all, until they complied with directions to read on my instruments. They never succeeded in reading anything on the relay, and all that they did read before the introduction of my instruments was, I believe, on a common telegraph galvanometer.

I have given these details with more minuteness than would otherwise have been necessary, because I think you may have seen incorrect statements on the subject, which appeared in the newspapers during Mr. Whitehouse's unfortunate discussion with the Board. . . .

The theory of induction in a submarine conductor led me, as early as 1854, by the aid of the analysis of your immortal countryman Fourier, to give a complete mathematical expression of the circumstances (*Proceedings R. Society*, May 1855). The numerical data which I subsequently obtained from Weber's measurement of the electric conductivity of copper in absolute electro-magnetic units, and his comparison between electro-magnetic and electrostatic units, enabled me to estimate the actual amounts of retardation to be experienced in telegraphing through 2400 miles of just such a cable as has since been constructed for the Atlantic Telegraph (see my letter in *Athenæum*, October 6, 1858), and now confirmed by observations on the cable both before and after submergence.

In the bitter polemics in the press between Whitehouse and the Directors Thomson took no part until, on October 14th, he wrote (from Arran) to the *Morning Chronicle* to correct an opinion falsely attributed to him as to the position and nature of the fault in the cable. To this he added a brief postscript :—" I refrain for the present from making any reply to the gross perversions and misrepresentations in Mr. Whitehouse's article regarding the instruments by which the messages were received at Valencia. The Directors' statement requires no defence against his attacks. It is with the deepest regret that I find myself in any way compelled to enter personally into this controversy."

On October 25th, Thomson wrote to his brother :—

INVERCLOY, *Oct.* 25, 1858.

MY DEAR JAMES—I got your note at the Gresham, but was very sorry not to be able to take the opportunity of going to see you, as we had just time to drive direct from the hotel to the steamer. As it turned out she did not sail till three-quarters of an hour after her time, but we could not trust to that, although if I had known there would be so much time I should have tried to get you down to the steamer by a message.

I need not tell you about doings at Valencia, as you will have heard enough about it in the newspapers, and had some details, no doubt, supplied by W. Bottomley. You would see that after all a message came through last Wednesday, I rather think it was in reality by no extraordinary or unsafe power, but simply by the application of my battery, which from the beginning, or from before the beginning, I have always advocated as the best means of working the telegraph. The chief danger to be apprehended now from battery power of any brand is, I think, that it is certain to eat away the wire when it is exposed by positive currents out through the leak. I have therefore enforced the use of "negative currents" as much as possible, that is to say, of such electrical arrangements at both ends as shall keep the wire throughout negative relatively to the water and earth beside it. The wire itself would be preserved indefinitely by this (and might be eaten through in a few days or weeks by the reverse), but the leak may get even worse, whatever we may do. As it is, no good work can be expected, I believe, from this wire, and the best, if not the only chance, is by lifting it as far as the fault. The most probable distance to this along the line is 254 statute miles; that of the "bank" 242 (also measured along the line). The estimate for the fault may be 15 miles wrong either in excess or defect, and it is possible, therefore, that it may be found in shallow water. Margaret has not been so well as she was in summer The journey to Valencia and stay there were not fortunate, but she is now rather better.—Your affectionate brother, W THOMSON.

The last spark of life had flickered out from the cable on October 20th. There is no doubt that from the first it suffered from defects caused by the damage received on board during the storm; but its actual failure was due to Whitehouse's bungling use of induction coils—some five feet long—working at some 2000 volts.

A year later, when a Departmental Committee[1] inquired into the circumstances, Thomson gave evidence which throws some light on the facts, and on the self-repression which he exercised during this trying time. The following are extracts :—

There was a very great omission in the apparatus on board in the want of standard resistance coils. I had urged on the electrician of the Company, as early as the month of May 1857, the very high importance of having a set of resistance coils properly made, giving a resistance at least equal to the resistance of the whole cable, and admitting of variations to the smallest measurable quantity. I urged this strongly, but the electrician of the Company had his own system of testing, which he considered satisfactory. The great want in our system of testing was a good constant battery and a set of resistance coils. The constant battery I supplied, as far as I could, from the resources which I, for another reason, had provided. A sufficient set of resistance coils could not at the time be extemporised, and, accordingly, much of the testing was necessarily mere guess work.

I had the very strongest misgivings as to the condition of the cable from the Monday forenoon[2] of the laying.

From the handing over of the cable to White-

[1] Report of the Joint Committee appointed by the Lords of the Committee of Privy Council for Trade, and the Atlantic Telegraph Company, to inquire into the Construction of Submarine Telegraph Cables, together with the Minutes of Evidence and Appendix, 1861. The members of this Committee were R. Stephenson, Wheatstone, Fairbairn, and Bidder, for the Board of Trade ; and Edwin Clark, C. F. Varley, L. Clark, and G. Saward for the Company.

[2] A few months later Thomson wrote :—"Those who were engaged in the undertaking were fully convinced that the cable they were laying was in perfect condition. It was not until rather more than half the cable had been laid that signs of defective insulation manifested themselves. When the ships had come to anchor on the two sides of the Atlantic, the signals which passed between them were of so satisfactory a character that none of us had then any doubt as to the complete and final success of the undertaking."

house on the forenoon of August 5 till August 9 near midnight, they never received anything that they could be quite sure was a signal current. On that night there were good signals; the first words being "Please repeat slower." Asked about the underrunning of the cable to the harbour mouth with a view to the discovery of a supposed fault, Thomson replied that it had been done "in opposition to the orders of the Board of Directors, and against my strongly expressed advice, and during my temporary absence." After Thomson left Valencia, on August 10, Whitehouse put in his own patented recording devices, by which there were "some indications of legible signals recorded, but not one complete sentence." "I have," said Thomson, "had all the slips, which were preserved by the Board, to examine, but these contain very little of the work of the first four days, the slips corresponding to those days having been all abstracted from the possession of the Company. The longest word I find correctly given is the word 'be.'"

Official documents show that those slips had been taken away by Whitehouse in October 1858 from the office of the Company in the absence of the Secretary, and were shown about by him at the Royal Institution as the performances of "his" inventions

Writing in 1860, in his article "Telegraph, Electric," in the *Encyclopædia Britannica* (eighth edition), Thomson says: "It was only in consequence of fallacious interpretation of experiments on the relative capabilities of 'battery' and 'induction coils' that the latter were ever introduced into the service of the Atlantic Telegraph." He added the following note :—

The induction coils were superseded by Daniell's battery at Valencia, after a few days' trial through the rapidly failing line had seemed to prove them incapable of giving intelligible signals to the Newfoundland station ; but, owing to the immediate introduction and continued use of an entirely new kind of receiving instrument—the mirror galvanometer, introduced for long submarine telegraphs by the writer—at Valencia, the signals from the Newfoundland coils were found sufficient during the three weeks of successful working of the cable. It is quite certain that, with a properly adjusted mirror galvanometer as receiving instrument at each end, twenty cells of Daniell's battery would have done all the work that was done, and at even a high speed, if worked by a key devised for diminishing inductive embarrassment, according to the indications of the mathematical theory ; and the writer, with the knowledge derived from disastrous experience, has now little doubt but that, if such had been the arrangement from the beginning, if no induction coils and no battery power, either positive or negative, exceeding twenty cells of Daniell's had ever been applied to the cable since the landing of its ends, imperfect as it then was, it would be now in full work day and night, with no prospect or probability of failure.

In 1863 a very unworthy attempt to revive an old sore was made by an anonymous article in *The Electrician*, vol. iv. p. 109, of July 10, purporting to quote a letter of Cyrus Field's of September 8, 1858, containing a statement that Thomson's system was "regarded by all practical telegraphers as perfectly childish." In the issue for July 17, 1863, p. 132, Thomson replied that this alleged letter of Field's was a forgery long ago disclaimed by Field; and in the number for July 31, p. 153, Field himself wrote, giving extracts from the *New York Tribune* of September 10, 1858, containing his repudiation.

Thomson returned to his winter's work at Glasgow worn but not disheartened. Scotland welcomed him with no uncertain voice. Professor George Wilson of Edinburgh, lecturing on the Telegraph on November 6th, spoke of Thomson as the Columbus of this voyage, and compared him watching the quivering magnet of his galvanometer on board the *Agamemnon* with Columbus watching the compass needle on his ship four centuries before.

In acknowledgment of Professor Thomson's services in connection with the Atlantic Telegraph a banquet was given to him by his fellow-citizens in the Queen's Rooms, Glasgow, the Lord Provost in the chair, on January 20th, 1859; Professors Rankine, Rainy, and others, the Parliamentary representatives, and many prominent local leaders being present. The Lord Provost, in proposing the toast of his health, spoke of the event as a public proof of the esteem with which he was valued as Professor in the College, and as the most efficient scientific promoter of the great Transatlantic undertaking. Professor Thomson had consented, along with Sir J. Anderson, M.P., and Mr. W. Logie, to act as Directors representing on the London Board the Scottish shareholders, and it would not be forgotten how considerable a share in the accomplishment of the great work he had had, as he was the only one of the Directors who had shared the dangers and hardships of the expeditions, the last of which had achieved success amid such universal enthusiasm as probably was never before called forth by anything short of a great national victory. He mentioned how, amongst the messages delivered before the cable broke down, there was one assuring the Cunard Company of the safety of the ship *Arabia* after collision off the banks of Newfoundland. He also read letters of regret from Mr. Stuart Wortley, and from Sir David Brewster, who was prevented from accepting the invitation to assist "in paying honour to one of the most distinguished philosophers that Scotland has produced, and one to whom those who are about to quit the field of labour willingly confide the scientific reputation of their country."

Professor Thomson in replying to the toast said:—

In returning you my thanks for the honour you have done me this evening, my sense of what I owe to your kindness cannot but be intensified by the consideration that while in general success is the criterion which determines the approbation of the world, you have taken the rare and generous part of attributing merit to efforts which have resulted in failure. After the harassments and disappointments of a year, when wealth and labour, care and anxiety, skill and invention might appear to have been absolutely thrown away, and to have gone to swell the vast amount of profitless labour which is done under the sun, it is no small solace to meet with such sympathy as you now manifest. For the too generous construction which you have placed on my own part in these events I thank you with the deepest gratitude. At the same time, impressed as I am with a sense of the kindness which you have shown to me, I cannot but be aware that a feeling beyond anything personal to myself has influenced you, and that by your presence here this evening you show an interest in the great undertaking, animated by a conviction that the foundation of a real and lasting success is securely laid upon the ruins which alone are apparent as the results of the work hitherto accomplished. (Cheers.) Am I right in supposing that you entertain such a conviction? That you do entertain it, and that you have good reason for entertaining it, I firmly believe. What has been done can be done again, as marvellous as it is; improbable, impossible as it seemed only six months ago—chimerical and merely visionary as such a project seemed ten short years earlier—instantaneous communication between the Old and the New Worlds is now a fact. It has been attained. What has been done will be done again. *The loss of a position gained is an event unknown in the history of man's struggle with the forces of inanimate Nature.* If it will not be considered that I am trespassing on your patience too much, I shall endeavour to explain, in as few words as I can, something of the great physical problem which now stands solved before the world and ready for application to promote the national, commercial, and social interests of the countries on the two sides of the Atlantic. The difficulties to be overcome in establishing telegraphic communication with America have not been magnified in the popular imagination. It was indeed among men, the most profoundly versed in mechanical and electrical science, and men of the greatest experience in nautical, in engineering, and in telegraphic enterprises, that those difficulties were considered most formidable. The forces tending to break the cable in laying it across an ocean of such depth—even in the most favourable circumstances of weather—seemed to present an almost insurmountable obstacle to the project. What is known of hydrostatics and of the laws of fluid friction gave but little encouragement, and to get the cable laid unbroken in a dead calm was about as much as the most sanguine dared to hope for. But we were not all sure of a dead calm over the whole Atlantic every day of June. So we went out, praying for fair weather, to lay the cable. And after nearly losing it in a storm, before any attempt could be made to lay it, we had as fair weather as could be desired. We did at last recognise that promised feature of the Atlantic which is so often, by a slight stretch of imagination, compared to a mill-pond. But a smooth sea did not bring safety. The cable, running out under the most favourable circumstances, broke away once from the *Niagara* and once from the *Agamemnon.* Another time it was quite away from each ship in consequence of a sudden and total loss of electric communication in the part paid out. After laying and losing 500 miles of cable in three successive trials the ships returned to Queenstown harbour. Permission having been granted by the Admiralty and the Board of Directors to renew the attempt, which seemed hopeless to all except those who had gained the experience by failures, the expedition again put to sea bound for the mid-ocean rendezvous. On 29th July the *Agamemnon* and *Niagara* lost sight of one another, steaming west and east with a thread between them, which was never broken, and now joins the Old and New Worlds. (Cheers.) Many important experiments, many exciting incidents, many periods of intense anxiety, some long intervals scarcely illuminated by a ray of hope, there were, which can never be forgotten by those who took part in the operations of the three Atlantic cruises of the telegraph ships. The raising of the splice from three miles' depth in the Bay of Biscay was an unparalleled achievement of nautical mechanics. The *Agamemnon*, moving round within her own length, more like a living animal than a ship 270 feet long, did what could scarcely be expected even of "Sir Edmund Lyons' brougham," and during this evolution the passing of the cable from her stern to her bow, while it hung down with a strain of more than two tons, completed a performance which is not marvellous only to those ignorant of the difficulties attending such manœuvres. But the story of the *Agamemnon* at sea has been so well told by the graphic pen of *The Times* correspondent that I need not weary you by a weak repetition of any part of it. I cannot, however, refrain from alluding to that hopeless period when at the height of the gale in June the main coil—a thousand miles of cable—began to move in the *Agamemnon's* hold. Each day the evil increased, and when the storm began to subside the precious freight seemed irretrievably ruined. Now it was that unflinching determination was tested, and Sir Charles Bright and Mr. Canning had a task before them such as few engineers have ever had to deal with, and boldly they faced it without a shadow of encouragement, except from their own investigations and their own conclusions as to the state of the unseen part of the cable. They began their labours while the storm in the hold was still exhausting its fury on the luckless wire, while the ship was still rolling so that men could scarcely stand; and almost as soon as the gale had subsided, "the Company's men," assisted by as many blue-jackets as could be spared from the still more important work on deck and aloft, had, not without danger and hard knocks, and even broken limbs, pulled out from the tangled mass one hundred miles of cable, weighing as many tons, and deposited it in clear coils, wherever in any part of the ship space and support could be had for such a load. By these efforts the cable was saved and a year gained, I believe, if only a year, in the first attainment of telegraphic communication with America. But looking under the all ruling Power to human agency, we must regard the skill, constancy, and unflinching performance of duty manifested by Captain Preedy, his officers and men, as having saved the ship and cable and all on board. (Great cheering.) I cannot pass from the *Agamemnon* without expressing the sense I entertain of the value of the assistance and support rendered by Captain Preedy to the engineers of the Company in some of their most difficult mechanical operations. The part taken in the actual laying of the cable by the officers and men under his command was, I believe, the most essential contribution towards the success which was achieved. The electrical conditions of the grand problem strike the imagination even more than the mechanical difficulties thus so perfectly overcome. This line of metal, stretching away two thousand miles under the Atlantic, must convey the subtle influence. Touch our European end: a magnet must instantaneously move in America. Timed repetitions of the simple signal compose letters, words, and sentences, till ideas flow through the wire. Astonishing result of science, from which no degree of familiarity can remove wonder! The speed with which these operations may be performed, each giving its own distinct effect through ordinary land telegraphs suspended on poles in the air, or the shorter submarine lines, has no limits yet discovered; and signals succeeding each other more rapidly than the hand can send or the eye follow, can be turned to account by using the proper mechanism at each end. But if "quick as thought" is an inadequate expression for common telegraphic operations, even lightning becomes slow through the Atlantic telegraph. The most sudden electric shock at one end gives a sluggish, long protracted current through the other which, after a quarter of a minute, is still working its way feebly out of the wire. That such would be the action through a telegraphic cable of ordinary construction, connecting Britain and America, was pointed out by Faraday long before the existence of the Atlantic Telegraph Company. An exact mathematical investigation of the circumstances showed that a sufficiently large-sized conductor and insulating coat would entirely remedy the anticipated embarrassment. But a cable of such dimensions as the calculations showed to be required for signalling through two thousand miles at the rapid rate of our ordinary telegraph would be too unwieldy and too costly to be thought of in a first attempt. Therefore the Atlantic Telegraph Company, in adopting the improvement which I suggested, did not carry it further than to making the quantities of copper and of gutta-percha in any part of their cable nearly double of those in an equal length of any previously constructed telegraph line, so far at least as their first cable was concerned, and prudently, in my opinion, they left the further mitigation of the anticipated slowness to be worked out by improvements in the methods and instruments to be used for the transmission and the receipt of messages. When the *Niagara* and *Agamemnon*, bearing the two halves of the cable from Birkenhead and Greenwich, met in Queenstown harbour on 29th July 1857, trials were made through the whole length of 2500 miles united in one conducting line, and the most unpalatable warnings of the mathematical theory were too surely fulfilled to the letter. This was only what was to be expected, but the result in reality was most satisfactory. Messages were transmitted with accuracy through the whole length by means of instruments constructed by Mr. Whitehouse for the Company, although not at the rate anticipated by those who ignored reason and trusted to fallacious experiments. Even the one word a minute which those trials seemed to promise is not to be regarded as a small result. Of what value would not sixty words every hour be, continued night and day between the Old and New Worlds? Who can put a limit to the value of two words transmitted in any two minutes when an ocean flows between? It is true that a sevenfold higher speed had been confidently promised, but that more than one word a minute was not to be had with the ordinary receiving instruments was finally learned by trials continued until the ships left Keyham Dockyard in May 1858. During the last four weeks of that period a new kind of receiving instrument and a new mode of working gave promise of a double or triple speed, and fulfilled theoretical estimates which had been published before the Atlantic Telegraph Company had commenced their undertaking. It would be unsuitable to the present occasion that I should give any minute or detailed statement regarding different systems of telegraphic operations; but the general character of the difficulty at first denied, and the solution at last brought into practice with reference to the Atlantic Telegraph will be sufficiently understood if I am allowed to tax your patience with a few words of explanation. The long continued effect received at one end, which I have already described as the result of a sharp signal at the other, when the communication is through a submarine line of great length and of ordinary lateral dimensions, renders it necessary that time enough be given from signal to signal to allow each to show its effect distinctly at the remote end. Ordinary receiving instruments can only show fresh signals after being almost perfectly relieved from the residual effects of previous operations. An instrument capable of distinctly marking a new signal when still under the influence of ten-fold or twenty-fold accumulations or undischarged residues from currents which have already told their tale must obviously give a higher speed of working. It is thus that messages at the rate of two fully spelled words a minute without a doubtful letter were received through the submerged and failing cable. Another way of increasing the speed has been found by making each signal in such a manner that its effects may subside very rapidly after the first indication has been received at the remote end. By a combination of these two principles a considerably more rapid rate was obtained in the transmission of messages through three thousand miles of cable on board the ships at Devonport than afterwards, by the application of the former alone, was realised between Valencia and Newfoundland. There can be no doubt, however, but that the combined use of both principles will be found perfectly practicable whenever a cable in fair condition is available for telegraphic communication with America. But I have forgotten how little these facts and these anticipations can interest you when no Atlantic cable exists from which useful work can be reasonably expected. When approaching Valencia on board the *Agamemnon* the last critical operation of changing from coil to coil was made, and the bight of cable came up easily and straightened itself quietly till the rope began running out with perfect smoothness in its new course through the ship. The enthusiastic clapping of hands and the three hearty cheers which resounded on all sides from the crowd of sailors and marines standing round anticipated the expressions of feeling which the electric nerves of Europe and America called forth on the following day, the ever memorable 5th of August 1858. (Cheers.) How much pleasure would have been lost if it had been known that after the 2nd of September the cable for all practical purposes would be worthless. "Prudens futuri temporis exitum; caliginosa nocte premit Deus." The bewildering feeling of having done that which the sanguine scarcely hoped to do, and which nobody on shore considered possible; the excitement of bringing the end to land, and the delight of seeing the first words thrown from the needle in the Valencia instrument room—though these were only, "Please repeat slower"—were soon followed by the most harrowing anxiety. The signals had been becoming very weak and vague during the days preceding that on which words began to come—so much so that the *Niagara's* end run short and landed somewhere without battery or instruments was the only explanation suggested to obviate the conclusion that the insulation of the cable, which the tests had shown to be imperfect during the three days before the landing, must be rapidly deteriorating. When the signal currents came only one-thousandth of the strength which the battery power employed would have sent through a well-insulated line of the same length, they became lost in the earth currents; and disturbances owing to chemical action at the leak, and vague fluctuations of the needle (watched day and night by weary eyes in the now cheerless station-house at Valencia and the dismal swamp at the head of Bull's Arm Bay), never even told whether electrical tides were ruled by the moon or sun. Sooner or later we all believe another Atlantic cable will be laid, but will it last any longer? Will it do any more work than the one which has raised and cast down so many hopes? Will another experiment be another gigantic failure? And will material locomotion be again fallen back upon as the only means of communication between the Old and New Worlds? That the next trial will not be a failure no man living can say is more than probable; although on no one point can it be said that there is any insurmountable difficulty, and least of all on that which has proved the cause of ruin in the present case. Increased caution in the manufacture and preservation, even without improvements in the material of the insulating cover, with a more searching system of detection for faults, might, I believe, make very sure against any recurrences of such a failure. The great risk which the enterprise undeniably involves depends not on any one source of danger, but on the multiplied chances of accident inseparable from the exposure of so great a length of cable to so great a variety of contingencies. You have spoken of the sacrifices I made to give my assistance in the undertaking. If I have made sacrifices, so have all the Atlantic Telegraph Company. The cheerful and uncomplaining spirit with which an apparently total loss of interest and capital has been met by the subscribers to the Atlantic Telegraph most forcibly demonstrates the high character of the motives with which they have entered upon an undertaking of which the success would have been of such large benefit to mankind. It is true that sacrifices have been made by those who have been most closely connected with the work. I shall always esteem it as an honour to have belonged to the Board of Directors, and in this capacity to have acted along with men who have devoted themselves with so much earnestness and at so great personal sacrifices to carry into execution a project of such high national importance. Under circumstances of the greatest discouragement the Directors have persevered with an untiring energy and undaunted resolution worthy of the great enterprise. Many among them continue day after day, and month after month, sacrificing their own convenience and setting aside their own business avocations to promote the interests of the Company. The unremitting daily attention given by the Executive Committee, the zealous co-operation of the general manager, Mr. Cyrus W. Field, and the able and energetic services of the secretary, Mr. Saward, may well be remembered with gratitude both now and when the final and complete success is attained. Nor in speaking of an undertaking in which all mankind are interested, would I omit to claim for our own city its contribution to the general resources; to the assistance of students of the University of Glasgow, and to the high ability and energy of Glasgow instrument makers, by whom scientific principles, novel in conception, were understood and carried out with extraordinary promptitude, we are indebted for the realisation of ideas which, without the aid of the practical element, must have remained powerless to achieve material results. In conclusion, Professor Thomson again returned thanks for the honour which they had done him. It was an honour which he did not expect, and which he did not think he was entitled to receive. The learned Professor sat down amid loud cheering.—*From the "Glasgow Herald," 21st January* 1859.

"That is a noble speech of William Thomson's, so like himself. I have sent it to Thackeray. You may well be all proud of him." So wrote Dr. John Brown to James Crum, on January 24. What Thackeray's comment was is unknown; but we have heard already, p. 324, what a high opinion of Thomson he entertained.

CHAPTER IX

STRENUOUS YEARS

NOT even the absorbing interests of the Atlantic cable could withdraw Thomson from studies of a more abstract nature or from his laboratory work.

In April 1857 an exceedingly interesting discussion arose at the Institution of Civil Engineers upon a paper by Mr. Robert Hunt on "Electromagnetism as a Motive Power." In this paper the possibility of economically driving electromagnetic engines—that is, electric motors—by currents derived from voltaic batteries, was discussed in the light of Jacobi's discovery of the counter-electromotive force in these machines, and of Joule and Scoresby's measurements of the mechanical equivalents of the battery power. Using Daniell's cells, they had found that for every grain of zinc consumed in the battery the motor performed a duty equivalent to raising 80 lbs. 1 foot high against gravity. In the Cornish engine, doing its best duty, 1 grain of coal was equivalent to a duty of raising 143 lbs. 1 foot high. Putting the price of zinc at £35 per ton, as compared with coal at less than £1 per ton, he argued that electric power would be about 60 times as dear as steam power, and concluded that it would be far more economical to burn the zinc under a boiler, and use it for generating "steam" power, than to consume zinc in a battery for generating "electro-magnetic" power. The debate which followed, and in which Grove, Tyndall, Smee, Bidder, and Robert Stephenson took part, makes strange reading now. One and all, these most eminent leaders condemned the idea of electric motive power as unpractical, and impossible commercially. To the debate Thomson sent a contribution in writing, which shows that he, at least, understood the true theory of electric motors, and knew that their intrinsic efficiency might far exceed that of the best thermal engines. He set forth in their proper bearing the results of Joule and Scoresby, and ended thus:—

> These facts are of the highest importance in estimating the applicability of electromagnetism, as a motive power, in practice; and, indeed, the researches alluded to render the theory of the duty of electromagnetic engines as complete as that of the duty of water-wheels is generally admitted to be. Among other conclusions which may be drawn from these experiments is this: that, until some mode is found of producing electricity as many times cheaper than that of an ordinary galvanic battery as coal is cheaper than zinc, electromagnetic engines cannot supersede the steam-engine.

In other words, Faraday's great discovery of 1831 notwithstanding, the real significance of the dynamo had not yet dawned upon the keenest minds of the time; and the most Thomson could say was that until some such source of electric supply was available there could be no economical use of electric motors.

In June 1857 Thomson read to the Royal Society a note on the alterations of temperature accompanying changes of pressure in fluids, and in January 1858 some remarks on the interior melting of ice. This was followed in February by a third, on the thermal effect of drawing out a film of liquid, and in April a fourth on the stratification of ice by pressure.

As a result, perhaps, of his visit to Dellmann, he paid much attention to the electricity of the atmosphere, a subject on which he had already worked with Joule. He devised an improved collector of atmospheric electricity, which he described in a letter to Helmholtz.

LARGS, *January* 2, 1859.

MY DEAR HELMHOLTZ—I was very sorry by your last letter to have so bad an account of your wife's health. I can fully sympathise with you in your anxiety. Mrs. Thomson wishes to be kindly remembered to you, and bids me say she has felt for you very much in this trial, and hopes that you have now less cause for anxiety.

The information you gave me regarding Kirchhoff's investigation of solar chemistry interested me greatly. I had just the week before been telling my students how other ingredients of the solar atmosphere besides sodium, which was already proved, were to be tested, and I therefore took your letter with me to my lecture and read the part of it which had reference to that subject. I enclose a memorandum of ideas I have had as to solar and stellar chemistry from a conversation I had with Stokes a long time ago, which you may perhaps think it worth while to show to Professor Kirchhoff.

My water-dropping *collector*, or more properly *discharger* for atmospheric observation has led me naturally to a self-acting *condenser*, in which electrified drops of water give up their electricity to an insulated metal funnel upon which they fall in a fine spray. This funnel is attached to a metal tube open below, as shown at the bottom of the sketch, so that large drops falling from the mouth of the funnel are each as much as possible quite unelectrified, even although a considerable charge may have accumulated on the outer surface of the funnel and tube. The fine drops originate in a stream issuing from a small aperture B at the end of a tapering tube A. An insulated tube CDEF surrounding this stream fixes by "induction" the electrical condition of the stream and drops breaking from it. Thus if the tube AB proceed from an uninsulated cistern, and if the metal of CDEF be the same as that of AB, an extremely minute charge of electricity communicated to this insulated inductor of which CDEF is a part, gives rise to a continued accumulation of the opposite kind of electricity on the lower insulated conductor. If this last be tested with my divided ring electrometer, a very strong effect is shown when $\frac{1}{10}$ of the electromotive force of a single element of Daniell's battery is applied to maintain a difference of potentials between G and H connected respectively with the tapering tube and the wide tube surrounding it. If on the other hand a metallic connection between G and H be established, and if AB be a different metal (say copper) from CDEF (zinc), the electrometer shows strong negative.

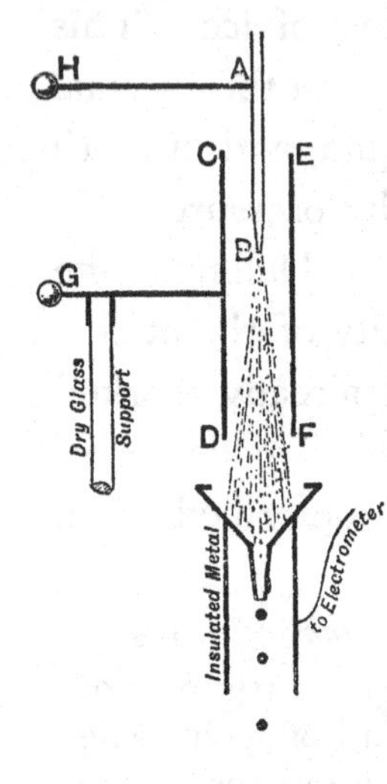

I have been greatly troubled in these experiments by phenomena of "polarisation," of which as yet I cannot make much, and I have been forced, too, into very difficult considerations regarding Volta's fundamental experiments. I hope for more light soon, as I have been much perplexed. Of one thing, however, I have I believe perfect experimental evidence, that a vitreously electrified body in the air over a metallic mass of zinc and copper, in contact with one another, experiences a force tending on the whole from the zinc and towards the copper. From this it follows, as indeed

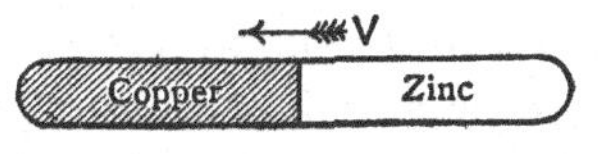

from Volta's experiment, that a sheet of zinc and a sheet of copper parallel to one another and connected by a metallic arc, attract one another; and when we understand the whole sufficiently I believe we shall see

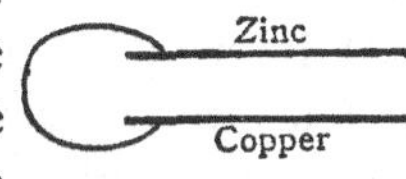

it simply as chemical affinity (another word for electric force) acting at a distance.

I shall be glad to hear from you again when you have time to write, and I hope you will have a much better account to give of your wife's health.—Believe me, yours always truly, WILLIAM THOMSON.

Thomson was not present at the British Association meeting of 1858, but at that of 1859 at Aberdeen he read two papers on atmospheric electricity: one on the false discharge from a coiled cable, and another on the periodical variations of underground temperature. He had hoped that Helmholtz would have come, and wrote to invite him:—

BIRKENSHAW COTTAGE, GLASGOW,
May 12, 1859.

MY DEAR HELMHOLTZ—At the request of Mr. James Crum, an uncle of my wife's, I write to ask if you will take up your quarters in his house during the approaching meeting of the British Association to be held at Aberdeen. It is anticipated that it will be a *very good meeting*, because the Queen and Prince Albert are to be present. Unfortunately, for the same reason the time cannot be precisely fixed yet, but it will in all probability be the middle of August or at the end of September, which I think would suit your University vacation. I hope you may be induced to come, as it is a long time since we have met, and there will be much interesting matter for conversation between us. If Mrs. Thomson is feeling well enough at the time she will accompany me, but in any case I hope you will give us a visit in Arran before or after the meeting.

On my return from Valencia last October I found a copy of your paper on rotatory motion in fluids you were so good as to send me, which I read with very great interest. I intended to write to you regarding it, but fell into the vortex of my winter's work before doing so, and have had little chance to think or to put pen to paper on any subject except that of the day from that time till this. As I hope there is now a prospect of seeing you before long I shall keep the discussion of the Ring-Wirbelfaden till then.

Since this time last year I have managed little in the way of experimenting, except in connection with telegraphic instruments, and I have now got a set of marine and land reflection galvanometers and resistance standards, which will very much shorten work in various more purely scientific investigations which I hope now to be able to proceed with. I think you would feel some interest in some of those instruments, which I shall be happy to show you if you come, although I do not mention them now as even the smallest inducement.

I shall be obliged by a line from you at your convenience to enable me to reply to Mr. Crum. If you write next week address care of Dr. J. P. Joule, Oakfield, Moss Side, Manchester, where I shall be till about the end of the month to experiment on the thermal effect of air in motion.—Believe me, yours always truly,

WILLIAM THOMSON.

Unfortunately Helmholtz was kept at home by the illness of his wife. Joule was present, and experimented with Thomson on atmospheric electricity on the Links. After the meeting Thomson wrote again :—

INVERCLOY, ISLE OF ARRAN,
Oct. 6, 1859.

MY DEAR HELMHOLTZ—We looked forward with much pleasure to seeing you here a month ago, and were disappointed on receiving your letter of the 30th Aug. to find that you were prevented from carrying out your intention. I hope by this time you have a better account to give regarding the health of your wife.

We should have had a strong scientific party in Mr. Crum's house at Aberdeen if you had been with us. We had Faraday, Joule, and Thomas Graham, besides my father-in-law, who is known as a chemist. The meeting "went off" on the whole very well. It was unusually large, owing no doubt to the attraction of the Prince as President. There was a great press to hear his opening address, and for the first time, I believe, in the history of the Association the issuing of tickets had to be stopped in consequence of insufficient accommodation. Upwards of two thousand, I believe, were present. The address was in general very much approved of, and seemed to be quite original. I succeeded in getting the subject of electric observation taken up effectively—at least in a manner which I hope will prove effective. I have been appointed a "committee" to prepare self-recording instruments which are to be set up and used at the Kew meteorological observatory, and I hope very soon to have the system in action. I have now got an extremely simple collector, and have tried it at the window of the house here, with perfectly satisfactory effect. It consists of an insulated can of water inside the house, discharging by a fine orifice at the end of a thin tube projecting about a yard from the wall. The effect is in about 20 sec. (or in a much shorter time when a higher head of water and a small aperture are used) to reduce the "potential" of the whole insulated conductor, including the electrometer, to (sensibly) the same as the atmospheric potential at the point where the stream of water breaks into drops. The amount which I have found since yesterday has varied from about 35 to 70, the unit being a single zinc-copper water cell. If the weather permits I hope in a few days to have simultaneous observations made by the fixed apparatus here, and a portable apparatus at various points on the hills, with a view to examining the influence of electrified clouds or masses of air in the vicinity.

The pressure of work connected with the Association (from which I only got free two days ago), and the time taken up by it, have prevented me from sooner writing to you in reply to your question regarding Stokes' investigation of effects of friction on the motion of fluids. Stokes was not at the meeting, being away in the south of Ireland for the sake of his health. I have, however, written to him for full information such as you desire. I wish he may be able to send you separate copies of his papers, but I fear he will have given all away before this time. His first paper on the subject, entitled "On the Friction of Fluids and Motion and the Equilibrium and Motion of elastic Solids," was read at the Cambridge Philosophical Society, April 14, 1845, and was unfortunately buried in the *Transactions* of that society, vol. viii. part 3. It contains the hydrodynamical equations with friction taken into account, which, so far as I can recollect, are identical with those you have found.

His second paper on the same subject was published some years later in the *Transactions* of either the Camb. Phil. Soc. or the Royal Society, probably the former, from your not having been able to meet with it. It contains solutions of the equations for various practical cases, especially the motion of a ball pendulum and the descent of a globule of water (a particle of a cloud, for instance) through air. By the former application he explains (I believe perfectly) the discrepancies which experiments on pendulums by Bailey and others showed from the ordinary hydrostatical and hydrodynamic theory. In the latter application he justifies perfectly the idea generally entertained regarding the support of clouds.

If Stokes has no copy to send you, and if his paper does not otherwise fall in your way, I shall send you some extracts, should you wish to have particular information, especially as to the numerical value of the coeff$^{t.}$ of fric$^{tn.}$ in air or water.

I wish you would send me a short account of your investigation of vowel sounds for publication either in the *Proceedings of the Manchester Literary and Philosophical Society*, from which it would be copied in the *Philosophical Magazine* or for the Royal Society, London. If you send it in German I shall have it correctly translated.—Yours very truly, W. THOMSON.

Mrs. Helmholtz died on December 28, and Thomson on receipt of the news addressed condolences in the following letter:—

GLASGOW, *Jan.* 22, 1860.

MY DEAR HELMHOLTZ—I was very much grieved to learn of the great loss you have suffered. Mrs. Thomson begs to join with me in expressing deepest sympathy. We were not unmindful of your anxiety, as you will see by the letter which I enclose. I had written it at Largs and brought it up to Glasgow with me to post, when I was much shocked to find waiting my arrival the intimation you had sent me. I would not send it to you now to trouble you when your mind must be so painfully occupied, but I wish you to know that I did not neglect the short period of leisure I had to write in reply to your previous letter.—Believe me, my dear Helmholtz, always yours sincerely, WILLIAM THOMSON.

One of the recurring scares of foreign invasion led in the autumn of 1859 to the Volunteer Rifle Corps movement. It found in Thomson an enthusiastic supporter,[1] and a College Company was raised in the original corps of the 1st Lanarkshire Volunteers in 1860. Rankine was captain of the 1st

[1] "That was in the days of what Lord Palmerston called the 'rifle fever,' and I was touched a little with it at the time."—*Popular Lectures*, vol. i. p. 448.

Company and Thomson of the 2nd Company. He took an active part until the accident which befell him at Christmas (see p. 412), and which left him permanently lame. His resignation he used to relate in the following terms:—

"It was this way, you know. I was all right standing at ease and at manual exercises, but when it came to evolutions, the more the officer commanding ordered us to 'march,' the more I 'halted.' So I resigned."

In 1863 a new battalion was formed, and Thomson was elected captain of the 1st Company of the 2nd Battalion, and held that rank till he resigned in May 1867.

Years afterwards, on the formation of the Volunteer Corps of Electrical Engineers (Royal Engineers) in 1897, Lord Kelvin accepted the position of Honorary Colonel.

That winter Thomson read several papers to the Glasgow Philosophical Society, and early in 1860 communicated two important papers to the Royal Society. One of these was on the "Measurement of the Electrostatic Force produced by a Daniell's Battery." Using one of his own absolute electrometers he weighed the mutual attraction exerted between the two metal plates connected to the respective poles of the battery, finding 3·57 (later corrected to 5·7) grammes as the equivalent force of the two plates, each a square decimetre in area, and one millimetre apart, when charged by 1000 Daniell's cells. The other paper was on the

"Electro-motive Force required to produce a Spark in Air between two Metal Plates"; the result was that to produce a spark between plates at $\frac{1}{8}$ of a centimetre apart would require a battery of 5510 Daniell's cells.

For some five years Thomson had been devising electrometers of various patterns, and had applied them to the study of atmospheric electricity. In October 1859 he had written to Joule describing the early instrument known as the divided ring electrometer—a forerunner of the more sensitive quadrant instrument—and a water-dropping collector for observation of the state of electrification of the atmosphere. He also mentioned a portable electrometer which he had taken to the top of Goatfell to get readings to compare with those at Invercloy (Brodick) at the foot of the mountain. He was invited to lecture on the subject at the Royal Institution, and gave the Friday evening discourse on May 18, 1860.

The following note of May 14 shows him preparing:—

LABORATORY R.I., *Monday* 4 P.M.

MY DEAR FARADAY—Many thanks for your kind note. I hope by Wednesday to be ready with my instruments, and when you come I shall have to ask your advice on many points connected with the arrangements for my lecture.

With Anderson's assistance I have been at work all day putting up one of my electrometers, and I shall probably have nearly as much work to-morrow with another. I find a great deal of trouble in making the glass fibre suspension, having very little skill of hand, so that what would be easy and short to others costs me a great deal of time and trouble.—Believe me, yours very truly, WILLIAM THOMSON.

In this lecture he gave a popular account of his instruments and the results of observations. The peroration of this lecture, which was in praise of Faraday's conceptions of the nature of electric forces, was striking and suggestive.

Earlier in the year Forbes had retired from the Edinburgh chair of Natural Philosophy, and after some delay Tait was appointed to succeed him. This event led to the subsequent collaboration of Thomson and Tait in their famous *Treatise*. The following letter alludes to the occurrence :—

GLASGOW, *Feb.* 14, 1860.

. . . . . .

MY DEAR JAMES—Did I tell you that a Mr. F Jenkin, whose experiments on cables were communicated to the British Association, has joined me in my patent, I having assigned him a charge, and that we have made joint proposals to the Red Sea Directors which I think have a good chance of being accepted? I am still working hard at the electrometer, and I hope at last have something convenient for general use.

I expected to see Prof. Tait before this on his way to or from Edinburgh. I was very much disgusted, but not excessively annoyed, to hear the other day that it is supposed a Mr. S—— (a mere nobody) has a good chance for the chair vacated by Forbes. I hope, however, this is not true. Margaret is on the whole better.—Your affectionate brother, WILLIAM THOMSON.

In R. L. Stevenson's *Memoir* of Fleeming

Jenkin,[1] p. clv., there is a note contributed by Sir William Thomson narrating how in the beginning of 1859 Lewis Gordon came to Glasgow to see Thomson's cable instruments, and said to him, "I would like to show this to a young man of remarkable ability at present engaged in our works at Birkenhead." "Fleeming Jenkin was accordingly telegraphed for, and appeared next morning in Glasgow. He remained for a week, spending the whole day in my class-room and laboratory, and thus pleasantly began our life-long acquaintance. . . . When he returned from Glasgow to Birkenhead a correspondence commenced between us, which was continued without intermission up to the last days of his life. . . . He was the very first to introduce systematically into practice the grand system of absolute measurement founded in Germany by Gauss and Weber."

Thomson was still at times occupied with thermodynamics; and in June 1860, with Joule, communicated to the Royal Society two papers on the "Thermal Effects of Fluids in Motion."

At the British Association meeting in June 1860, at Oxford (memorable for the famous encounter

[1] "It was at this time besides that he made the acquaintance of Professor, now Sir William, Thomson. To describe the part played by these two in each other's lives would lie out of my way. They worked together on the Committee on Electrical Standards; they served together at the laying down or repair of many deep-sea cables; and Sir William was regarded by Fleeming, not only with the 'worship' (the word is his own) due to great scientific gifts, but with an ardour of personal friendship not frequently excelled. To their association Fleeming brought the valuable element of a practical understanding; but he never thought or spoke of himself when Sir William was in question; and I recall, quite in his last days, a singular instance of this modest loyalty to one whom he admired and loved."—*Memoir*, p. lxi.

where Huxley met—and silenced—Wilberforce with his famous *argumentum ad episcopum*), Thomson had some further notes on atmospheric electricity, recounting extended observations made by students in his laboratory. Amongst other points noted was that the steam from the funnel of a locomotive engine is always negatively, while that escaping from the safety-valve is always positively electrified. He had been writing just before to Faraday on his observations.

THORNLIEBANK, GLASGOW,
June 12, 1860.

MY DEAR FARADAY—The insurance trial regarding the Atlantic Telegraph, for which I expected to be summoned to London before this time, has been put off until the end of the month, and I expect to be required to attend between the 25th and the 1st of July. I shall call to see you if you are in London, and to ask if you are disposed to come out to Kew along with me, where I shall have to go to look after the recording atmospheric electrometer. I have made several attempts to discover, if possible, indications of electric force in the air over the surface of two liquids, such as sulphate of zinc and sulphate of copper, separated by a porous partition, but as yet with no result. I think there must be something to be found ; and probably strong in such a case as caustic potash and nitric acid, since these two liquids when substituted for acidulated water next the zinc and platinum of a galvanic element increase its electro-motive force very largely. I yesterday had an opportunity of observing something with my portable electrometer during thunder. No lightning was visible, but I could perceive the instants of the discharges that gave rise to audible sound by sudden motions of the needle. The thunder came about 20 seconds later than an impulse of this kind, several times, from which I judged that it was about 5 miles distant.

The motion of the needle was more sudden than that which takes place when the conductor with the match burning is suddenly insulated. When this is done the needle gradually deflects without vibration, and shows nearly the full effect in 5 or 6 seconds. The changes yesterday were so sudden as to leave the needle vibrating, and were therefore *inductive* beginnings of the electric change in the conductor which the burning match completes. Besides the larger impulses which I was able to connect with the thunder, there was a constant flickering of the needle, which seemed to show that between flash and flash sufficient to make audible thunder, there were countless smaller discharges. On a small scale the same thing is produced and is indicated by the needle in the same way when shreds or fibre assist disruption of the air in any "field of electric force" in connection with experimental apparatus.

The ordinary atmospheric changes, although sometimes very rapid, for instance doubling the force in a minute or less, are not instantaneous, and show their effect by a gradual motion of the needle without vibrations. It seems certain that such changes are produced by motions of electrified air, while those I observed yesterday must have been due to discharges.—Believe me, my dear Faraday, ever truly yours, WILLIAM THOMSON.

In this year, too, Thomson wrote for Nichol's *Cyclopædia* several articles on "Atmospheric Electricity," "Velocity of Electricity," "Telegraph," and "Thermomagnetism."

Helmholtz visited Thomson again in Arran in August 1860.

With the close of the year came a great disaster. Thomson, who, while staying at Largs, engaged in the favourite pastime of "curling" on the ice, had the misfortune to fall and break his left leg. This was on Saturday, December 22, 1860. There was some doubt at first whether there was actual fracture or not, and not till the following Friday, after a visit from Dr. Syme of Edinburgh, and removal of the sufferer to another house, was the leg finally set. He was in great suffering, and for some days under continual opiates. The following letter from Mrs. Thomson narrates the events :—

AUCHINEAN, LARGS,
11 *Jan.* 1861.

MY DEAR PROFESSOR HELMHOLTZ—I am sure you will be surprised when you read the date of this letter, and I indeed little expected to be here at this time. We are detained by a most unfortunate accident which happened to Mr. Thomson three weeks ago to-morrow. He came down here on the 21st to spend the Xmas holidays with our friends Mr. and Mrs. Lang, I having come down a week sooner. On the 22nd he went to a frozen pond a mile or two distant to curl along with some friends. They were very late of returning, and at last Mr. Lang came to tell us my husband had fallen and hurt himself, but would soon be home. He was standing on a board underneath which was a narrow piece of wood, the board swung round with him and he was thrown with great violence on the ice. He attempted to rise, but fell again immediately, and had to be carried home in great pain. It was necessary to give him chloroform in order to examine the limb, after doing which Dr. Kirkwood concluded there was fracture of the neck of the thigh bone, but wished to have further advice. We sent to Glasgow for the professor of surgery, but he being in London another physician came. Chloroform was again administered, and the opinion of the Glasgow doctor was that there was not fracture, and that we were to continue fomenting with hot water cloths. It was not till a week after the accident that Dr. Kirkwood (who is very skilful) told me that he had thought all along that there was fracture, and was still more convinced that it was so, also that it was high time that the limb should be set, as he added that frequent lameness was the consequence of this fracture, and also as it would lay Mr. Thomson up eight weeks from his class (shd. it be set for fracture). We decided on having Professor Syme from Edin^b. He would not say decidedly whether there was fracture, but said it must be treated as such—the injury being very severe. Being most unwilling to subject Mr. and Mrs. Lang to the penalties of a long illness in their house, we had this one (my father's) prepared, and my husband was moved over on a litter, after which his limb was set with "the long splints," and there he must remain for several weeks. He never showed any tendency to fever, but suffered much from the effects of the chloroform, of which it required very large doses to make him insensible. He suffers much uneasiness and considerable pain. The pain caused by setting the limb was very great. It was found to be an inch and a half shorter than the other, and had to be stretched out to the same length. Our chief anxiety is as to whether bad effects will be left. Some physicians are of opinion that "boney union" never takes place at the part where this injury is (I am told the blow was on the apex of the trochanter major), but then this is an injury that seldom happens except to old people. . . . I fear there is considerable risk of shortening of the limb, and I shall be very uneasy till he is able to use his limbs again, when I *hope* all may be found right. I do not think we shall be here less than 3 months. William was very unhappy until something was settled about carrying on his class. Two of the professors have undertaken to divide the work, and he has slept better since this anxiety was removed. They kindly came here and discussed what was to be done with him. He is in a very helpless position, and cannot even raise his head, but he is very patient and bears it wonderfully. We have got a nurse to attend upon him through the nights, and my sister

Elizabeth is with me. Jessie came the first fortnight as Mary had a cold. . . . I have written to you minutely as you understand these subjects. W^{m} bids me say he was just about to write to you before this happened, to tell you he had read yr. paper with experiments to the Philosophical Society, and that it had excited much interest and been listened to most attentively.

Helmholtz, replying on January 16 to Mrs. Thomson, to express his sympathy with the sufferer, sends him messages about his own work :—

I have penetrated a long way into the Theory of Music with my physical theories—much farther than I dared to hope at the outset—and the work has amused me considerably. In developing the consequences of any valid general principle in individual cases, one constantly comes on new and quite unexpected surprises. And as the consequences are not arbitrary, nor contingent on the caprice of the author, but develop according to their own laws, I often have the impression that it is not my own work that I am writing out, but some one else's. Mr. Thomson must have found the same thing in his own work on the mechanical theory of heat.

On February 21, Mrs. Thomson again wrote to acknowledge a letter in which Helmholtz had announced his impending marriage with Anna von Mohl, and to send congratulations. She added :—

Mr. Thomson has made little apparent progress since I last wrote to you. . . .

I begin to fear that he will not be able to return to his work this Session. He eats and sleeps pretty well, but is exceedingly thin, and is easily fatigued by any mental exertion.

It was at this time, when compelled for many weeks to lie on his back, that Thomson began to use the famous green-backed note-books which ever afterwards he carried about with him. They were of quarto size, with detachable leaves. In these, in all odd moments, when travelling or waiting for a train, he would jot down as they occurred to him suggestions for experiments, calculations, diagrams, draft paragraphs of scientific papers, all dated punctiliously, and often cross-referenced. His *green-book* became his inseparable companion, and the series now preserved of over one hundred volumes is a witness to the extraordinary fertility and bewildering variety of his genius.

One of the first letters he wrote—a long pencil scrawl—was to his sister.

AUCHINEAN, LARGS,
AYRSHIRE, *January* 28, 1861.

MY DEAR ELIZABETH—I have been keeping in a very uniform state since you heard last, and do not expect to have any *news* of my progress to communicate for some time yet. The object of the splint is to protect the joint as perfectly as possible from motion, and there is nothing for it but to wait the required time to allow the union to take place. Dr. Kirkwood is most attentive and unremitting. He comes every morning and evening to see that all is going right, and to readjust the splint if it has shifted at all. Once a week he takes it down completely and sponges the limb, which has just been done to-day and given me great relief. Since the last time I had suffered a great deal of uneasiness occasionally, but I feel now as if I were not likely to be troubled so again soon. My general health has been very satisfactory, and all the circumstances have been as favourable as possible. At best, however, my recovery must be rather slow, and I have no hope held out of being able to return to Glasgow and take up work again till some time in March. It has been a great relief to me to have my class in such good hands in my absence.

Margaret has been sometimes very well, and able to walk much more and with greater ease than ever since her illness. At other times she has been not so well, and has suffered a good deal of pain and uneasiness. . . . She is, however, feeling better to-day and is out having a walk which I hope will not prove too much for her. All the work she has had for me, and the writing of letters, and attending to troublesome business, done with so much difficulty by post, have been very severe.—Your affect. brother, W. T.

At Easter, 1861, Helmholtz went to London to give two lectures at the Royal Institution on the Physiological Theory of Music. Faraday induced him to give also a Friday evening discourse on "The Application of the Law of Conservation of Force to Organic Nature." Thomson, who was now beginning to limp on crutches, wanted Helmholtz to come to Largs, but he was not able. On April 4th, Mrs. Thomson wrote:—

William desires me to say he wishes he could have been present at your lectures both for pleasure and instruction. Also I am to ask your opinion about applying electricity to his limb. Dr. Kirkwood has proposed this, but would like to know what you think of it. The limb is still quite stiff. It can move a little up and down at the knee, but not one inch to right or left.

The injury left him permanently lame. He used to walk fast, limping along with his left hand pressed against the hip. When in the summer they returned to their favourite resort at Kilmichael, in Glen Cloy, Arran, he wistfully remarked that he would never again be able to climb Goatfell—and he never did, though his activity in after-years was wonderful.

Writing in March to Dr. King, Thomson says:—

I am glad you were interested in some parts of my paper on "Atmospheric Electricity." It is merely an abstract, and necessarily very condensed, but, so far as the details of instruments and electrical observations are concerned, it will be not the less acceptable on this account. I am afraid you say more than is deserved as to the style; but one thing is certainly true, that *I in general get on at a very slow rate in writing for the press.*

At the close of the session in May 1861, he sent his colleagues, Professors Grant and Rogers, a hundred guineas each, as an acknowledgment for their having carried on his classes during the time when he was incapacitated.

In June Thomson gave to the Royal Society a paper, on the Measurement of Electric Resistance, in which he described the double-bridge, and introduced the non-inductive method of winding resistance coils.

This year the British Association met in Manchester. Thomson presented a paper on The Possible Age of the Sun's Heat (see 537), and joined with Joule in another on the Thermal Effects of Elastic Fluids (see p. 292). This meeting was notable for the appointment of the Electrical Standards Committee, which after six years of labour fixed the principal features of the now international System of Electrical Units. A paper had been read[1] by Sir Charles Bright and

[1] See the *Electrician*, vol. i. p. 3, November 9, 1861.

Mr. Latimer Clark proposing certain units and suggesting names for them: "galvat" for current, "ohma" for electromotive force, "farad" for quantity, and "volt" for resistance. The matter was keenly taken up, and on the motion of Thomson a committee was appointed, consisting of Professors Williamson, Wheatstone, Thomson, and W. H. Miller, with Matthiessen and Fleeming Jenkin, to report on the subject. They were reappointed next year, with additional members, Clerk Maxwell, C. W. Siemens, Balfour Stewart, Joule, Varley, and Sir C. Bright, and later with Carey Foster, L. Clark, and Hockin, and published, for several years, a series of reports. Thomson had so far back as 1851 been using absolute units, and now urged their adoption. After several meetings the committee agreed upon the main resolution, namely, to adopt as unit of resistance one based on the metre and second in the electromagnetic system of Weber, since one of the decimal sub-multiples of the metre-second system agreed within a few per cent with the arbitrary mercury standard suggested by Siemens. Thomson prepared some new resistance coils and sent them to Weber to be checked. He also devised a new method of determining resistance by means of a revolving coil; but owing to close occupation of himself and Jenkin as jurors at the International Exhibition in the summer of 1862, their experiments were incomplete when the committee had to present its first report. During the second year of the committee's existence great advances were made.

Maxwell, Balfour Stewart, and Jenkin, at King's College, carried out the classical experiments with the spinning coil to determine the value of the unit of resistance; Thomson was the designer of this apparatus, which was constructed under his eye by James White of Glasgow, expressly to carry out the method devised by him. He not only improved his method during the progress of the experiments, but devised an absolute electrometer, and himself drafted the main report, while Maxwell wrote the famous appendix on the relations between Electrical Measurements, in which the Theory of "Dimensions" was expounded with a masterly hand. The third report, in 1864, resulted in the issue of the standard of resistance called the "B.A. unit," or "ohmad," since shortened to "ohm." During the fourth year the committee went on with tests and measurements on units of resistance to confirm their constancy, and Thomson prepared new apparatus for absolute measurement, but had to suspend his work because of his absorbing engagements upon the Atlantic cable. In the fifth report, presented at Dundee in 1867, the chief features were a long account of electrometers by Thomson, and a new determination by Joule of the mechanical equivalent of heat, by electrical methods which would afford a check as to the absolute value of the standards of electrical resistance. Thomson also constructed at his own expense a new electro-dynamometer. The sixth and final report of 1869 contained the experiments of Thomson and those of Maxwell on the ratio of the units, called "*v*," while Jenkin added a report on the new standard, the "ohm." Throughout all its protracted labours Thomson was the inspiring force in the committee, incessant in interest, and fertile in suggestion.

Vexations came to Thomson at the end of 1861, when new statutes, promulgated by the University Commissioners, deprived the Professor of Natural Philosophy of certain graduation fees out of which hitherto he had paid the expenses of his laboratory and the salary of £60 to £70 to his laboratory assistant. It was some years before this matter was righted. In the meantime the work went on with unabated activity.

Old Donald MacFarlane, the "official assistant" of the Professor during many years, was a marked figure[1] amongst successive generations of students. Devoted to his chief, for whom he made many experimental researches, he was never happier than when busy over the working out of some abstruse calculation of spherical astronomy, and his laborious task of looking through the answers of students to the problems set in the weekly examinations of the class was fulfilled with a regularity and zest that was astonishing.

Thomson was also busying himself with investigations on the rigidity of the earth, on its secular cooling,

[1] The following is one of the many stories told of this faithful assistant:—

The father of a new student when bringing him to the University, after calling to see the Professor, drew his assistant on one side and besought him to tell him what his son must do that he might stand well with the Professor. "You want your son to stand weel with the Profeessorr?" asked MacFarlane. "Yes." "Weel, then, he must just have a guid bellyful o' mathematics"!

and other questions affecting its geological history. Of these some account is given in Chapter XIII., below. He was still collaborating with Joule on the thermal effects of elastic fluids. He was busied with kinematical and dynamical problems of bodies in revolution. His laboratory corps was kept busy with experimental work. Moreover, he was examining candidates for the India Telegraph Service, and sending weekly problems and exercises to Mr. (now Sir William) Preece for them to work. And as if there was not enough to fill his days, he projected the writing of an extended treatise on Natural Philosophy with his friend Professor Tait. The following letter to Dr. David King mentions the matter:—

THORNLIEBANK, GLASGOW, *Jan.* 8/62.

MY DEAR DAVID—I think it possible that the sun's heat, which we know to be radiating away dissipated through space, may have the effect of evaporating matter in very distant regions, and preparing a suitable medium or *atmosphere* (for I think the word atmosphere, or interplanetary air, is quite as appropriate as ether, which is in fact *aer*, or a luminiferous medium, by which the fluid occupying the region in space through which the earth moves is more commonly called) for the propagation of light, and generally for the requisites of a "world."

I send (if possible by this post) a very rough draft of my paper on "The Possible Age of the Sun's Heat," wh. was read at the British Assoc^{n} for me by Prof. Rogers. A delusive and altered abstract appeared in the *Athenæum*, but the paper itself will soon appear in *Macmillan's Magazine*—next month, perhaps. In the meantime I think you will make out enough of its general tenor by the MSS. I send.

If you are not using the abst. of my Roy. Inst. lect.

on atmosph. electy. which I sent you last spring, will you put a band round it and address it by post to Prof. Tait, 6 Greenhill Gardens, Edinburgh. I have been projecting a book on Natural Philosophy (elementary and non-mathematical) along with him, and as he has very great executive energy and facility in writing, I hope we may soon get a vol. i. out.

With love to Elizabeth and the children,—I am, yours affectionately, WILLIAM THOMSON.

P.S.—My paper for *Macmillan* has been in type some time, and I presume will appear in next number. I shall send you a copy.

The next letter mentions his change of view as to the origin of the sun's heat:—

THORNLIEBANK, *Feb.* 3/62.

MY DEAR DAVID—As to falling stars, you will remember (as I think I told you, and as the rough MS. draft in your hand shows), I have formally abandoned the hypothesis that they continue at present falling into the sun abundantly enough to compensate him for the heat he radiates away from year to year, and I suppose him at present to be a liquid mass cooling. But it seems highly probable that the heat he is now radiating away was generated in ancient times by the falling in of meteors.

I feel very strongly the difficulty you state as to shapeless detached stones being a primitive form of matter. It was put very strongly to me about two years ago by Bishop Ewing (" Argyle and the Isles "), and I have felt it ever since; but, indeed, it always seemed to me, much as it does now, a very improbable supposition. We know, however, that shapeless fragments of matter are actually met in millions by the earth in its course through space. In a theory accounting for solar heat by such masses having at one time fallen into the sun, or having fallen together and built up the sun, hot and ready for his appointed task, we are not called upon to go a step farther back and discover, or guess, an antecedent condition of those fragments or masses. Some suppose them to have arisen from the disruption of more dignified masses, but this is a mere hypothesis. What is *large* and what is *small*, even to our ideas, enlarged and enlightened by science? We are equally far from comprehending an act of creation out of nothing, whether it be of matter in a finished and approximately round globe like the earth, or in small solid fragments, or in a general diffused medium, although perhaps the last may seem the most probable to us in our present state of feeble enlightenment.

But without attempting anything so much beyond our powers as the discovery of *the* primitive condition, of matter, we successfully investigate the present condition, and argue from analogies and from strict dynamical reasoning, what must have been the antecedent condition, with more or less of detail, back to more or less ancient times. Looking at the present conditions and functions of the sun, I argue back on this principle to the probable supposition that he has been built up by the falling together of smaller masses.

Actual energy wasted. I do take into account that water by friction will "heat its channel as well as itself." The whole heat generated, *if it were confined to the water*, would heat the water by the stated amount. The heating actually experienced by the water is that which the portion of the whole generated heat which does not go to the solid channel effects in the water. Joule in his fundamental experiments, stirring water in a cubical box by means of a paddle, always allowed for the heat that went to the metal of the box and the paddle. . . .

With love to all, I remain, yours affectionately,

WILLIAM THOMSON.

In the spring of 1862 some letters passed between William and James Thomson on the question of vital forces in relation to the conservation of energy (compare p. 289). In June the jury work at the Exhibition claimed his energies. The British

Association that year met at Cambridge on October 1, but Thomson did not attend it.

Later in the year Thomson addressed to Helmholtz a letter of twenty-four octavo pages, the bulk of which is taken up with a mathematical disquisition on the distribution of potentials in the neighbourhood of the edge of a solid conducting body, a problem which he treated by the method of electric images. Apart from this, the letter is as follows:—

THORNLIEBANK, GLASGOW, *Nov.* 23, 1862.

MY DEAR HELMHOLTZ—I owe you many apologies for having so long neglected to answer your letter[1] received, I am ashamed to say, some six months ago. It came to me when I was hard at work as a juror in the Exhibition. This lasted till the end of June and kept me, as you may imagine, in very incessant occupation; but I used to say every day "heute will ich schreiben," and every day passed without my being able to find time. I wanted to write something that might possibly be useful to you in reply to your mathematical question; but had I known that I should not manage to do so even by this time, I should at once have written to thank you for your letter at least, and for the tidings it contained regarding yourself, which interested me much. I hope all has gone well with you and your family since that time, and I shall hope to hear again from you soon, unless you are disposed to reward me according to my own conduct in the matter of letter-writing.

Ever since I got away from the Exhibition (whence after spending nearly two months there we went to Arran for the rest of the summer), I have been very busy, first, with two papers, one on the "Secular Cooling of the Earth" (now published *Trans. Roy. Soc. Edin.*, I shall

[1] This was a letter of May 27, stating a mathematical difficulty about the distribution of electricity at the edge of a circular tube or at the straight edge of a conducting body, and asking for a solution which Thomson had announced but not published.

send you a copy as soon as I get one [one sent Dec. 1], and communicated but not published R. S. Lond.), the other on the "Deformations of Elastic Spherical Shells," and then preparing for a book on "Natural Philosophy," which, along with Tait, Professor in the same department in Edinburgh, I am going to bring out for the use of our students. I have long found the want of such a book, but the labour to prepare has seemed too formidable until I found a most energetic and able coadjutor in my colleague of Edinburgh. We have one sheet in type now ("Kinematics") only so far, and we hope to have a first of two good-sized volumes out by summer. "Sound" is to be our last chapter of vol. i., and any suggestions, contributions, or references from you (if, assuming that we know all that is published in your book, you have anything more to tell of), will be thankfully received. Is your book on Acoustics now finished? I suppose it is from what you told me in your letter; and if it is to be had, we shall get it immediately to help us with what we have to do. . . .

The next time you come to Glasgow, which I hope is not very long in the future, you will find a great improvement in my working place. From the beginning of this session (a month ago) I have had a really convenient and sufficient laboratory for students. Out of about 90 who attend my lectures, about 30 have applied for admission to the laboratory, and of these 20 or 25 will work fairly. I hope I may have half a dozen who will do good work. Some of them are at work at present on new electrometers which you would not recognise. The old (mammoth) species of portable electrometer, which you know, is extinct, and has been succeeded by one of which some individuals do not exceed $3\frac{1}{2}$ inches in any dimension, and yet are more sensitive and more easily managed than their progenitors. I have also a much-improved mirror-electrometer retaining some of the same organs as the original species, but so much altered that you would scarcely know them to be the same, and some new organs; also (owing to the habits of the instrument makers), some of the old organs retained but abortive.

I hope this next time you fix upon the seaside for your holidays it will be Arran that you will choose. It will be a pleasure to us, too, to make acquaintance with your wife, if you will come and bring her with you.

Mine sends you her kindest regards, and I remain yours always truly, W. THOMSON.

In December 1862 Mrs. Thomson met with a bad accident when at Largs, being thrown out of a dog-cart. Her health, which was never robust, suffered severely from the shock.

The following letter to Professor T. Andrews, of Belfast, is of interest in relation to the state of electrical engineering at this date :—

GLASGOW COLLEGE, *March* 4/63.

MY DEAR ANDREWS—Tait asks me to write to you and tell dimensions, etc., of the circuit in Holmes' electric light (marine) apparatus. This I cannot do, but I took a note of the particulars of the French apparatus (Nollet), which I hope may answer for your purpose :—

$4 \times 6 = 24$ bobbins, each 16 metres of eight-fold wire, each wire 1 mm. diam. 24 *large steel magnets*, six plates each. 16 *small steel magnets*, three plates each. Weight of each large magnet 21 kilo, "carries from 60-70 units." Weight of each small magnet 11 kilo, "carries from 35-40 units." 300-400 turns per minute. One machine requires $1\frac{1}{4}$ horse power to drive it. "Light of one machine = 125 carcel."

Height of machine (stated with a view to convenience on board ship), 1.5 metre. Breadth a little less.

Ganot gives a drawing of one, but of somewhat different proportions. Each machine costs £400.

There, I have made a clean breast of it. I know nothing more of the matter *except that Nollet does not reverse his connections, and therefore does have alternately reversed current in his flame* ; whereas Holmes does reverse and does not leave reversals in the flame. Thus Nollet escapes the commutator, a *great evil*, and gets a flame which does not burn one of the points faster than the other—a small but sensible benefit. The reverse of each proposition applies to Holmes. *The commutator is a frightful thing.* I don't mean that Holmes' is bad, because I do not know it, and it may be very good; but the thing to be done at the requisite speed is appalling. However, Holmes does it successfully. But I believe it cannot be done except theoretically without great waste of energy and consequent burning of contact surfaces.

I was assured that the reversing flame is in every respect as good as the flame of constant electro-polar direction, and I have no reason to believe it should not be so. Now I have told you rather more than all I know about it.

But I believe a large voltaic battery will be more economical than any electromagnetic machine. I am not quite confident about this, but shall be so soon, as I am getting a large voltaic (120 elements), crude carbon (zinc and one liquid, *i.e.* dilute sulphuric acid), and I shall soon learn how expensive its habits are, and multiply by the number required for a lighthouse. Serrin supplies an electric lantern with movement for the unequal burning of the points. (This movement adjustable to equality for the Nollet apparatus.) As far as I could gather at the Exhibition, Serrin's was the best, and I accordingly ordered one which I have recently received, and is to be tried to-night with a view to making a bright spot on the top of the observatory on the evening[1] of the 10th. If you care to hear how it acts or how my battery works I shall be glad to write when I know.—Yours very truly,

WILLIAM THOMSON.

In 1863 Helmholtz was invited to London to deliver the Croonian Lecture at the Royal Society,

[1] March 10th, 1863, the occasion of the wedding of H.R.H. the Prince of Wales to Princess Alexandra.

and six lectures on the Conservation of Energy at the Royal Institution. In anticipation of this visit Thomson wrote him :—

2 COLLEGE, GLASGOW,
March 16, '63.

MY DEAR HELMHOLTZ—I am very glad to hear you are soon to be in England, to give some lectures at the Royal Institution, and I write to say that we hope you will be permitted to come as far north as this, and give us a visit in Glasgow before you return to Germany. We are now living in our own house in the College, and it will be a great pleasure to Mrs. Thomson and myself to see you. I have got a great improvement in my laboratory recently, which gives me, what I never had before, space for allowing the students to work in a systematic manner. I have a few new things, both electrical and others, which may be some slight inducement to you to come, should you think it worth while to make so long a journey to see your friends in Scotland. But above all, I want to have a great deal of conversation with you on many subjects.

The book of Tait and myself is dragging along very slowly. We have about 400 pages in type, only, but hope to have a volume of 700 pages (including preliminary matter, Dynamics, and Properties of Matter) published by the middle of June. Some time ago I sent you, at the request of the author, an article which appeared in the February number of the *North British Review*. It is only the first of two articles, of which the second is to contain electric, chemical, and magnetic thermodynamics. The author would be glad to have your opinion of it as a whole or on any points, especially any *objections*, if any occur to you against it, as he is very anxious to be as correct and just as possible, and would have an opportunity in the second article of repairing to some extent errors or omissions of the first. Should any remarks occur to you, I should therefore be much obliged by your telling me them, that I may communicate them to the author.

My paper on the "Rigidity of the Earth" is only now completely in type, and finally corrected for print. I hope soon to be able to send you a copy. Mrs. Thomson joins with me in kind regards, and I remain yours very truly,
W. THOMSON.

P.S.—If you come here, as I hope you will, you will see three clocks regulated and a separate pendulum kept going constantly by electric currents from our University observatory three miles distant. I have an object glass on the pendulum, on your method, as a preliminary to determining the force of gravity. I can also show you several new electrometers, some highly sensitive, and the experiments on contact electricity regarding which I wrote to you a long time ago; also ironclad galvanometers.

Mrs. Thomson followed this up with a further note:—

2 THE COLLEGE, 25*th March* [1863].

MY DEAR MR. HELMHOLTZ—It will give us much pleasure to see you on Monday, the 28th, and I hope you will be able to come. Please let me have a word as soon as you know, to say if we may expect you, and at what hour you will arrive. It is very kind of you to come so far to see us,—In great haste, yours very truly,
M. THOMSON.

Helmholtz's visit to England lasted four weeks. "I always look," he wrote, "on a journey to England as a kind of intellectual 'cure,' which shakes one out of the comfortable indolence of dear old Germany into more active life." He visited Oxford, Manchester, and Glasgow. He sent to his wife a full report of his doings.

My journey to Glasgow went off very well. The Thomsons have lately moved to live in the University buildings [the old College]; formerly they spent more time in the country. He takes no holiday at Easter, but his brother James, Professor of Engineering at Belfast, and a nephew who is a student there, were with him. The former is a level-headed fellow, full of good ideas, but cares for nothing except engineering, and talks about it ceaselessly all day and all night, so that nothing else can be got in when he is present. It is really comic to see how the two brothers talk at one another, and neither listens, and each holds forth about quite different matters. But the engineer is the most stubborn, and generally gets through with his subject. In the intervals I have seen a quantity of new and most ingenious apparatus and experiments of W. Thomson, which made the two days very interesting. He thinks so rapidly, however, that one has to get at the necessary information about the make of the instruments, etc., by a long string of questions, which he shies at. How his students understand him, without keeping him as strictly to the subject as I ventured to do, is a puzzle to me; still, there were numbers of students in the laboratory hard at work, and apparently quite understanding what they were about. Thomson's experiments, however, did for my new hat. He had thrown a heavy metal disk into very rapid rotation; and it was revolving on a point. In order to show me how rigid it became in its rotation, he hit it with an iron hammer, but the disk resented this, and it flew off in one direction, and the iron foot on which it was revolving in another, carrying my hat away with it and ripping it up.

After he left, the following note was sent him by Mrs. Thomson :—

2 THE COLLEGE, GLASGOW,
19th April [1863].

MY DEAR PROFESSOR HELMHOLTZ—Will you be so kind as take the charge of a small parcel, which will be left for you at the Royal Institution, containing a book, which I hope your daughter will accept with my kind regards. It is Longfellow's Poems which I have chosen, as being more easily understood by a foreigner than some of our other poets. I wished to send it by you when here, but I was prevented going out to choose it.

Your hat is here, and if you have not left London the first week in May, I think we may find an opportunity of sending it to you. I was very sorry when I heard of your accident in the Laboratory, which I did not until you had left us. I trembled to think what it *might* have been, and I am distressed that you should have been exposed to such a danger with us. I hope that you are enjoying your stay in London, and that you have good accounts from home, and with kind regards I remain,—Dear Professor Helmholtz, yours very truly,

M. THOMSON.

In this spring Thomson delivered a course of three lectures on Electric Telegraphs at the Royal Institution. The syllabus of these shows no feature of special novelty; and no report of them is known.

At the British Association meeting of 1863 Thomson read a paper on the result of his self-recording electrometer at Kew. He was continuing experiments on the electrostatic capacities in connection with submarine cable work, when, one evening in October, he noticed the peculiar circumstance that a condenser, formed of two metal plates separated by a film of air, emits at the moment of its discharge an appreciable sound. An account of this discovery, written to Professor Tait, was published in the *Philosophical Magazine.*

The Atlantic telegraph project was now being revived, and in the autumn Thomson spent some time in London to advise upon the preliminaries.

Early in 1864 he gave to the Royal Society of Edinburgh several papers, including one on the influence of dew in protecting vegetation from destructive cold at night.

In May Mrs. Thomson wrote to Mrs. James Thomson :—

William, I am sorry to say, is still very lame, and able to walk almost none, without pain. He *will* not rest as much as he ought to do, although he abstains from walking. It is most disappointing, after he had been so well, and I begin much to fear that he will never be as free from lameness as he was some weeks ago.

In the spring Mrs. Thomson's health was again cause of anxiety. They went in May to Creuznach, whence, on June 2, he wrote to Mrs. King :—

Margaret has removed to a house nearer the spring, where she goes to drink the waters. . . . There are very pretty drives and walks in the neighbourhood. . . . Tell George that his cheroots are very good, and will be most useful here.

Later he wrote to Helmholtz :—

WILDBAD, *July* 31, 1864.

MY DEAR HELMHOLTZ—The amount of *Cur* prescribed for me is 21 baths, which will be completed on Friday next, and as we leave immediately after, there is no chance, I fear, of our seeing you here. We think on returning by Strasburg and Paris. We like the place very much (a great deal better than Creuznach), and Dr. Berckhardt says the baths are sure to do me some good, but that I am not to expect too much (which I have certainly never been disposed to do). I shall not forget to return your Kirchhoff on Plates, which has been very useful to me. I now see quite distinctly both how there are only two, and how there might be supposed to be three equations for the boundary. It is certainly a great card of Kirchhoff's, to have set this matter right after such great people as Mlle. Sophie Germain and Poisson got wrong on it, and Lagrange had it in hand without doing it. The full working out of the solution, too, for the circular plate, shows no small amount of courage, skill, and *well-spent* labour. Oh! that the CAYLEYS would devote what skill they have to such things instead of to pieces of algebra which possibly interest four people in the world, certainly not more, and possibly also only the one person who works. It is really too bad that they don't take their part in the advancement of the world, and leave the labour of mathematical solutions for people who would spend their time so much more usefully in experimenting.

I have got the theory of an elastic plane plate now I think on a very simple foundation. The only result I have taken from the general theory of an elastic solid is that the couples of forces required to bend a plate depend solely on the curvature, provided the displacement is at no point more than an infinitely small fraction of the thickness.

. . . .

I hope you will not forget the promise you kindly made, to order your apparatus of tuning-forks, and a harmonic syren, for me. If any improvement in either occurs to you, do not hesitate on account of expense to have it introduced. Shall I be able to have either or both by November? I should be glad at all events to have them, if possible, before the end of the year, so that I may be able to use them for my lectures this session. We hope your little boy is getting better. My wife joins me in kind regards to yourself and Mrs. Helmholtz.—Yours always truly, W. THOMSON.

P.S.—We have a great deal of amusement talking with people of all nations here, Germans, Russians, Poles, and English; no French or Prussians. The last are much abused by the first.

In October he wrote again:—

We have been in Arran since our return until a few

days ago, and remain here till November, when the College session commences. The book has been dragging its slow length along more slowly than you could conceive. Still I hope by Christmas to be able to send you a volume of 700 pages.

To Professor Andrews, who had presented him with that optical curiosity a Barton's button, he wrote on January 28, 1865:—

MY DEAR ANDREWS—The Barton is beautiful, and will be a great acquisition for my lectures. I have long wished both to see one myself and to be able to show it to my students.

I am extremely glad to hear that you think of resuming your experiments on the relations between the gaseous, liquid, and transition states. I am sure you will get most valuable additional results, but I hope you will not delay later than you can conveniently to write out an account of what you have done already, so giving it to the world. The experiments you showed me on carbonic acid seemed to me quite complete in themselves and ready for publication. They throw a perfectly new and most unexpected light on the subject. I have no very satisfactory way of showing the heat generated in air rushing into a vacuum. I merely show, by aid of a thermo-electric junction, that there is a heating effect. I should be greatly obliged by your describing to me the arrangement you have made for this purpose, according to your kind offer, and any of the other class experiments on heat to which you refer, if you can do so without too much trouble. But I feel quite ashamed even to suggest this, knowing how valuable your time is; and I must beg that you will not scruple to postpone indefinitely, unless you find some convenient moment to write to me about them.

Joule has been with us for a week, having come to Scotland to lecture at Greenock in commemoration of Watt, on the occasion of his birthday, observed as an anniversary there.

I have thus been very much engaged, over and above my ordinary class work, and obliged to postpone work on *the book*. Tait and the printers had just begun to fret, but they ought to be in better humour now.—Yours always truly, W. THOMSON.

With the absorbing work of experimenting in connection with the Atlantic cable, and with frequent visits to the factory in London, the winter session of 1864-65 was one of extreme pressure. But the laboratory continued in full swing, and the study of the properties of matter had now taken the turn of inquiries into the elasticity and viscosity of metals. In May 1865 a memoir on this subject was presented to the Royal Society. Thomson had observed that by vibrating a spring alternately in the air and in the exhausted receiver of an air-pump there was an internal resistance to its motions—as attested by the rate of subsidence of its vibration—immensely greater than the resistance offered to it by the air. This internal molecular resistance is of a frictional nature, and is a part of the phenomenon of viscosity. To test this viscosity, Thomson, with the assistance of Mr. Donald MacFarlane, hung wires of different metals from a rigid support and submitted them to torsional vibrations by means of heavy vibrators attached to their lower ends. Determinations of the values of Young's modulus were made also on wires 80 feet long hung in the College Tower, and the rigidities were measured. Rigidity was found to be reduced by longitudinal traction. In this paper Thomson adopted metric units; and he appended in a footnote the following caustic comment on "the British no-system": —

> It is a remarkable phenomenon, belonging rather to moral and social than to physical science, that a people tending naturally to be regulated by common sense should voluntarily condemn themselves, as the British have so long done, to unnecessary hard labour in every action of common business or scientific work related to measurement, from which all the other nations of Europe have emancipated themselves.

In May 1865 Thomson paid a flying visit to Cambridge—the first since his marriage in 1852. With the summer came the cable-laying expedition narrated in Chapter XI. The autumn found him as devoted as ever to pure science, and able to detach his thoughts from the absorbing topic of the day. Three papers were read by him to the Edinburgh Royal Society on December 18, 1865. Of these, the first, an address delivered at the request of the Council, was on the forces concerned in the laying and lifting of deep-sea cables—an importunate topic, since the broken cable of 1865 was then lying at the bottom of the Atlantic, and anxious engineers were cogitating how to raise it in the coming year. The second was a note on the dynamical theory of heat. The third, consisting of but four emphatic sentences, was entitled the "Doctrine of Uniformity in Geology briefly Refuted." It is considered in Chapter XIII. (p. 540). A month later he was giving to the Glasgow Philosophical Society a discourse on electrically-impelled and electrically-controlled clocks, the inventions of Mr. Robert L. Jones of Chester, which then were attracting much attention, and exhibiting a couple of new standard electrometers. But he published little in 1866 : the cable work was too intense.

For many years Thomson had been a stranger to Cambridge, but now the University offered him the compliment of an Honorary Degree of Doctor of Laws. It was conferred on Tuesday, May 22, 1866 ; but, unfortunately, the Latin speech of the orator in presenting him to the Vice-Chancellor has not been preserved.

The following day Thomson delivered the Rede Lecture for that year. As this was never published, the following abstract from the *Cambridge Chronicle* of May 26, 1866, from a copy corrected by the author's own hand, will be of interest.

THE REDE LECTURE

On Wednesday afternoon last, Professor Thomson, of Glasgow, delivered Sir Robert Rede's lecture in the Senate-house, in the presence of a large and highly-distinguished audience, including many ladies. The subject was "The Dissipation of Energy." We make the following abstract:—The great principle of the conservation of energy teaches us that the material universe moves as a frictionless machine. *Vis viva*, or, as we now call it, Kinetic Energy, is never lost or gained. If its amount becomes less in any portion of matter, an equivalent of work is done, and remains ready able, "potential," to generate the same quantity of kinetic energy anew. Or if kinetic energy increases, it is necessarily at the expense of potential energy drawn upon for the work by which it is generated. Until thirty years ago naturalists did not require any explanation of the apparent losses of energy manifested in every movement. But Davy and Rumford, at the end of last century, in concluding from their experiments that heat is a state of motion, had prepared the way for the great generalization which marks the fourth decade of the nineteenth century as an era in Natural Philosophy. They had not made this generalization, nor quite proved that they had even imagined it. Davy, when he said that the communication of heat follows the laws of the communication of motion, did not suggest the idea that in the generation of this kind of motion there may be no loss of energy by frictions and impacts as there always is in the communication of visible palpable motions. But Rumford's merciful treatment of his mill horses is described in language too suggestive of a direct relation between work spent and whole amount of heat generated, to allow us to suppose that he had not a very distinct idea of mechanical equivalent between them in his mind. When he finds that nine wax candles all burning at once generate heat as fast as a single horse working hard driving a cannon-boring machine, he gives us a reckoning in horse-power to measure the activity of a fire. And when he tells us that in no case can it be economical to keep horses for generating heat by friction, BECAUSE *more heat could be obtained by burning their food*, he anticipates, in all but the number, Joule's discovery that the heat of combustion of a horse's food is from four to six times that obtainable through friction from a horse's work, and comes very near to that deepest part of Joule's and Mayer's philosophy in which it is concluded that animal energy and heat together make up an exact equivalent to the heat that would be generated by the chemical actions in the living body if these were allowed to take place without any performance of mechanical work. Joule himself has deduced a fair approximation of his own directly measured dynamic value of the thermal unit, from the record which the *Transactions of the Royal Society of London* contain of Rumford's experiments in the Military Arsenal of Munich on the heat of friction. The first step towards the establishment beyond the range of abstract dynamics of the great law of Nature now known as the Conservation of Energy, was made by Joule in his investigation of the heat generated by electric currents, described in an article communicated to the Royal Society, December 17, 1840, under the title "On the Production of Heat by Voltaic Electricity." About the same time many naturalists in many countries began to feel strongly the want of some general principle to account for the effects of work done when seemingly lost in friction. Séguin in France, Grove in England, Mayer in Germany, Colding in Denmark, all speculated of a comprehensive theory in which the kinetic character of heat, established by Davy and Rumford, should show how it is that energy is not annihilated when work is done against friction. The first published distinct mention of a "mechanical equivalent" of heat was, I believe, that in 1839 by Séguin, for which he refers back to his uncle Mongolfier. But the first solid ground gained in advance of that occupied by Davy and Rumford was conquered by as thorough and determined sapping as has ever won a fortress for a stubborn army. The genius to plan, the courage to undertake, the marvellous ability to execute, and the keen perseverance to carry through to the end, the great series of experimental investigations (from 1830-1849) by which Joule discovered and proved the conservation of energy in electric, electro-magnetic, and electro-chemical actions, and in the friction and impact of solids, and measured accurately, by means of the friction of fluids, the mechanical equivalent of heat, cannot be generally and thoroughly understood at present. Indeed it is all the scientific world can do just now in this subject to learn gradually the new knowledge gained. Many of Joule's subtlest discoveries and richest mines of future research are still almost unknown. I refer particularly to his paper on the "Heat of Electrolysis" (1842, Manchester Literary and Philosophical Society). The greatness of a Swiss mountain is scarcely discovered when looked up to from the adjoining valley, and is only appreciated when distance shows outlines in their true proportions. A hundred years hence Joule's massive work of 1840 to 1850 will be seen not only towering above that of all contemporaneous experimenters except Faraday, but stretching across a range of physical science which includes regions little known and rarely visited at present. The lecturer then explained the Dissipation of Energy, and showed that the molecular motions which constitute heat, though containing a true dynamical equivalent for the work spent in generating it, are not mechanically equivalent to it, inasmuch as only a small part of it can ever be "reconverted into potential energy" (that is to say, applied to raise weights). The subtle but profoundly practical reasoning of the French republican war-minister's son, Carnot ("Motive Power of Fire," Paris, 1824), was alluded to as explaining the conditions under which heat must be presented in order that the motive power, as in the steam-engine or air-engine, may be got from it. The heat radiated from the sun was referred to as a case of dissipation in which energy is lost from every square foot at the rate of 7000 horse-power, and ultimately applied to warm the air (or æther, as some call it) of space, except such small parts of it as, falling upon the earth, have given the energy of coal, and keep giving the energy of growing combustibles and fuel. There is dissipation from the earth in the combustion of coal, at a rate so rapid as thoroughly to require anxious attention in this country at the present time. And by the conduction outwards of heat from within there is dissipation at an average rate of one thirty-millionth of a horse-power per square foot all round, or at 180 million horse-power from the whole surface. The friction of waters flowing by the tides gives rise to dissipation of energy of the earth's rotation, the ultimate tendency of which is to make the earth, sun, and moon turn like parts of one rigid body. The effect of the tides is to produce retardation of the earth's rotation, which might possibly amount to one or two hundred seconds of time in a century, four times as much in two centuries, and so on. In the present very imperfect state of clock making, which does not produce an astronomical clock more than two or three times as accurate as a good pocket watch, the only body by which the accuracy of the earth's rate can be compared is the moon, and it seems, so far as the physical astronomers Adams and Delaunay have been able to correct her great irregularities, that the earth has lost on her by some ten seconds of time in the last hundred years. This may well be due to the tides, but it would be rash to conclude that it is so, as the earth cannot be trusted as an accurate time-keeper for two-tenths of a second in a year (or 365 times as accurate as a pocket watch good for two-tenths of a second per day). Snow melting any year, or succession of years, from the polar regions and raising the average sea level by the almost undiscoverable difference of one inch and a half all over, would make the earth go slower by two-tenths of a second per year. The direct effect of the moon on the tides in the Thames from London Bridge to the Nore Light is to retard the rotation of the earth; from the Nore all through the English Channel as far as Exmouth and Jersey the direct influence is to accelerate the rotation. West of this along the English coast to Land's End, the south and west coasts of Ireland, and the west coast of France and the Peninsula, the direct effect is to retard the earth. All over the oceans the action is either accelerating or retarding, but the fact that there is loss of energy by fluid friction makes it certain that the whole sum of retarding effects due to one set of patches or spaces of ocean, exceeds the sum of accelerating effects from the other spaces. The lecturer concluded by summing up in the following terms [quoted from the paper of 1852, see p. 290]:—

(1) There is at present in the material world a universal tendency to the dissipation of energy.

(2) Any *restoration* of energy, without more than an equivalent of dissipation, is impossible in inanimate material processes, and is probably never effected by means of organised matter, either endowed with vegetable life or subjected to the will of an animated creature.

(3) Within a finite period of time past the earth must have been, and within a finite period of time to come, the earth must again be unfit for the habitation of man as at present constituted, unless operations have been or are to be performed, which are impossible under the laws to which the known operations going on at present in the material world are subject.

One who was present declares[1] that Thomson's multitudinous enthusiasm and dissipative energy was so great that any really adequate or representative report of the lecture was impossible. A small portion of it appeared in the *Philosophical Magazine* (vol. xxxi. pp. 533-537; Supplement, 1866) under the title, "On the Observations and Calculations required to find the Tidal Retardation of the Earth's Rotation," reprinted in *Popular Lectures*, vol. ii. p. 65.

Writing to Helmholtz in December 1872, Sir William Thomson stated that he "did not succeed in getting it written out, and it has not been published." But the subject still engaged his thoughts. He had first announced the discovery in 1852 (see p. 290, *ante*), and he returned to it in February 1874 before the Royal Society of Edinburgh; and in March 1892 he wrote a popular article on the Dissipation of Energy in the *Fortnightly Review*

[1] The thread of the discourse was incessantly interrupted by digressions. As soon as he had got a little under weigh, something apparently unconnected with his subject would occur to his mind, and he would remark: "When you meet with a fallacy in vogue it is well not to leave it alone without having a rap at it" (or something to that effect), and thereupon proceeded to demolish the fallacy. That done, when both he and his audience had forgotten where he left off, he resumed the old track, but not quite where he had left it, and by the time he and his audience had got together again, some new hare crossed his track, and we were all scattered in pursuit of it; and so it went on, the time running out, and the short-cuts to regain the track becoming more and more difficult to follow, until the end. The hunt pleasant enough to those who were nimble enough, but the game killed mice and rats—and no hare.

prevalent in the Scottish Universities, reverently recite the morning prayer. In his case the prayer—always the same—was chosen from the English Church Service, the third Collect for Morning Prayer: "Almighty and Everlasting God, who hast safely brought us to the beginning of this day. . . ." Then he would take a hurried glance at the apparatus arranged on the table by his assistant, and plunge into the topic of the lecture.

In an article on "Kelvin in the 'Sixties,"[1] the late Professor Ayrton, who was one of Thomson's students in these years, has given an inimitable account of his master, whom he describes as not only a giant mentally, but of extraordinary physical activity:—

> When he came into his class-room—a room festooned with wires and spiral springs hanging from the ceiling like the rigging of a ship—he had hardly given a thought to what he was going to talk about. If it were Monday morning, he had just returned from staying the week-end with Tait at Edinburgh, and he gave us an enthusiastic account of their talk, bubbled over with what they had been doing, was full of suggestions about it, told us how the manuscript of *Natural Philosophy* was progressing. We felt that we also had been discussing these points with Tait in his Edinburgh study, and listened with rapt attention to Thomson's narrative.
>
> In his mathematical physics lectures—aye, even in his elementary lectures—the suggestions that he poured forth were much above the heads of the ordinary undergraduates—over 100 in this class—and they gained little by coming to them, except a register of their attendance

[1] *Times Engineering Supplement*, Jan. 8, 1908, and *Popular Science Monthly*, March 1908.

necessary for their degrees. For as soon as he turned round to write on the blackboard, the students row by row began to creep out of the lecture-room through a back door behind the benches, and steal downstairs, their bodily presence following their mental presence, which had left as soon as the reading of the roll-call was finished. From time to time Thomson put up his eye-glass, peered at the growing empty space, and remarked on the curious gradual diminution of *density* in the upper part of the lecture-room.

This class consisted mainly of divinity, medical, and law students, who, of course, should have been taught the elements of natural philosophy by some assistant provided by the University. To waste the time, energy, and extraordinary original power of a genius like Thomson on such teaching was like using a razor to chop firewood. The junior clerks in Downing Street require instruction, but the prime minister is not expected to personally hold daily classes for them. And yet, during the past eighty years, there have been many prime ministers, but only one William Thomson.

But to those, like myself, who, after receiving some scientific training, had come from other countries, to hear Thomson's talk, his suggestions, his buoyancy, were like the rays of brilliant May sunshine following April showers.

The ideas of those students sprouted as never had they done before. The more thoughtful gazed with eyes of wonder at Thomson developing an original paper during a lecture on anything he might be talking about, well knowing that any notes or calculations that he might previously have made were on the back of some old envelope, and left probably with his greatcoat in the hall.

Delight and wonder, says another of Thomson's old students, were ever daily companions during those strenuous years.

Emery Walker Ph. sc.

William Thomson

[1870]

CHAPTER X

THE EPOCH-MAKING TREATISE

"Les causes primordiales ne nous sont point connues ; mais elles sont assujetties à des lois simples et constantes, que l'on peut découvrir par l'observation, et dont l'étude est l'objet de la philosophie naturelle."—FOURIER.

TIME and chance happen unto all; but it is not all who are able to turn time and chance to account for the benefit of posterity. Friendships between scientific men are not uncommon, but rarely does friendship bear fruit so direct, so permanent, as did the friendship between William Thomson and Peter Guthrie Tait. Their collaboration in the production of the *Treatise of Natural Philosophy* in 1867 marks an epoch in the teaching of the foundations of physical science scarcely less important than that in which Newton produced his immortal *Principia.*

Early in his career as Professor at Glasgow, Thomson had been confronted with the lack of adequate text-books to place in the hands of his students. Books, indeed, there were of the feebly descriptive sort that one associates with the name of Lardner; also powerful mathematical treatises like the *Mécanique analytique* and the *Mécanique céleste.* Books, too, to carry the perspiring candidate through the mental gymnastics then in vogue for passing examinations in Theoretical and Applied Mechanics. But none of these were written from the modern standpoint of energy; they were mostly loose in their phraseologies, confounding acceleration with force, and force with work; and, where mathematical, were too often apt to lose sight of the physical meaning in pursuit of the mathematical interest. "Books?" he used to say to his students when they asked him in what books to find what he had been telling them in his lectures. "Books? I am telling you what is *not* in books." But books are necessary for deliberate study; and are all the more wanted when the master, inspired by his own internal and vehement force, leads his students with almost volcanic energy and erratic genius to the confines of knowledge in the making. Thomson, nurtured under the mature yet original genius of his father—a great teacher, if ever there was one—was deeply conscious of the needs of the students struggling to follow his own inspiring but discursive lectures. More than once he had set himself to recast in simple form the doctrines he was discovering: witness the second paper on the "Theory of Electricity," of 1848 (p. 142), and other publications, in which he deliberately chose elementary methods of exposition. But it was more particularly in the fundamental mechanical principles underlying all physics that the need was sorest and the provision least adequate. We have seen

(p. 190) how he discriminated between the "natural history," or descriptive stage, and the "natural philosophy," or correlational and deductive stage of physical science; and as professor *philosophiae naturalis*, he wanted to place in the hands of his students an exposition of the laws and relations discoverable by experiment, and of their logical development into rational theories. For fourteen years he had held the chair with peculiar distinction; but, almost overwhelmed with the new discoveries that were so largely products of his own genius, he had never methodized his teaching, nor found any comrade capable of entering into the task.

With the election of Tait to the chair of Natural Philosophy at Edinburgh the chance came. Tait was a Scotchman, reared in Auld Reekie, a schoolfellow of Maxwell, a Peterhouse student, and, like Thomson, a pupil of Hopkins. He was Senior Wrangler in 1852, and was in many points a disciple of Thomson, though personally unknown to him prior to his election in 1860 as successor of Forbes. Himself a man of remarkable individuality and singular accomplishments, a first-rate mathematician and a skilled experimentalist, Tait possessed qualities that differed widely from those of Thomson. Tait was methodical and restrained where Thomson was discursive and vehement. Controversy, which was ever distasteful to Thomson, was to Tait almost the breath of life. Tait loved a neat mathematical demonstration for its own sake;

Thomson cared little for the mathematical form, or its deduction, provided it expressed the physical relations he sought to convey. Professor Andrew Gray, so long assistant to Thomson, has stated another point of contrast. "Tait's professorial lectures were always models of clear and logical arrangement. Every statement bore on the business in hand; the experimental illustrations, always carefully prepared beforehand, were called for at the proper time, and were invariably successful. With Thomson it was otherwise: his digressions, though sometimes inspired and inspiring, were fatal to the success of the utmost efforts of his assistants to make his lectures successful systematic expositions of the facts and principles of elementary physics." Thomson, always striving to set himself free from hypotheses of a doubtful sort, kept his physical conceptions clear-cut, and was almost pedantic in his efforts at precise definition of the terms he used, objecting to all metaphysical glosses or subtleties that would take him away from the concrete. Tait, too, was precise and clear, but in a fashion of his own. Tait who had early imbibed a passion for the quaternion method of treating directed quantities, such as velocities and forces, in accordance with the ideas of Sir W. Rowan Hamilton, advocated the quaternion analysis as particularly adapted to the problems of physics. Clerk Maxwell, though he himself employed ordinary analysis in his physical investigations, adopted the language of quaternions in his great treatise. But

Thomson to his dying day would have none of these things, and even grew to hate the name of *vector*. In spite of such divergences of view, the two men worked in the most harmonious association in their undertaking, and each supplied to the other something necessary for the achievement of the task. They thoroughly enjoyed one another's society, and the intimate discussions renewed week by week, enlivened as these often were by quips at one another's expense. With them and their friends the book was familiarly spoken of as "T and T′," a notation which passed into their letters; for Tait, instead of heading his epistles "Dear Thomson," substituted "O T," while Thomson's reply, usually scribbled on the margin of some printer's proof, would begin "O T′"; and this logogram, being non-epistolary in form, saved a halfpenny in postage!

The origin of their collaboration is told in the following letters:—

PROFESSOR CHRYSTAL to LORD KELVIN

5 BELGRAVE CRESCENT, EDINBURGH,
12*th July* 1901.

MY DEAR LORD KELVIN—I am, in default of a better, engaged on a short obituary notice of Tait for *Nature*. There is one point of great interest on which Crum Brown is unable to inform me.

I should like to know when Tait first became personally intimate with yourself, and what led to the auspicious conjunction of T and T′, which I have always regarded as one of the most important scientific events of the Victorian era. Also when and under whose influence (if under any particular influence) did Tait first enter what I may call the "Energetic" school of Natural Philosophy. Joule's earliest papers on Energy date from about 1842; your own, I think, from 1850 or 1851.

Tait appears as a fully indoctrinated disciple in 1862, and had apparently fully declared his position and views in his introductory lecture in Edinburgh in 1860. Before that I have no trace of him.

I should esteem it a great favour if you could send me a few words on these two points which I could use in *Nature*. I should, of course, be very careful not to mix up any views or statements of my own with anything given on your authority. . . .—Yours sincerely,

G. CHRYSTAL.

LORD KELVIN to PROFESSOR CHRYSTAL

15 EATON PLACE, S.W., *July* 13, 1901.

DEAR CHRYSTAL—I first became personally acquainted with Tait a short time before he was elected Professor in Edinburgh; but, I believe, not before he became a candidate for the chair. It must have been either before his election or very soon after it that we entered on the project of a joint treatise on Natural Philosophy. He was then strongly impressed with the fundamental importance of Joule's work, and was full of vivid interest in all that he had learned from, and worked at with, Andrews. We incessantly talked over the mode of dealing with energy which we adopted in the book, and we went most cordially together in the whole affair. He gave me a free hand in respect to new names, and warmly welcomed nearly all of them.

We have had a thirty-eight years' war over quaternions. He had been captivated by the originality and extraordinary beauty of Hamilton's genius in this respect; and had accepted, I believe, definitely from Hamilton to take charge of quaternions after his death, which he has most loyally executed. Times without number I offered to let quaternions into Thomson and Tait if he could only show that in any case our work would be helped by their use. You will see that from beginning to end they were never introduced.

Excuse haste, as I am leaving London this instant for three days in the country. I am exceedingly glad you are writing the article for *Nature*.—Yours very truly, for Lord Kelvin, WM. ANDERSON, *Secy.*

The earliest reference found in Lord Kelvin's papers is the following letter from Tait, dated from Edinburgh, December 12, 1861:—

MY DEAR THOMSON—I have great pleasure in accepting your brother-in-law's kind invitation, and shall be at your laboratory about 4 h. 30 m. on Friday week. . . .

I have not yet heard definitely from Macmillan about the treatise, but there is one point on which I feel very strongly, and on which therefore I am desirous to talk with you. I mention it now, that *you* may remember it, if *I* forget it, when we meet. Look at Jamin's book—it is a storehouse of valuable details for a man like Regnault, or some one who is setting up as a thorough experimentalist *with verbal instruction.* Now it seems to me that such detail in modes of avoiding every sort of error, and availing oneself of every nicety, is NOT the thing our students, or the general public, want. Explanation, as thorough as possible, but *not* elaborate detail, seems to me the proper line to take.

Let us be very full on the mathematical part—but not spend, as Jamin does, nearly $\frac{4}{5}$ths of each of his volumes in explaining the precautions necessary in the laboratory. Any student who wants these can get them from his teacher or read them up in the French.

If this be agreeable to you, I fancy that we might easily give in three moderate volumes a far more complete course of Physics, Experimental and Mathematical, than exists (to my knowledge) either in French or German. As to English, there are NONE. . . .

I am myself a good example of the want of such a book as we contemplate, having got all my information bit by bit from scattered sources, which often contained more error than truth. The next generation will thank us.

P. G. T.

The next letter preserved drafts out the skeleton of the treatise :—

6 GREENHILL GARDENS, EDINB.,
Dec. 25th, 1861.

MY DEAR THOMSON — I wrote to Macmillan on Monday last. Here are the *Postulates* :—

I. *At least two* vols. Experimental.
II. Illustrations *in the text.*
III. Authors retain copyright.
IV. A *definite* number only to be printed. *New* arrangements for subsequent editions.
V. If these agreed on, it is desirable to commence at once.
VI. That there will probably be *two* vols. of Mathematical —and, as their sale is not likely to be so speedy, that McM. submit his terms for *them.*
VII. That, as we are certain to be pirated in America and translated abroad, if no precautions be taken—the publisher do all that is requisite in that way.

I shall send you his reply as soon as I get it. Meanwhile, I have been trying my hand at a programme for the Experimental part. I wish you would try also before looking at mine, and see whether we are driving on the same road or not :—

I. General reflections on Matter, Force, Motion, Measures, Energy, Work, and Experiment.
II. Ordinary Statics.
III. Kinematics and Dynamics.
IV. Hydrostatics, Pneumatics, and Hydrodynamics.
V. Properties of Matter.
VI. Sound.
VII. Light.
VIII. Heat.
IX. Magnetism.
X. Electricity.
XI. Electro-dynamics.
XII. Conservation of Energy.

Please correct this list by your own, or, if yours be

entirely different, let me have it. We may then take (say) alternate *chapters*, *not subjects*, and each submit his MSS. to the other, who will forward it to the Pitt Press, after making any remarks. The proof-sheets in the same way will pass from one to the other before returning to Cambridge. I think, if my scheme be anything like the thing, that the first seven articles may be the first vol., and the others the second. However, we are not tied down to *two*. When you have made all the remarks that may appear necessary, I wish you would return the green proof-sheets of my article "Force." Also, I am *particularly* desirous that you should *annotate very freely* my notes for lecture on "Properties of Matter." You must feel how much we shall gain (in working together) by making the most abundant comments on each other's work. I am in no hurry for the MSS., but I should like to have the proof-sheets some time next week.

I got the plateau, but nothing else. I expect the other instruments as soon as White can send them. I asked MacFarlane to tell him to give me notice of their coming. —Yours truly, P. G. TAIT.

26/12/61.

P.S.—I wrote you the enclosed late last night, and luckily had not posted it when yours arrived this morning.

First, then, about the proof-sheets. I quite agree to all your comments, most of which, I am happy to see, are on points on which I felt I had no satisfactory information, and, indeed, had told you so—but it is amusing to see how definitely you go into the case of conception and treatment of the continuous uniform medium in which atoms (or at all events matter) are supposed to float. I am quite willing to adopt your views, but I should like you to send me, as soon as you have leisure, a little sketch of your proposed mathematical treatment of such a fluid or solid—or refer me to the work, Stokes' or others', in which it is found, if already in print.

I quite agree with you that conductivities of all kinds should be put under properties of matter, but I think *merely mentioned*, as they cannot be properly explained till we come to Heat and Electricity.

I should imagine that we ought to give the whole (save the Mathematics, of course) of Elasticity, both of form and volume, under Properties of Matter. You must remember that my Notes for Lectures are *mere headings*, and were intended for very great expansion when I contemplated writing alone.

As I have already said in last night's part of this letter, the more remarks you pen on these the better shall I be pleased, for we shall thus have an opportunity of *blending* our styles from the outset, and without that our volumes will be patchwork, of excellent materials, no doubt, but awfully inartistic—a *mixture*, not a *chemical combination.*

Give me your idea on the general table of contents which I have drawn up. I made it for my last year's course, nearly as it stands, and have put in a few corrections here and there from the results of experience in a year and a half's teaching. But I have *not* put in Astronomy, nor Meteorology, as I have, up to the present time, only examined, not lectured, on these great branches. Of course, all the Chemistry we do we can pop into Properties of Matter. We *must* do Physical Astronomy in our Mathematical part. I am quite game for the Lunar and Planetary part; but I shall hand over to you (as you seem to be well up in them) Precession, Figure of Earth, Tides, and Effect of $\oplus$'s oblateness on moon's motion, at least if they are to be treated to a formidable extent. I can also venture on Capillarity, Magnetism, Static Electricity, Electrodynamics, Undulatory Theory, Sound, and Conduction of Heat (and all about Ohm); but you must furnish Elasticity (a very long article), Dynamic Electricity, Induction, Induced Magnetism, Dynamic Theory of Heat, and Thermo-electricity, and perhaps other things, which, as I write in haste, I may have omitted. So you see we have no light undertaking before us.

As to time, I don't think that there is anything to

hinder us from having a volume ready (or at all events in a state in which I can manage it alone) by May, if we can only agree within the next three weeks to our general plan, and settle the order and nature of the contents of the first volume. We have all our matter (for the experimental volumes) at our finger-ends—let us apportion our work, and fall to. An average of three or four (or less) hours a day would give us the volume in six weeks in such a state as to require little correction in the present state of the science, for we have *both* been, and shall be, talking every day on the subject. As I said before, I firmly believe that all we want is order and arrangement. That being *very carefully* attended to before beginning—the rest is easy. In fact the arrangement of our matter is our difficulty, for it is hardly possible to take any part of the subject up without referring to some other which has not yet been explained. In fact, put them as you like, the cart *must* inevitably come before the horse in some cases. . . .—Yours truly,

P. G. TAIT.

Three days later Tait writes on the suggestion that a shorter elementary text-book should be first completed, the more mathematical developments being reserved for the treatise. On January 8 Tait sends Thomson a further draft for commentary, and acknowledges the receipt of Thomson's "Introductory Lecture" (see p. 239), which was suggested as basis for a preface. On January 11 Tait urges that the illustrations in the text must be very copious. He acknowledges the receipt of a draft of a chapter on Properties of Matter, a schedule of contents of Vol. I., and fourteen pages of comments. They certainly lost no time in sketching out the work.

On January 15 Tait writes:—

I sent you yesterday a meagre attempt at our Preface; it wants completeness, and also grinding down in its many obvious asperities, but I think it has the ring of true metal. I fear, however, it will come back shortly shorn of its beams in a way in which (to my great surprise) you did *not* treat my Prop. of Matter.

As to the latter, I will shortly send you the *revised* readings, that you may see whether they correspond with your ideas, which I confess I have but vaguely gleaned from your notes. However, we shall see that next week.

I am now engaged with the Abstract Mechanics wh. you wish me to do, and you shall have them in a few days. I wish a few hours' contemplation of a subject, with a pot of beer, and the lurid glare from my pipe showing in the darkness; then I can sit down and write you off a chapter in double quick time. I find there is no use whatever in putting pen to paper till the subject is carefully thought over—and when that is done, writing is almost mechanical.—Yours truly,

PETER G. TAIT.

P.S.—You have evidently thought more deeply about matter than I have, but I can scarcely admit the *ultimate* compressibility of its molecules.

The next day brought a most characteristic epistle.

6 GREENHILL GARDENS, 16.1.62.

MY DEAR THOMSON—§ Ω. 0. Of course the Preface will have to be drawn mild, but I am glad you think it contains the proper elements. As you say, it may be dismissed at once from our minds till the first vol. is printed off, but I can't help telling you that in conversation with Simpson (who is, of course, a capital authority), I told him of the nature of the performance, and, though he was amused, he seriously said he did *not* dislike the idea of showing up the critics, *especially* the *Athenæum*. What do you say to a shot at the critic of Faraday?

Such as I have hinted (in an interlineation) without mentioning names, about that philosopher who is *so* great that other men's discoveries become his as soon as he repeats their experiments.

§ Ω. 1. I return by this post your Introductory Chapter, with a few MSS. notes of my own. I fancy you ought to cut out Neptune, and add a good deal at the end. Such being done, it will be the proper complement of Chap. I., Division I., as by our recent arrangement.

§ Ω. 2. I think as we are going ahead so rapidly it will be useful to § our lucubrations as I have done this note, taking some outlandish letter as our starting point for each chapter. Say you take Hebrew and I Greek. The foreign letter means the absolute number (to be afterwards found on putting all together) of the first § of a chapter, and § Ω 72 will thus mean (if Ω be, say, 19,385) § 19,457. This will give us great facilities for reference as we go on with the writing. It is to be understood that

$$\S\Omega = \S\Omega\ 0.$$

§ Ω. 3. I send you also my first approx^{n} to Div. II., Chap I., "Explanation of Abstract Mechanics," wh. you wished me to try. I have put some bosh towards the end which, of course, I will rewrite, but I want to know whether you think the *object* (at wh. said bosh is aimed) to be properly cognizable in such a place.

§ Ω. 4. I have just received the beginning of your "Laws of Motion," of which (of course) I shall say nothing till I see the whole; except that you are welcome to write what you like, so long as you save me from such an abominable subject. . . .—Yours truly,

P. G. Tait.

The correspondence continues :—

Edin., 20/1/62.

My dear Thomson—§ Δ. 0. I am very glad you are pleased with my chapter on Abstract Mechanics.

As to the title of the whole, I think there are great advantages in using "*Dynamics*" instead of Mechanics. *First*, that in reality there is no such thing as Statics—only dynamical equilibrium. *Secondly*, and very happily, *Dynamics* really means the science of Force or Power, and is erroneously used as a contrast to Statics. I am perfectly willing to drop Mechanics entirely, and make Dynamics the general title. What would you propose as a substitute for the phrases *mechanical equivalent of heat*, etc? THIS QUESTION IS IMPORTANT AT THE OUTSET.

Δ 1. I sent you on Saturday (to the College, but I shall remember your hint in future) a sort of index to Section III., Chap. I. I don't think I have quite understood you about *Porosity*, but I think most of the other things are pretty well. I wish you would send it back speedily, and I will get *that* chapter off my mind before the month is out if possible. . . .

Δ. 8. As to the *Yoke*, I shall bear it with perfect equanimity, feeling assured that if I be galled with it (wh. I don't expect) in the present book, YOU, too, will have your fair share of abrasion when our *Mathematical* volumes are being got ready. Think of that and don't prematurely waste your stock of sympathy.

Δ. 9. Andrews has just sent me all his papers on Heat of Combination, and I intend to avenge him on F. and S., who (if not your friends) are simply THIEVES, at least as far as I can see.—Yours truly,

P. G. TAIT.

6 GREENHILL GARDENS,
EDIN[R]., 23/1/62.

MY DEAR THOMSON—I feared the Porosity wouldn't suit you. I am sorry I didn't send a scheme of some other chapter which might have passed muster more easily. . . .

I have *done* nothing more in the *Book* line, except receive, and answer, as you find enclosed, another note from Macmillan. Make your comments on both, and send *on*, or *back*, as you see fit. . . .

I think we should put in parabolic, circular, etc., motions in Division I., Chap. V., Kinematics (I don't see the force of cycle or Socrates—what is the us auv K in an Alfabet at aul if C duz as ouel?) and I should also there go into Velocity and Acceleration. I am like M. Scot's Demon and *want work*, anything short of ropes of sea-sand; and even *that* is not impracticable, for the new dodge for submarine cables is founded on glass-thread. Shall I try Chap. VI. "Experience," or II. ii. "Statics," or what? . . . *I want work* (for I *must* say it again) and I will take it out either in writing the Book, or in reading what may bear upon the above question—since I can't get sunlight for my own investigations. . . . Tell me then by the beginning of next week (Monday is a holiday, and I shall golf all day at Musselburgh if it is fine) what to attack, and to what extent, and I shall set to with vigour. My Quaternion Paper and my article on Heat are both despatched to the printers; so that my hands are (with the exception of Examination Papers and Answers) almost free in the evenings, especially as I have given up the idea that there is *any* pleasure (even) to be obtained at a dinner or evening party—the only profitable conversation being that at which either science or tobacco is freely admitted—their union being the nearest sublunary approximation to perfection. . . .—Yours truly,

P. G. TAIT.

W. Thomson, Esq.

6 G. G., E., 28/1/62.

MY DEAR THOMSON—The Demon will probably have his hands full enough this week, as he has one of his general Ex^{ns} on Saturday, and a Medical d^{o} soon, for both of which papers have to be made and printed, and, moreover, he has two papers to read to the R.S.E. on Monday night, and a batch of 113 exercises to examine before Friday first. However, he is "bon diable au fond," and may perhaps manage 20 or 30 pages of Properties of Matter before the week is out. . . .

Your article on the Laws of Motion pleased me very

much, so far as it went, save that on thinking it over after reading I felt that at the beginning and end it was not *expanded* enough to be at once intelligible to an average student. But *that* is easily supplied, and is even possible on a proof-sheet, where I daresay much of our work will undergo a little emery or filing down. It is rather curious to observe that Newton's first Law defines time to be an independent variable. For my own part I own that I think this a very satisfactory way of putting the whole; for I don't believe in *time* at all, only in order of succession, and intervals of d° measured as above.

By all means drop the finer Book if you like. I put it in that sort of way to Macmillan, as I told him we had only adopted it to satisfy what we imagined would be his peculiar pride. But the Cambridge books are ridiculously dear, and I think he may learn a useful lesson on that point by being made to do the affair cheap for once. *We* can afford to do the humility dodge, for no one will suspect us of the "darling sin" so neatly described by Coleridge. If we can do the two volumes under 20/-, I think we may consider ourselves to have succeeded. Future editions, as requiring few new cuts, etc., will, of course, be producible more cheaply.

Well, be it Cinematics (do you propose to say *S*inematics?), I wonder where we got kine, and kyloes, and kangaroos, besides kirks and kirn, etc., etc. But I can't help looking on Cinématique and Conductibilité as being equally French and equally erroneous. Macdonall is not a man to swear by on points like these; but still, though wrong, there is a dignity about them which the sibilants possess not—they delight in the dust, grovelling. So, in haste,—Yours truly, P. G. TAIT.

Prof. W. Thomson.

6 G. G., E., 30/1/62.

MY DEAR THOMSON—I wish you would send back my sketch of the Chap. on Prop. of Matter, with your amendments, etc., and I will have it written as soon as is consistent with care and completeness. I cannot commence without my skeleton (why not *sceleton? or if you like French*, SQUELETON??). . . .

Jan. 31, 1862.

. . . I have been so harassed by unavoidable work that I have done nothing of the book this week, besides, I had not my *sceleton* of the Prop. of Matter. . . . Also tell me soon what you have done in the Macmillan business, AT ALL EVENTS ACT SPEEDILY, so that there may be no delay in the appearance of our advertisement. If Mac. is satisfied that his honour is safe in bringing out a cheap book, I am sure *we* won't object, as the condition of our students is becoming from year to year more lamentable as regards text-books. If we go in for a cheap one we may print a larger edition—say 2000 as a lower limit. . . .

6 GREENHILL GARDENS, *Feb.* 5, 1862.

I hope you have got the letter off, and so got the advertisement started. As the French say, *je m'ennuie d'elle, i.e.*, I long to see it in print.

I have received with great delight your letters on Capillarity, especially the foam question, which latter is one that I have very often studied in my summer musings on a clear glass bottle just emptied of its frothy contents direct from Burton. . . .

I return your Axiomata with a few notes. I am now busily at work with "Prop. of Matter," of which by the middle of next week I will send for an MS. volume, the reverse of each leaf being left blank for interpolations from either of us. When the latter have been carried to their limit I shall recopy and remould the whole, and it will be ready for the printer.

Feb. 6, 1862.

. . . I promised to write you last night about Prop. Matter. . . . And, indeed, having just now reperused your letters of New Year time, I think I see my way to a great part of that chapter without giving you any further trouble till you see it in MSS.

P.S.—Six hours' work done since writing the above, and I have got well on with the chapter. Having come to *Inertia*, I shall go to bed forthwith.

Feb. 17, 1862.

. . . . I am glad you are going to do the Cinematomorphology, or whatever you call it, it will help me greatly in writing about Elasticity, etc. (which I am now at), to have an ideal chapter of yours to refer to, and I shall send you, when you wish it, a copy of your sketch of your treatment of it. And I am most thankful for the data for Friction and Cohesion, whose turn is about to come. . . .

6 GREENHILL GARDENS, EDINR.,
4/3/'62.

MY DEAR THOMSON—I have just received "Elasticity." The first part of it should, I think, be dispatched in "Sinematics" (this is the true way to soften a K). The latter portion, with a few references and dates, and with a condensation of your observations on Faraday and others, will slip very nicely into the Properties of Matter which I am at work at. I shall, therefore, in what I sent to you as a first essay, leave out what you have just written, as, indeed, I have done with those parts of Capillarity, Cohesion, and Friction which you sent me. And I have followed the same rule with all my own work, namely, writing in red the title of anything to be afterwards put in in copying for press, if that were a matter of such simplicity that one could scribble it at a second's notice. All the really cranky bits I have written in full, in order that you may have a fair shot at them even at this early stage. I intend to remain some time in Edin^r^ *after* the session is over, and I shall devote that to incorporating and completing that most important chapter.

Sound and Light can cause us no trouble whatever in writing, and may be left for leisure time of w^h^ we have at least six months in prospect.

I think the Dynamical Top might be done more briefly while quite as lucidly as in Maxwell's paper, but that's no reason for your not getting a copy from him for the paste and scissors process. I have a copy (also bound up), which I will consult carefully to-morrow. . . .

You are a little hard on the Macmillanites; but the article is a most interesting one, and the paste and scissors will be at IT for our final chapter—"Conservation of Energy." . . .—Yours truly, P. G. TAIT.

W. Thomson, Esq.

Thomson had told his brother-in-law Dr. King, at the beginning of 1862, of the projected book, which was to be "elementary and non-mathematical." His letter to Helmholtz (see p. 425), ten months later, lets us know that by that time one sheet on "Kinematics" was in type, and that they hoped to have the first volume (ending with a chapter on "Sound") out by the summer of 1863.

Helmholtz replied:—

Your undertaking to write a text-book of Natural Philosophy is very praiseworthy, but will be exceedingly tedious. At the same time, I hope it will suggest ideas to you for much valuable work. It is in writing a book like that that one best appreciates the gaps still left in science.

By March 1863, 400 pages were in type, and they still hoped (see p. 428) to have a volume of 700 pages, including "Dynamics" and "Properties of Matter," published by the middle of June.

The *Glasgow University Calendar* for 1863-64, announcing the Natural Philosophy classes, states:—

The text-books to be used are: *Elements of Dynamics* (first part now ready), printed by George Richardson,

University Printer; *Elements of Natural Philosophy*, by Professor W. Thomson and P. G. Tait (two Treatises to be published before November: Macmillan).

The *Elements of Dynamics* (1863) mentioned above was a small volume of eighty-one pages, compiled by Mr. (now Professor) John Ferguson, then a member of Thomson's laboratory corps, partly from the sheets already completed and partly from Thomson's lectures. It was hurried into existence to meet the pressing needs of students, but dropped when the maturer work appeared, and used to be referred to as "the Glasgow pamphlet." It was reprinted at least once, and in 1869 had been slightly enlarged. As for the *Treatise*, it grew and grew under its authors' hands, and its appearance—alas! for the vanity of human wishes—was delayed until 1867, when Vol. I. appeared. It was printed, not as originally intended as a private enterprise, but at the charges of the delegates of the Clarendon Press.

The late Professor Ayrton[1] tells this recollection of his student days:—

At that time the advanced proofs of only a fragment of that book had been printed off for the class. We saw the book grow, we felt pride in its growth, we almost felt that we were helping that growth. That book by "T and T'," as is well known, consists of chapters which are more original than the papers usually read before scientific societies. Only one volume has ever appeared —the second, alas! alas! never will now.

To test the power of the Clarendon Press to publish such

[1] "Kelvin in the 'Sixties," *Popular Science Monthly*, March 1908.

a book, Tait and he wrote down at random complicated equations, lines of wholly unintelligible reasoning, and then thought it would be a good joke to send out the proofs—as copies of an original paper—to various of their friends. And one day Thomson told me, with a twinkle in his eye, "Nobody has yet found any mistakes in that paper."

With the long delays over the proof-sheets Tait became very impatient. To Thomson, then at Creuznach, he wrote in June 1864:—

I wish you would go ahead. I am getting quite sick of the Great Book, for I see plainly that if you don't *send immediately* the whole mass of your self-imposed contribution to Elastic Bodies we shall not get out in September. . . . I am particularly anxious that my time should be employed here, as I must stay here; but if you send only scraps, and these at rare intervals, what can I do? You have not given me even a hint as to what you want done in our present chapter about Statics of Liquids and Gases! I have Kinetics of a particle almost ready. . . . I sent you a great bundle of proof-sheets nearly ten days ago, but you have taken no notice of these whatever.

Actual publication of the book took place in October 1867, the same month that witnessed the publication of Tait's *Quaternions*. It bore the title:—

A Treatise on Natural Philosophy. By Sir William Thomson, LL.D., D.C.L., F.R.S., and P. G. Tait, M.A., vol. i. Clarendon Press Series. Oxford and London: Macmillan & Co., 1867.

It consisted of 727 pages, and professed to be the first volume of four, of which the complete treatise was to consist. Immediately on its appearance it was greeted with a chorus of approval, and recognized as marking an advance over any and every text-book of Natural Philosophy that had hitherto been written. Any one who might doubt the enormous and enduring service rendered to science by this work has only to compare any recent treatise with the best of those in existence before 1867, to be convinced on the point. The forty-two years that have elapsed since its publication have made the work so familiar to every student of Natural Philosophy that any analysis of its contents would be an impertinence here. But attention may be drawn to one or two features that are characteristic.

The Preface opened with a quotation from Fourier, which has been transferred to the heading of the present chapter.

The authors then begin :—

The term Natural Philosophy was used by NEWTON, and is still used in British Universities, to denote the investigation of laws in the material world, and the deduction of results not directly observed. Observation, classification, and description of phenomena precede Natural Philosophy in every department of natural science. The earlier stage is, in some branches, commonly called Natural History; and it might with equal propriety be so called in all others.

They then state the plan of construction of the text:—

Our object is twofold: to give a tolerably complete account of what is now known of Natural Philosophy, in language adapted to the non-mathematical reader; and to furnish, to those who have the privilege which high mathematical acquirements confer, a connected outline of the analytical processes by which the greater part of that knowledge has been extended into regions as yet unexplored by experiment.

To effect this double object the text was arranged in numbered paragraphs in two sizes of type: the large type being the non-mathematical part for general readers; the small type being reserved for the more mathematical portions. Not infrequently a proposition is stated first in the large type, generally, and often with geometrical illustration, then restated with due proof in mathematical form in a paragraph, or several paragraphs, of smaller type. If the result seems inelegant to a typographical connoisseur, its essential convenience for readers of different classes is incontestable. Geometrical diagrams were added unstintingly to the earlier sections.

For the benefit of mathematicians the authors added this significant hint:—

> We believe that the mathematical reader will especially profit by a perusal of the large type portion of this volume; as he will thus be forced to think out for himself what he has been too often accustomed to reach by a mere mechanical application of analysis.

Also this warning:—

> Nothing can be more fatal to progress than a too confident reliance on mathematical symbols; for the student is only too apt to take the easier course, and consider the *formula* and not the *fact* as the physical reality.

Another pregnant sentence of the Preface runs:—

> One object which we have constantly kept in view is

the grand principle of the *Conservation of Energy*. According to modern experimental results, especially those of JOULE, Energy is as real and as indestructible as *Matter*.

Thus for the first time the co-ordinating principle of energy was made the basis of a systematic treatise. Then again the authors, discerning that the geometry of motion can be treated apart from the forces that produce the motion, introduced kinematics as a definite study prior to dynamics. This enabled them to introduce at an early stage the analysis by the method of Fourier of periodic motions, the use of generalized co-ordinates, and the spherical harmonic expansion of arbitrary functions, branches of the higher mathematics much needed for the problems of dynamics, but previously ignored by the writers of text-books and never before reduced to a simple and relatively comprehensible form. Amongst the novelties were the principle of varying action, the ignoration of co-ordinates, and a discussion of kinetic stability, also such matters of practical bearing as the degrees of freedom of a mechanism, the design of geometric slides, and a host of beautiful and important things about the gyroscope and the dynamics of rotation. The second part of the volume dealt with abstract dynamics, the laws of attraction and potential theory, the statics of solids and fluids, including a very beautiful discussion of torsional rigidity, and a good deal of geodetic matter regarding the figure of the earth, and other applications of the theory of elasticity. The work included a vast quantity of original matter, some of it gathered up from Thomson's fragmentary investigations, some of it written specially for the book, and much added while it was going through the press, to the dismay of his colleague and the discomfiture of the printers.

A review in the *Scotsman* of November 6, 1868, evidently from the pen of one who knew the authors personally, says :—

Two authors better qualified for a great joint work have seldom or never attempted it. . . . They are to a certain extent a happy complement of each other—the one being deeply speculative, but slightly nebulous in the utterance of his original thoughts, as often happens with profound thinkers ; the other, though not deficient in originality, being clear, dashing, direct, and practical. They are both honest and candid, free from that solemn humbug which has been known to hang as a sort of sacred curtain about professors even of Natural Philosophy.

The world of which they give the Natural Philosophy is not the abstract world of Cambridge examination papers—in which matter is perfectly homogeneous, pulleys perfectly smooth, strings perfectly elastic, liquids perfectly incompressible—but it is the concrete world of the senses, which approximates to, but always falls short alike of the ideal of the mathematical as of the poetic imagination. No iron beam is there met with so rigid as not to bend, no sphere of metal equally tense in its parts, no body that does not yield so much as to be incapable of having a fixed centre of gravity. Nowhere is there actual rest ; nowhere is there perfect smoothness ; nowhere motion without friction.

Four years later a German translation by Helmholtz and Wertheim was issued, when Helmholtz took the opportunity in a Preface to point out the remarkable features of the book. Amid so much original matter he had encountered new words difficult of translation. Thomson was a precisian as to language, and shared with his brother James a passion for creating new names where needed for the connotation of new ideas. Never before had dynamics been treated with so clear a grasp of physical principles. To give clear expression to clear ideas new definitions were often necessary. But the fixed determination of the authors to keep in the forefront the physical side in all their mathematical demonstrations made the work seem scrappy and tentative to those who looked for a formal and consistent analytical exposition. Helmholtz, far from condemning this characteristic, hailed it as leading, in spite of the inevitable lacunae and disjointed transitions, to a wider outlook than the formal treatises of earlier and less original writers. He told the reader who would not grudge the effort to master the work that he would reap an ample reward. He expressed the gratitude of the scientific world to Sir William Thomson, one of the most inventive and penetrating of thinkers, for admitting us to the laboratory of his thoughts and unravelling for us the clues which had helped him in controlling and ordering the entangled and refractory materials with which he had to deal. He pointed out that in this work actuality, consistency to physical fact, was preferred to elegance of mathematical method. "Perhaps when science is perfected physical and mathematical order may coincide." Helmholtz, who himself in his lectures on mathematical physics treated mathematics "as the means, and not as the end," could not but hail a work conceived in this spirit.

The book sold rapidly, and for a time went out of print. It brought, however, a very inadequate remuneration to its authors. Indeed, in March 1869 the delegates of the Oxford Press wrote to Tait that the book was still so much in debt to the Press in the actual outlay that there was no balance payable to the authors. In January 1870, after accounts had been presented, Tait wrote to Thomson, "We do look like a couple of sold gorillas!" The dispute was only settled in March 1872 by the friendly intervention of Mr. Alexander Macmillan.

To meet the immediate needs of students the authors in 1867 printed a small octavo volume of 120 pages, consisting of the first two chapters of their then unpublished *Elements*. The title-page styles the work *Elementary Dynamics*, by Sir W. Thomson and P. G. Tait. It was printed by the Clarendon Press, and is now exceedingly scarce. Another private edition, slightly enlarged, appeared in 1868.

In 1873 there was published by the Clarendon Press the *Elements of Natural Philosophy*, Part I. This book, of 279 pages, of large octavo size, was partly reprinted from the *Elementary Dynamics* (as is shown by retention of certain misprints), enlarged in certain paragraphs, but with the addition of Chapters V. and VI., remodelled from the larger *Treatise*, and also of Chapter VII., "Statics of Solids and Fluids," taken from the previously mentioned Glasgow pamphlet.

The second edition of the *Treatise* was not ready till 1879, when Part I. appeared, revised and considerably extended, the more important additions being an Appendix of 47 pages on Laplace's coefficients, and another of 30 pages on the calculating machines invented in the intervening years by Thomson and his brother, also new sections on degrees of freedom and geometrical slides, on motions of cycloidal systems, and on gyrostatics. The proofs of this edition were read by Mr. W. Burnside[1] and Professor Chrystal.

Part II., in much enlarged form, came out in 1883. The authors then announced that their intention of proceeding with the other volumes was "now definitely abandoned." The most important share in editing of this part was taken by Mr. (now Sir) George Darwin, who himself added numerous valuable sections on the figure of the earth, elastic tides, and an Appendix on tidal friction. Thomson himself rewrote much and added much on the theory of elasticity, and the equilibrium of rotating masses. With regard to this last matter, the authors make the remark: "Year after year, questions of the multiplicity of possible figures of revolution have been almost incessantly before us, and yet it is only now, under the compulsion of finishing this second part of our first volume, with hope for a second volume abandoned, that we have succeeded in finding anything approaching full light on the subject." Three Appendices, On the Secular Cooling of the

[1] Now Professor at the Royal Naval College, Greenwich.

Earth, On the Age of the Sun's Heat, and On the Size of Atoms, were inserted. They are noticed elsewhere (pp. 537, 538, and 566).

It is possible from scattered references to infer the proposed contents of the other volumes that were never published. Sound was to have been included in Vol. I. Vol. II. was to have dealt with Optics; but later this was changed, and Vol. II. was to include waves and vibrations in spherical masses of fluid or of elastic solids, and the velocity of long free waves. There was to be a volume on Heat, containing a chapter on the various forms of Energy. There was to be a volume on Electricity. At least a dozen references are made in Vol. I. to the projected volume on the *Properties of Matter*, which looms large in prospect. In it there was to be a discussion of the question: What is Matter? and another on the subjectivity of Force. Newton's experiments on the pendulum and the relations between mass and weight were apparently to lead to a discussion of gravity as the action of parallel forces on all the molecules of a body, and the non-interference with gravity by the interposition of other matter. There was to be an exposition of spherical harmonics as applied to cylindrical problems corresponding to the spherical problems applied in Vol. I. to solid and hollow elastic bodies. Forces of elasticity, and of imperfect elasticity, and the viscosity and compressibility of liquids were to be dealt with; and the work done in a fluid by distorting stress was to be considered. There were to be chapters on Friction and on the coarse-grainedness of Matter. Doubtless all this would have been worked up with the illuminating penetration that marked the one completed volume. But it was not to be. The little volume which Tait, in 1885, brought out under the same title, excellent as it is, was a purely elementary work. Professor Gray, commenting on the keen regret that all physicists feel at the non-fulfilment of the original plan, singles out the "Properties of Matter" as the loss to be most lamented. "No one," he says, "can ever write it as Thomson would have written it. His students obtained in his lectures glimpses of the things it might have contained, and it was most eagerly looked for. If that chapter only had been given, the loss caused by the disappearance of the book would not have been so irreparable."

A story was current in Scotland thirty years ago that part of the Optics had been written, but that the manuscript had been lost by Thomson. He certainly lost, while travelling, in the year 1873, a bag containing papers, but the actual contents are not known. After the main labour of collaboration in the preparation of the second edition of the great treatise was over, and the authors had agreed that no further volumes would appear, each seems for a time to have devoted himself to the subjects which most interested himself. I asked Lord Kelvin once why no more than Vol. I. had been given to the world. His reply was that the ground they had proposed to cover had in the years that followed been largely covered by such books as Rayleigh's *Sound* (1873), Maxwell's *Electricity and Magnetism* (1873), and Lamb's *Hydrodynamics* (1879), and that he and Tait had so much specialized work of their own that they felt they had better frankly abandon the idea of proceeding further.

The articles on "Elasticity" and "Heat," which Thomson wrote for the *Encyclopædia Britannica* in 1878, doubtless contain matter that would have appeared had the volumes on these subjects of the Treatise been composed. The article on "Elasticity" includes much that was already contained in Vol. I.; that on "Heat" is noticed elsewhere (see p. 688).

The following letter to Professor Simon Newcomb gives Thomson's own statement of the abandonment of the completion of the Treatise.

26th April, 1881.

DEAR PROF. NEWCOMB,—. . . I am working hard now at reprint of Vol. I., Part II., of Thomson and Tait's *Natural Philosophy* which will contain some considerable additions. Alas, alas! for vols. ii., iii., and iv.; *Ars longa; vita brevis.* I am afraid neither of us will live to see them. We are both working hard in different branches of our study, and I hope there is some good work in both of us yet before we die. I am bringing out a reprint of all my papers already published. About 70 octavo pages are already in print. It will fill three or four octavo volumes, and will, in occasional different papers, bring out a great deal that I would have written for "T and T′," volumes ii., iii., and iv.

I look forward also to possibly a separate publication on "Hydrodynamics," and on the "Equilibrium and Motion of Elastic Solids." Alas! however; I have been absolutely stopped, for three or four months now, in the work on "T and T'," Part II., Vol. I., and in the reprint of my own papers, on account of incessant and pressing engagements both here and in London. To-morrow I become freed from my University duties; and the day after I hope to take refuge in the *Lalla Rookh*, where very soon I shall get to work, at least on my reprint.—Believe me, with kind regards, yours very truly,

WILLIAM THOMSON.

After the death of Professor Tait in the autumn of 1901, Lord Kelvin gave to the Royal Society of Edinburgh the following appreciation of his illustrious colleague, as reported in the *Scotsman* of December 3:—

The President read an appreciation of the late Professor Tait. After referring to his early life, he said: In 1860 Tait was elected to succeed Forbes as Professor of Natural Philosophy in the University of Edinburgh. It was then that I became acquainted with him, and we quickly resolved to join in writing a book on Natural Philosophy, beginning with a purely geometrical preliminary chapter on Kinematics, and going on thence instantly to Dynamics, the science of Force, as foundation of all that was to follow. I found him full of reverence for Andrews and Hamilton, and enthusiasm for science. Nothing else worth living for, he said; with heartfelt sincerity, I believe, though his life belied the saying, as no one ever was more thorough in public duty or more devoted to family and friends. His two years as "don" of Peterhouse, and six of professorial gravity in Belfast, had not wholly polished down the rough gaiety nor dulled in the slightest degree the cheerful humour of his student days; and this was a large factor in the success of our alliance for heavy work, in which we persevered for eighteen years. "A merry heart goes half the day, a sad one tires in a mile O!" The making of the first part of "T and T′" was treated as a perpetual joke, in respect to the irksome details of interchange of drafts for "copy," amendments in type, and final corrections of proofs. It was lightened by interchange of visits between Greenhill Gardens, or Drummond Place, or George Square, and Largs, or Arran, or the old or new College of Glasgow; but of necessity it was largely carried on by post. Even the postman laughed when he delivered one of our missives, about the size of a postage stamp, out of a pocket handkerchief in which he had tied it, to make sure of not dropping it on the way. . . . About 1878 we got to the end of our "Division II." on "Abstract Dynamics"; and, according to our initial programme, should then have gone on to "Properties of Matter," "Heat," "Light," "Electricity," "Magnetism." Instead of this we agreed that for the future we could each work more conveniently and on more varied subjects, without the constraint of joint effort to produce as much as we could of an all-comprehensive text-book of Natural Philosophy. Thus our book came to an end with only a foundation laid for our originally intended structure. . . . After enjoying eighteen years' joint work with Tait on our book, twenty-three years without this tie have given me undiminished pleasure in all my intercourse with him. I cannot say that our meetings were never unruffled. We had keen differences (more frequent agreements) on every conceivable subject—quaternions, energy, the daily news, politics, *quicquid agunt homines*, etc. etc. We never agreed to differ, always fought it out. But it was almost as great a pleasure to fight with Tait as to agree with him. His death is a loss to me which cannot, as long as I live, be replaced.

In 1875 appeared anonymously a work of speculative theology called *The Unseen Universe*, the authorship of which was afterwards acknowledged by Professors Balfour Stewart and Tait. On its publication a Glasgow bookseller announced it as a new volume by Thomson and Tait. Sir William Thomson did not like the book, but he did not even trouble to put out a formal contradiction.

One other piece of literary work belongs to this period. This was the issue in 1871, by Sir William Thomson and Professor Hugh Blackburn, of a fine reprint of Newton's *Principia.* They found that all editions of the *Principia* were out of print, and so, partly in order to have available copies to award as prizes to University students, they had a handsome large-type reprint made of Newton's last Latin edition, without note or comment, unaltered save for corrigenda or typographical errors.

CHAPTER XI

THE ATLANTIC TELEGRAPH—SUCCESS

Happily the story of the Atlantic Cable does not end with the disappointment of 1858. Public opinion was, however, adverse. Mr. Robert Stephenson, in July 1859, declared that no cable was sufficiently durable to render it a satisfactory or remunerative speculation. In a debate at the Civil Engineers in November 1860, the president, Mr. G. P. Bidder, stated that submarine telegraph engineering was "a branch of the profession the practice of which, up to the present time, had been signally unsuccessful. Upwards of 9000 miles of submarine telegraph cable had been laid down, of which not more than 3000 miles could be said to be in working order. . . . Patents had proved the curse of telegraphy." In these circumstances it required much faith to revive the project of an Atlantic telegraph. After several years of delay fresh capital was, however, raised, largely by the assistance of Mr. Brassey and Mr. (afterwards Sir) John Pender; and with the co-operation of the contractors, who took nearly half the financial risk, £600,000 was available to construct and lay a new cable. Much experience had been gained since 1858 by the laying of other cables between Malta and Alexandria, in the Persian Gulf, and elsewhere. Further, the famous British Association Committee of 1861-62 had, on the suggestion of Bright and Latimer Clark, adopted names for the units of electrical measurement, and had taken a great step forward in perfecting means for electric testing. Cromwell Varley had devised the "artificial cable," with which it was possible to study in the laboratory the properties of cables in general, and had invented the use of the signalling condenser to sharpen the electric pulses, so augmenting the speed of working. A scientific committee, of which Thomson, Wheatstone, and Whitworth were members, was appointed by the Directors, and gave zealous and gratuitous assistance. They examined into the various types of cable, and then, out of 120 different specimens, selected a suitable construction. This, in accordance with the earlier advice of Bright and Thomson, was of a larger size, with 300 lb. of copper and 400 lb. of gutta-percha per nautical mile. The copper conductor was three times as thick as that in the first cable. With its armouring of manila-covered wire the cable weighed, per mile, 1·8 tons in air, or 0·7 tons in water, and could withstand a pull of 8 tons. At this time there existed but one ship capable of holding the entire cable, and this ship, the *Great Eastern*, of 22,500 tons, was lying idle, having been found unsuitable as a traffic boat. As she was furnished with both screw and paddles she was specially well adapted for manœuvring. She was chartered and equipped for cable service. The British Government, after protracted negotiations, guaranteed £20,000 a year and 8 per cent on the capital; the guarantee to be strictly conditional on success, and to be continued only during the working of the cable. By July 1865 the cable was shipped on board at Greenwich. De Sauty was able to send messages through the coiled-up cable at the rate of 3·8 words per minute. Thomson and Varley, using a new curb-signal key, found a speed of 6 words per minute practicable, and were confident of attaining 12 words when the cable should be uncoiled and laid. The ship was under the command of Captain (later Sir James) Anderson, with Canning and Clifford as engineers, and De Sauty as chief electrician. The staff and crew numbered nearly 500 persons. Varley and Thomson accompanied the expedition on behalf of the Atlantic Telegraph Company as consulting experts, Thomson being final referee in case of any difference of opinion between De Sauty and Varley, but he had no power of interference or control, and was simply to report on the testing and certify results. Willoughby Smith represented the contractors. With them was Dr. (later Sir W. Howard) Russell as newspaper correspondent. Staff-commander Moriarty (formerly of the *Agamemnon*) joined as navigation expert. At first some trouble was found with the ship's compasses, which were perturbed by the iron of the cable.

Thomson joined the ship on July 14th, which then steamed quietly to Valencia. On July 23rd, the heavy shore-end having been laid by the *Caroline*, the splice was made and the *Great Eastern* started westwards, accompanied by the *Terrible* and the *Sphinx*. Just before leaving, Thomson wrote to his brother-in-law, Dr. King:—

S.S. Great Eastern,
Monday, *July* 23 [1865], 2 P.M.

MY DEAR DAVID—One line to say all well, and that we are on the point of moving permanently westwards. The splice is now more than half made on board the *Caroline* (which, till Wed^y, was a great anxiety to us). The weather is beautiful, and all are in the highest spirits as to the undertaking. I would like to have written you a proper letter telling something of what we have been doing, but I have had a great deal of arrears of work to do at the few moments I have taken from open air and the deck, and my duties in the electric room. One thing, however, is that three days ago I went up Hungry Hill, 2050 feet high, with Cap. Anderson, Field, Varley, and others, and that, tho I was a little stiff the morning after, my leg, and body generally, have been much the better for it. We were out about 6 h. altogether, with plenty of wind and rain.

I have a very pleasant recollection of the few days we spent with you before leaving. Now I must stop, to get this off in time to go on shore by the *Hawk*.

Varley wrote immediately correcting the false statements of *The Times* article of yesterday week. We considered very carefully the terms in wh. he should write, and I *preferred* a very moderate statement. He is most useful—quite up to his share both in profit (prospective) and credit.

With love to E. and the children.—I am, your affectionate brother, W. THOMSON.

After 84 miles had been laid a fault was found, but speedily picked up and repaired. By July 26th they were over the bank where the sea-bed falls rapidly from 550 to over 1750 fathoms. Then it was discovered that there was no sounding apparatus on board either the *Great Eastern* or the *Terrible*! On July 29th, when 750 miles had been paid out, another fault was observed, and also repaired. In each case a bit of iron wire, of the armouring, had been, by accident, driven into the core of the cable as it was drawn out of the tanks, and was not discovered till submerged in the sea. By the morning of August 2nd, 1200 miles had been paid out, when at noon all signals suddenly ceased. Great was the consternation at Valencia where hitherto the operations had been followed with utmost satisfaction. Tests indicated a "dead earth" at a distance of 1250 miles, the inference being that the cable had parted at that distance and sunk to the bottom. A week passed and the cable remained silent. No news came from the ship. On August 10th, while still without news, the Directors called a meeting of the Company, and with quiet confidence passed resolutions for consolidating the stocks and for a fresh share issue to raise sufficient new capital to construct and lay an entire new cable. Not till the 16th, when the *Great Eastern* appeared at Valencia, was it known that the cable had indeed parted in mid-ocean during the turning of the ship after hauling back a short length of cable to recover a fault, and lay in some 2100 fathoms of water. But it was also learned that she had made attempts even in water of that depth to recover the cable by grappling-irons, and had thrice succeeded in finding and lifting it, but each time had again lost it through failure of the swivel-bolts in the gear used for raising the grapnels. They returned only when they had exhausted their supply of ropes. To a man the entire staff were unanimous as to the practicability of raising and completing the lost cable, but the Directors decided to wait till next year.

On board the *Great Eastern* some of the cheerful souls had amused themselves by producing daily in lithograph a lightsome journal somewhat in the manner of *Punch*. From this we extract two paragraphs as a sample:—

Wednesday, Aug. 2, 1865.

Professor Thomson gave a lecture on "Electric Continuity" before a select audience. The learned gentleman having arranged his apparatus, the chief object of which was a small brass pot, looking like a small lantern with a long wick sticking out at the top, spoke as follows:—"The lecture which I am about to give is on a subject which has ever been of great interest to the intellectual portion of mankind, and—." The luncheon bell ringing, the learned Professor was left speaking.

A novel and economical mode of illuminating London and other cities by electricity has been patented by Professor Thomson. It condenses the currents by the use of matches, cigars, cigar lights, lucifers, etc., in the streets, by means of a simple but beautiful adaptation of Ruhmkorff's Coil and Papin's Digester, and will soon, no doubt, supersede the ordinary gas-pipe and collector.

On reaching London to take counsel with the Directors, Thomson wrote to his sister:—

A. T. C. BOARD-ROOM, LONDON, *Tuesday, Aug.* 22, '65.

MY DEAR ELIZABETH—It is very kind of you to propose my staying at Colville Square, but I think for the very short time I am to be in London it is scarcely worth while to leave the hotel. I shall be going away either to-morrow night or Thursday night, and am here merely to report to the Boards what I have seen, so far as they want to know for deciding their course of action. The favourite idea is that capital for 1800 miles of new cable must be raised at once, and the *Great Eastern* go to sea next May with it, and both complete the present line and make a new one. I have no doubt but the end of the present cable *may* be recovered and the line finished. The probability of success depends chiefly on what arrangements are made, and if wise counsels prevail, I believe it will be done.

Meantime there is much to settle as to claims between the two companies, and until these matters are settled nothing can be done.

The electric tests were solely arranged by De Sauty, and we were never allowed to see them before going to sea. Nothing could well be conceived worse, and the last fault fatally exemplifies the chief defect of the plan, which probably was the chief cause of the loss of the cable. I am writing during a meeting, and have only time to close it now before going off with the Sec^y and Varley on an important appointment. I am glad you are all well, and that David is enjoying the sea-bathing. . . .—Your affectionate brother,

WILLIAM THOMSON.

On his return to Scotland Thomson wrote to the *Glasgow Daily Herald* :—

KILMICHAEL, BRODICK,
ISLE OF ARRAN, BY ARDROSSAN,
Sept. 13, 1865.

SIR—I enclose two articles[1] regarding the Atlantic Telegraph, which speak for themselves. I have only to

[1] The articles contained an authoritative declaration as to the improvement and the working of a cable when submerged, as to the excellent qualities of the *Great Eastern* as a cable-laying ship, and as to the certainty of grappling a cable to lift it from two miles' depth. They also gave a financial statement.

add that, should you think proper to publish them you may trust to the facts being correct on which the writer's conclusions are founded. I feel more than ever confident in the early and complete success of the undertaking, as the operations of this year, though not successful in immediate result, have proved some most important truths which could not have been learned from a simple success unimpeded by accident. What I have actually seen done by the contractors' engineers during the eight days following the accident of the 2nd August has convinced me that the attempt to recover the lost end next summer, and to complete the line will in all probability be successful. And as to the proposed new cable, notwithstanding the possibility of faults such as the three which appeared in the cable of this year during its submergence, there is every reason to expect confidently that it will be successfully laid, as the management required to cut out faults while laying a cable in deep water has been so well learned from the experience of the recent expedition as to do away with almost all danger for the future from this source.—I remain, yours very truly,

WILLIAM THOMSON.

Resting in the quiet of his house in Arran, Thomson resumed the writing of the great *Treatise*. Tait was expected, and Mr. Alexander Macmillan came over from Corrie twice to see him. In October Mrs. Thomson wrote:—

I never saw William better than he is just now, and I am so glad he is having a little rest before beginning his winter's work. I do not think we shall be here above a week or ten days, unless William succeeds in getting this house taken off his hands for next year (as he expects to be away), and then we should have a good deal to arrange about the furniture.

In December 1865 Thomson gave to the Royal

Society of Edinburgh an Address[1] on the forces concerned in the laying and lifting of deep-sea cables. After discussing the forces involved and the engineering conditions to be fulfilled, he expressed his absolute conviction that with strong enough tackle and a hauling-machine that was both strong enough and under perfect control, the lifting of a submarine cable, as good in mechanical quality as that of 1865, from the bottom at a depth of two miles was certainly practicable. If one attempt should fail another would succeed; their failures had taught them the road to assured success.

Thomson was kept busy for many months perfecting new instruments and testing-appliances in his laboratory, and running up to London from time to time to test the cable at Millwall during manufacture. Professor Ayrton has narrated that Thomson's secretary not infrequently used to be sent to the Glasgow railway station a few minutes before the mail train started, with this urgent message from Thomson: "I have gone to White's to hurry on an instrument. The London mail train must on no account start to-night until I come." And such was the national importance of the problem, and such the honour in which Thomson was held, that the station-master obeyed.

By March the manufacture had made good progress, and a new mode of continuously testing the insulation of a cable throughout the entire period of laying it had been devised by Mr. Willoughby

[1] Reprinted in *Popular Lectures*, vol. iii. p. 422.

Smith. On March 19th a meeting of Atlantic Telegraph shareholders was held in Glasgow, at which Thomson delivered a speech. He emphasized the advances in cable manufacture since 1858, the added experience with other cables, and the improvements in testing. He spoke with entire confidence as to the success of the project, and had not the slightest doubt that they would succeed in transmitting eight words a minute in regular work.

Early in the summer all was ready for the expedition. While the ship was lying at Millwall the public were admitted to view it at one shilling per head. Thomson wrote to a nephew, telling him not to ask for him when visiting it. "Every moment I am on board will be more than occupied, and I cannot even take friends into the electrical room. I am sure you know too much of my difficulties last year, and are too entirely interested not to understand this perfectly."

On July 3 Mrs. William Thomson wrote to her sister-in-law Mrs. James Thomson :—

25 COLVILLE SQUARE, *Sunday Eveng.*

Many thanks for your kind letter, which I should have answered sooner, but that the last few days have been very fully occupied in preparing for William's departure, which was expected to take place to-morrow. Only last night we learned that the *Great Eastern* will not sail till Saturday, and so, although Wm. must be on board to-morrow and the two following days, he will come back to London before starting.

I have had to get quite an outfit for him, as when we left Glasgow we had no idea that we should not be for some weeks at home before the voyage, and had brought very insufficient supplies for that, and a stay of perhaps two months at Newfoundland. We have had an exceedingly anxious time for some weeks, and William has been terribly worried and annoyed. Now, the agreement with the Board is signed, and I hope all will go on smoothly.

My heart fails at the thought of the parting on Saturday and the very anxious time that must follow. I would gladly have gone to Valencia, but there are to be no ladies admitted on board the *Gt. Eastern*, so the parting must be here. . . .

I am very sorry to hear that James has been suffering so much from anxiety. He is too conscientious—a rare fault, but one that I fear causes him suffering. I enclose a telegram from X., as James may like to see it. If the cable is laid, and Wm. makes anything by it, he says it will be his pleasure if James will allow him to take the full burden of X.'s debt. If it is *not* laid, I wish he may not be in debt himself! as in that case he accepts nothing for expenses even. . . .

The *Great Eastern*, laden with the new cable and the unlaid portion of the 1865 cable, reached Berehaven on Thursday, July 12th. The same day Thomson arrived on board H.M.S. *Racoon*. At noon a religious service was held, at which Mrs. Thomson and several of the Directors were present. On Friday the *Great Eastern* picked up the new shore-end which had been previously laid, spliced the new cable on to it, and proceeded on her western voyage. Thomson was on board, Varley remaining in charge at Valencia. The Thomson galvanometers used for testing and signalling were somewhat improved from those of previous expeditions. The mirror with its magnet now weighed only three-quarters of a grain, and was suspended on stretched silk fibres; the galvanometer being iron-clad to eliminate any magnetic disturbances from without. Before the *Racoon* returned to port Thomson wrote the following note to his sister, Mrs. King:—

Great Eastern, The Splice (at Sea),
July 13/66.

My dear Elizabeth—You may imagine I have been too busy for any writing. Margaret will tell you I got well off. I find a moment to get a note off from here. The splice is being made, and I am standing beside it looking on. All goes well. I hope every day you will have good news through the cable.—With love to all, I am, your affec^{te} brother,
W. T.

The weather was dull and wet, but the voyage comparatively uneventful. The "old coffee-mill," as the sailors called the paying-out gear, kept grinding

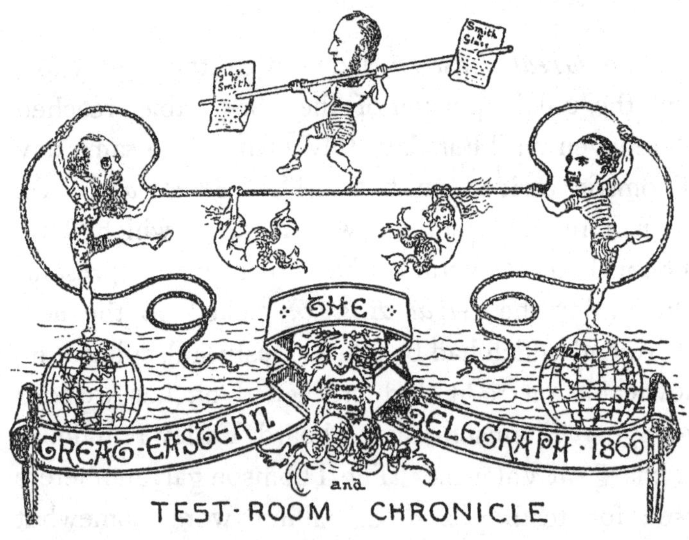

away unceasingly. News was received regularly on board from Europe, and printed in a lithographed journal, styled *The Great Eastern Telegraph and*

Test-Room Chronicle. The title-page of this publication bore a sketch, here reproduced, by the artist R. Dudley, depicting Thomson and Varley at the ends of the cable, on which Captain Anderson is balancing himself. Beyond a temporary tangle when six days out, there was not a single hitch in the laying, and Heart's Content Bay was reached on Saturday the 27th July. Early the next morning the work was completed by the landing of the end.

Communication with Ireland having been maintained throughout, public excitement rose daily as the successful issue grew into certainty. Within twenty-four hours the line was inundated with messages from Europe. The newspapers sang paeans of jubilation. In the *Glasgow Herald* Scotland acclaimed the achievement as her own. "Is Captain Anderson not one of the most gallant of her sea sons? Is Professor Thomson, the distinguished electrician, without whose inspiring genius this great business had not been so easily achieved, not a Glasgow man? And were the principal electrical instruments employed in testing and working the cable not manufactured by Mr. White, the optician of this city, though under Professor Thomson's directions?" It was undeniably so.

Thomson meanwhile quietly wrote the following simple letter as his official "report" to Mr. (later Sir Richard) Glass.

GREAT EASTERN, HEART'S CONTENT,
Aug. 6, 1866.

MY DEAR MR. GLASS—I have been thinking how many sheets of foolscap I ought to cover in my report to you, but really I cannot find that I have anything to say as to my department, except that everything went like clockwork in the testing, and I had the satisfaction of being "a cypher" from the time we got our arrangements completed before leaving Berehaven. I hope by about this time fortnight to be able again to make a similar report. But if, unhappily, there is trouble with faults I am very confident we shall be able to avoid confusion as far as can be done by speedy, and accurate enough, localization.

The cable, as you know, has far exceeded our expectations as to speed. So there will be ample time to arrange about curb keys and permanent plans of working when we are all home again. I hope before Christmas to be able to give you a recorder that will take messages through the cable with security at fifteen words per minute or more.

We are all very sorry to hear you have been so poorly. I hope and trust by the time this reaches you you will be perfectly well again.

I owe you many thanks for your great kindness about my messages to and from my wife. They have been a great comfort to me, although I feel distressed that the accounts of my sister-in-law's health have been so bad.—With kind regards to Mrs. Glass, I remain, yours very truly, WILLIAM THOMSON.

But there was more to come.

For nearly twelve months there had lain silent on the Atlantic bed those 1200 miles of cable. Night and day for the whole of those months had it been under observation by a telegraph clerk watching a galvanometer reflecting a spot of light for possible indications of any change in its electrical condition. None had been found: the cable, so far as it reached along the sea-bed, was believed to be in perfect condition, only awaiting the day when it should be raised. On August 9 the *Great Eastern* quitted Newfoundland along with the *Medway*, and preceded by the *Albany* and the *Terrible*, to attempt to recover the lost end. Their task was to grapple the cable at the depth of two miles, raise, splice, and complete it. On August 13 Thomson was able by means of a passing steamer to dispatch a letter to his wife. They expected to reach the grappling ground that night, and to get down the grapnels. He told her that in a week they might hear of their having found the end; then they would be three or four days in laying the cable to Newfoundland, where they would be two or three days; then ten or twelve more to Liverpool. He had "done a good deal of work at his book." The letter reached Mrs. Thomson on the 23rd. From August 15th to September 1st various attempts were made, with varying experience, resulting in the buoying of the cable after partial lifting at two different points; and finally, at noon on September 2nd, the cable was raised to the surface, brought aboard the *Great Eastern*, and connected to the instruments in the electrical cabin, where a few minutes sufficed for the tests that showed continuity to be perfect. At once they attempted to signal to Ireland.

It was at a quarter to six on Sunday morning, September 2, that the watcher at Valencia, Mr.

May, suddenly became aware that the spot of light showed unsteady flickerings. These speedily changed to coherent movements to left or right, as the cable began to speak. His signals in reply were received with enthusiastic joy by the workers on board the ship. That same afternoon the splice was made, and the *Great Eastern* once more steamed westward, paying out the remainder of the cable to Newfoundland.

Early the next day Mrs. Thomson, in Glasgow, received the following message from her husband in mid-Atlantic:—

VALENCIA, IRELAND,
September 3, 1866.

Mr. Graves has just been requested by Professor Thomson to write to Mrs. T., to inform her that he is well, and to request her to telegraph the latest news in full on Friday by means of the 1865 cable.

It is expected that the *Great Eastern* will arrive at Heart's Content on Saturday next. This marvellous achievement of science has put fresh life into every one.

The *Great Eastern* arrived at Newfoundland on September 8, thus putting the Company in possession of two perfect cables. It was found that their electrical qualities were improved by submersion, and the speed of signalling attained was thirteen or fourteen words per minute. To demonstrate the delicacy of the apparatus, Mr. Latimer Clark, at Valencia, on September 12, performed the experiment of joining up the two cables into a single circuit of 3700 miles, through which he readily transmitted intelligible signals, using only the power of a Liliputian voltaic cell consisting of a lady's silver thimble containing a few drops of acid, into which dipped a slip of zinc. A still more marvellous test of Thomson's instrument was the reception at Valencia of intelligible signals transmitted from Newfoundland by Mr. R. Collett, using as battery Dickerson's cell, a gun-cap containing one drop of acidulated water and a minute anode of zinc.

"I wish William were home," wrote Mrs. Thomson to Mrs. King on September 10. "I am anxious now on windy nights, and get more impatient as the time goes on." The *Great Eastern* left Newfoundland finally on September 10, and reached Liverpool on the 18th; and the day following Thomson returned to Glasgow. Sad news was awaiting him; his sister-in-law, Agnes Crum, had been for some months in a decline, and she passed away while he was returning. To his sister he wrote :—

THE ROUKEN, THORNLIEBANK, GLASGOW,
Mond. Sept. 24 [1866].

MY DEAR ELIZABETH—I was very glad to receive your letter this morning, and had intended to write to you to-day, just a few lines to say that I had got safely back. I was met, as you know, by very sad news on arriving in the *Great Eastern*, and I came here direct from Liverpool. It has been a great blow to us all, and will be, as you may conceive, a grief which will be felt not less strongly after the first shock passes.

I had heard through the cable just before leaving Heart's Content to return home that she might be spared only a few weeks, and it was a great pain to me when I arrived to find that it was too late to allow me to see her again. Margaret has had a great deal to tell me that was very beautiful about the way she passed away, which may make us feel that for her it is a happy change. Mr. and Mrs. Crum are well. Margaret sends you her kind love, and thanks you and Margaret for your letters. . . .

I hope you and David and all have been well. I have been as well as possible during the three months.—Your affectionate brother, WILLIAM THOMSON.

The Secretary of the Company wrote in terms worthy of record :—

ATLANTIC TELEGRAPH COMPANY,
12 ST. HELEN'S PLACE,
BISHOPSGATE STREET WITHIN,
LONDON, E.C., 18 *Sept.* 1866.

MY DEAR PROFESSOR THOMSON—Though prevented by important matters from being present to greet you in person upon your arrival once more in England, I cannot refrain from tendering to you my humble meed of thankfulness and admiration for the labour and perseverance which have resulted in so glorious a consummation of the work upon which your heart has been set during so many years.

I must repeat what I have often said before, that it is mainly due to your own courage and science that this work is so early brought to a practical result. Had it not been for your most valuable aid in 1857-8, whereby it was proved that messages could be sent across the Atlantic, I doubt very much whether there would have been any telegraphic communication with America for many years yet to come.

All distrust is now, however, finally set aside by a success as magnificent as it is complete, and I can only hope and sincerely wish that during a long life you may witness the great results of the work you have done so much to promote, and that for the part you have taken in it you may reap all the honour and profit you have so well deserved.— I am, my dear sir, sincerely yours

GEO. SAWARD.

Professor W. Thomson, LL.D., F.R.S.,
Great Eastern Steamship, Liverpool.

Great were the national expressions of joy at the successful ending to this great enterprise, and the feat of recovering the lost cable from the depths of the ocean was particularly the occasion of enthusiastic comment on all hands.

Then came a notable letter from the chairman of the Company, the Rt. Hon. J. Stuart Wortley, M.P.

Private. BROADSTAIRS, KENT,
30 *Sept.* 1866.

DEAR PROFESSOR THOMSON—In offering you my congratulations on your return home, it is most gratifying to be honoured by a commission from the Prime Minister to intimate to you that it is the gracious intention of Her Majesty the Queen to confer upon you the honour of knighthood in token of her appreciation of your services in connection with the Atlantic Telegraph, and in recognition of your high position in science and classical attainments.

You will believe that it is a great pleasure to all your friends upon our Board, as it is to myself, to know that your invaluable services in this great enterprise have attracted the attention of the Queen and her ministers, and procured you this honourable mark (the only one she has to bestow) of her Royal approbation.

Hoping that a title which your predecessor bore in Glasgow will be agreeable and acceptable to you,—I remain, dear Professor Thomson, yours very sincerely,

J. STUART WORTLEY.

Professor W. Thomson.

The reply was simple and sincere.

THE ROUKEN, THORNLIEBANK,
GLASGOW, *Oct.* 6, 1866.

MY DEAR MR. WORTLEY—I have only now, on my return from London, received your kind letter intimating to me that it is the gracious intention of the Queen to confer the honour of knighthood on me.

I accept with gratitude this mark of Her Majesty's favourable consideration of my humble efforts in connection with the Atlantic Telegraph.

I thank you very sincerely for the kind expressions with which you accompany the intimation. I wish at the same time to express my sympathy with my former colleagues, and other friends on the Atlantic Board, in the gratification they must now feel in the complete success of their great work, and to say that I shall retain a very pleasant recollection of the friendly manner in which they have always met me, not only when I was their colleague, but during the time which has passed since I ceased to be a member of the Board.—I remain, my dear Mr. Wortley, yours very sincerely,

WILLIAM THOMSON.

On October 2 there was held a public banquet at Liverpool to commemorate the event. The chair was taken by the president of the Board of Trade, Sir Stafford Northcote, and there were present Captain Anderson, Sir Charles Bright, Canning, Clifford, Latimer Clark, and Willoughby Smith. Thomson was not able to be present. At the conclusion of the chief toast the chairman read a letter from the Prime Minister, Lord Derby, announcing that Her Majesty, desirous of testifying her sense of the merits displayed in this great enterprise, had directed that the honour of knighthood should be conferred on Captain Anderson, Professor Thomson, Mr. Glass, and Mr. Canning; and baronetcies upon two of the directors, Mr. Lampson and Mr. Gooch. Honours were also offered to Mr. Stuart Wortley and Captain Hamilton, who declined them.

Not to be behindhand, the City of London gave a

banquet at the Mansion House on October 30, 1866. Cabinet ministers, bankers, merchant princes, and parliamentarians met to exchange congratulations with the successful heroes of the telegraph. The response of Sir William Thomson to the toast of "Science as applied to Telegraphy," is thus recorded in the *Morning Post* :—

Professor Sir W. Thomson, in reply, said that the toast of "Science as applied to Telegraphy" opened up an extensive subject, and he could not help feeling his own incompetence to do justice to it. It was impossible to treat it adequately in the limited space of time at his command ; but he would take that opportunity of saying that unless men of science pursued their studies out of a pure love of knowledge, or from an abstract desire to become acquainted with the laws of nature, they would seldom carry on their labours with success. But, at the same time, no greater reward could crown their investigations than when, as in the case of electric telegraphy, they were the means of conferring a practical service on mankind, the value and importance of which was admitted on all hands. The history of many of the most important discoveries showed that they were the result of investigations carried on from a pure love of science, and a desire to increase our acquaintance with the mysterious powers of nature. Who could have supposed that the years which Stephen Gray spent in discovering and defining the law of electrical conductivity as a distinct property of nature would lead to a discovery by which the most distant nations of the earth would be brought into immediate communication with each other, and which would be celcbrated in the middle of the nineteenth century by the representatives of the wealth and commerce of the English metropolis? Or, who would have supposed that the discovery of the relation between the laws of electricity and magnetism—or that Faraday's investigations into the properties of matter in reference to electricity, and his announcement that gutta-percha was the best insulator—would have borne fruit in the thousands of miles of electric cable that were now the means of conveying intelligence to and from every portion of the globe? His only object in these remarks was to point out that science, to be true to itself, must be followed for its own sake, and that all the most important services it had rendered to mankind had been the results of arduous investigations carried on by men animated with the hope of no other reward than that which awaited every sincere, industrious student of nature. In this he was sure that his learned friend, whose name was also associated with the toast (Professor Wheatstone), could bear him out, while he fully admitted, as he had said, that no greater reward could befall the scientific investigator than when his secret and silent labours resulted in some discovery of immediate practical benefit to the world.

The city of Glasgow, on November 1, conferred on Sir William Thomson the freedom of that City. The ceremony took place in the Corporation Galleries, and the Lord Provost in presenting the freedom referred to the circumstance that though his studies and researches had given to him a place amongst the most eminent scientific men of the age, he had yet remained comparatively unknown and unappreciated by the country generally until the practical application of his theories concerning submarine telegraphy attracted general attention. After quoting a eulogium by Sir David Brewster upon Thomson's scientific achievements, the Lord Provost referred to the practical part he had taken in devising his instruments and methods of testing, by which "Sir William Thomson has provided the world of thought with the finest instruments of observation and research, and the world of action with the means of carrying the messages of commerce and civilization which have yet to cross the uncabled oceans that separate the families of the earth." The burgess ticket presented contained a dedication thus recorded in the official roll.

1866, NOVEMBER 1.

SIR WILLIAM THOMSON, Knt., LL.D., etc.

In testimony of their high appreciation of his successful efforts to increase our knowledge of the natural laws of heat, magnetism, and electricity, the means of rendering their powers practically useful, and especially of his valuable services in connection with submarine telegraphy, and the now successful completion of the laying of the Atlantic Cable.

In acknowledging the presentation Sir William Thomson said :—

I find it impossible to give any adequate expression to my feelings in receiving this distinguished mark of the generous consideration with which my fellow-citizens have, from the commencement of the undertaking, regarded my connection with the Atlantic Telegraph. I can never forget that, at a time when failure and loss were the only material results of our labours, I was rewarded by citizens of Glasgow with a recognition which would have been ample and liberal had our work given satisfaction in full to all expectations. And now at last, when the long-looked for success has come, it is most gratifying to me to meet with the sympathy which you show me this day. How highly I value the privilege of being enrolled in the list of your burgesses I cannot find words to tell; but I can only say that, as long as I live, it will be a stimulus to me to endeavour so to act as to do no discredit to the ancient and famous city of Glasgow.

At the dinner which followed there was a toast to "Sir William Thomson, our youngest Burgess." The speech in reply contained several notable passages:—

I feel, indeed, that kindness has led my Lord Provost to say too much regarding my humble efforts in connection with the Atlantic Telegraph. I feel that so far as he has exalted what has been done by science he has not said too much; but that any small additions to electric science that I may have been fortunate enough to have made have had such great results, I think, must be viewed under much restriction, when you consider that the progress of science is sure and strong, and is not dependent on the weak powers of individuals. A few years sooner or a few years later such results as are now spoken of must have been achieved; there can be no doubt of that. Abstract science has tended very much to accelerate the results, and to give the world the benefit of those results earlier than it could have had them, if left to struggle for them and try for them by repeated efforts and repeated failures, unguided by such principles as can be evolved from the abstract investigations of science. . . .

It is curious that one of the earliest records of electrical phenomena that have come to us from the time of the Greeks is immediately connected with the electric telegraph. Pliny describes as known before his time that a certain fish had the wonderful power of killing its prey by some peculiar influence that it exercises upon it through the water; and further, that this influence can produce a torpor, or give a shock to any person at a distance who may touch the fish with a spear. Now, what is this shock communicated along the substance of the spear, and producing an effect on the hand of the person holding it but an electric shock? We now know that it is electricity. Faraday has proved to us what was anticipated by Cavendish; and, after all, this is a system of telegraphy in which the fish is the sender and the person holding the end of the spear the receiver of the message, and that message is *Noli me tangere.*

Sir William's speech continued with a reference to Stephen Gray's researches on conduction, to Faraday's investigation of the dielectric properties of gutta-percha, to the advantage of collaboration of practical engineers and manufacturers with those men whose occupation is with the abstract principles of science. After alluding to Varley's insistence on the necessity for using a larger conductor and a greater thickness of gutta-percha in cables for rapid signalling, as a vindication of the practical value of abstract science, he claimed for scientific institutions in general, and for the University of Glasgow in particular, the support and assistance of the Government and of the Town Council.

It was at Windsor Castle on November 10, 1866, that Queen Victoria conferred knighthood by accolade on Sir William Thomson. The event is thus recorded in the *Court Circular*:—

After the Council, Her Majesty, accompanied by Princess Christian, entered the White Drawing Room, when Mr. John Rolt, Attorney-General, Mr. W. Thomson, LL.D., Captain James Anderson, Mr. Samuel Canning, and Mr. Samuel White Baker, were introduced to Her Majesty by the Lord Chamberlain, and severally received the honour of knighthood.

The arms borne by Sir William Thomson were these:—

Argent, a roebuck's head, caboshed, gules; on a chief, azure, three mullets of the first. *Crest*—A cubit arm dexter, erect, vested; in the hand five ears of wheat, proper. *Motto*—Honesty is the best policy.

On November 22 another banquet was given in Dublin, when, in response to the toast of his health proposed by the Lord Mayor, Sir William Thomson spoke gratefully of the hospitality of Irish friends not only now in the hour of success, but also in the hour of failure in 1858.

On November 28, Sir William lectured on the Atlantic Telegraph at the Glasgow Athenæum, which was densely crowded, hundreds being turned away from the doors.

He traced the history of early land telegraphy, and remarked that the word electric current was an expression which might be used without its being easy to explain. That there was an electric fluid was probably a truth, or certainly half a truth, and that there was a fluid which actually flowed through the wire during the electric current was also probably true; but still he had to make the humiliating confession that when he spoke of a current flowing from left to right he really did not know whether it ran from left to right or from right to left. He spoke of the battery, and illustrated his remarks by sending currents through a portion of the old Atlantic cable of 1858 suspended round the room. He also described and illustrated the telegraphic use of the deflexions of the magnetic needle from its more primitive up to its comparatively perfect modern form. After explaining the causes of retardation which limited the working speed of cables, he described the mirror receiving instrument, in which there was a little needle not three-eighths of an inch long, attached to a mirror smaller than a fourpenny piece, and weighing about one-third of a grain, and which reflected a ray of light, an arm of infinite lightness. What was necessary was a receiving instrument which would be ready to superimpose upon the effects of a previous signal the additional, sharp, sudden effect produced by a new signal; hence the needle must be free to move, and must be controlled by a powerful magnet to make it execute the motions of the signal with sufficient rapidity. He alluded to the services of Mr. White, stating that it was he who first suggested the use of the microscope glass for the thin mirror. He then referred to Varley's improvements in signalling through condensers; the advantage of the curb key; the improvements in construction and insulation of cables; and Mr. Willoughby Smith's method of continuous testing during laying. It had been found possible to transmit as many as from thirteen to fourteen words a minute[1] with perfect security; and the operators, when signalling on private matters between themselves, had spoken as fast as eighteen or nineteen words a minute. The two cables could do far more than the land lines on the other side. Referring especially to the electrical department, Sir William said that by this organization one man's ideas, a second man's ideas, and a third man's ideas were all blended into a harmonious whole, and made ready to be brought into action. Of the organizing work of Glass, Canning, and Clifford, and of their skill in the operations of laying and picking up the cables, he spoke with enthusiasm. After explaining the grappling tackle, he observed that possibly fifty years might elapse before such a grappling rope was again needed. He did not know of any property of matter that indicated a termination to the good action of cables in deep water. He looked upon the Atlantic cable as property which would cost less in repairs, and have a less chance of being at any time found at zero in value than any railway property, house property, or ship property. He concluded with praise for the wonderful

[1] In a communication made by Sir William in 1890 to the Institution of Electrical Engineers he said:—"Varley, Jenkin, and I were more pleased than surprised when we found that, after having promised to the Company eight words per minute, we actually obtained by hand-signalling, with mirror-galvanometer as receiver, fifteen words per minute through each of the two cables, when the laying of both across the Atlantic was completed in 1866."

accuracy of seamanship which had enabled Captain Sir James Anderson and Captain Moriarty to find and hook up the lost cable.

The chairman, in moving a vote of thanks, observed that the picking up of the Atlantic cable was one of the greatest testimonies to the value of abstract science; and that while the audience could not fail to admire what they had heard of his fellow-labourers in this great work, they had heard nothing of what Sir William Thomson had himself done. He was certain that if any one of Sir William's fellow-workers had delivered an address, they would have heard a great deal of his labours amongst them.

To *Good Words* for January 1867, Sir William Thomson contributed an illustrated article on "The Atlantic Telegraph." It was translated into French, and appeared in the *Revue des cours scientifiques*. For some reason it was not reprinted in the volumes of *Popular Lectures*.

CHAPTER XII

LABOUR AND SORROW

RELEASED from the arduous labours and responsibilities of Atlantic telegraphy, Sir William Thomson resumed his quest into the properties of matter, his unremitting activity being, however, from time to time overshadowed by anxieties from the failing health of his wife. Many a time and oft his eager mind had turned to speculate on the intimate structure of matter, trying to conceive some rational explanation of the causes of hardness, plasticity, tenacity, crystalline structure, and other physical properties. So far back as 1856 he had worked out a mathematical theory of elasticity (p. 319), and his laboratory had been the scene of many experimental researches, with his faithful henchman, Donald MacFarlane, to carry out the details of the experiments. He had also made notable contributions to the theory of hydrodynamics, and was familiar with the problems of the kinetic theory of gases, to which, indeed, the work of Joule and himself had contributed. There is no evidence that in these twenty years of his activity he had busied himself much with the inner problems of the atomic constitution of matter; he had been dealing with molar rather than molecular dynamics. Apparently he had accepted the usual notions of the chemists as to the atomic nature of matter of different kinds, and with these notions the current conceptions as to the grouping of atoms into molecules, as, for example, the molecules of water consisting of atoms of hydrogen and oxygen united in some sort of bond. If he did not share the prevalent view that the atoms themselves might be regarded as excessively minute, indestructible, infinitely hard bodies, of probably spherical form, he had at least expressed no dissent, and had offered no theory as to how they came to possess the properties thus attributed to them. But now a new view was to present itself to him. Helmholtz had, in 1858, published in *Crelle's Journal* a short memoir of extraordinary power "On the Integrals of the Hydrodynamic Equations which express Vortex-motion." In almost all earlier investigations of the motion of fluids the difficulty and complexity of the subject had led to assumptions that seriously limited the scope of the theory. The mathematical treatment adopted had either assumed the absence of friction between the different parts of the fluid as they moved past one another, or else had admitted it only in a few simpler cases. Helmholtz abolished this limitation, and endeavoured to ascertain the forms of the motion which friction produces in fluids. Analysing the motion into three constituents,—a simple translation, a rotation about an instantaneous axis, and a voluminal expansion,—and postulating that in a frictionless fluid such particles as have no initial rotation do not acquire rotation at any subsequent period, he showed that if lines—termed vortex lines—are drawn (in imagination) through the fluid, such that at every point of such line its direction coincides with the instantaneous axis of rotation of the element of the fluid at that point, the fluid along such a vortex line will remain composed permanently of the same elements of the fluid; and if they are being shifted in the fluid by any motion of translation, the vortex-line will be shifted with them. In other words, the vortex-filament, consisting of this indefinitely thin line of rotating fluid, will persist. A further deduction was that the product of the cross-section and angular velocity, taken over any cross-section of such a vortex-filament, is constant throughout the whole length of the filament, and is constant in time too if there is no frictional dissipation. But if it is constant all along the length of the filament, the filament must either be re-entrant on itself, as a ring or other closed figure, or else it must have its ends on the boundary-surfaces of the liquid. Helmholtz pointed out other consequences, including certain analogies with electric and magnetic problems. He also considered the reactions which two vortex-filaments in one mass of fluid would have exerted on one another. If the fluid were truly devoid of internal friction, it would be impossible to produce in it any such closed vortex-ring.

This memoir of Helmholtz, one of the most

wonderful pieces of mathematical physics ever written, was translated by Tait for the *Philosophical Magazine* in 1867, and Tait himself devised some extremely clever experiments to illustrate the points brought out in the theory, using vortex-rings of smoke produced in the air. He showed that they could rebound against one another, or against solid bodies, as if possessed of elasticity. Thomson, on being shown these experiments, was immensely struck with them, and seized with avidity upon the mathematical points in Helmholtz's memoir. And not only upon the mathematical points. With the lightning rapidity of thought, and the dominant passion for physical interpretation characteristic of his mind, Thomson, by a flash of inspiration, perceived in the smoke-ring a dynamical model of the atom of matter. If in a perfectly frictionless medium no such vortex can be artificially produced, it followed logically that if such a vortex existed it could not be destroyed. But being in motion, and possessing the inertia of rotation, it would have quasi-elasticity and kindred properties. Applying his fine mathematical powers to the results which Helmholtz had deduced by strict analysis, Thomson proceeded to develop the theory of vortex-motion. In the great memoir, which he subsequently presented to the Royal Society of Edinburgh, he showed that in a perfect medium vortex-rings are stable, and that in many respects they possess the qualities essential to the properties of material atoms,—permanence, elasticity, and the power to act on one another through the intervening medium. Thomson's visit to Tait, when he saw these smoke-rings, was in the middle of January 1867. His first paper on vortex-atoms was given to the Royal Society of Edinburgh on February 18th. But in the meantime he had written of them to Helmholtz :—

Jan. 22, 1867.

MY DEAR HELMHOLTZ—I have allowed too long a time to pass without thanking you for your kind letter. I need scarcely tell you that your congratulations were most acceptable and highly valued. It has been very satisfactory to all concerned in it to have succeeded at last so completely, and, above all, to have raised and completed the cable of last year.

From the beginning of this session I have been hard at work on various matters which I had to set aside on account of the Atlantic. Among other things, the book of Tait and myself has suffered great delay. The last sheets are now in the printer's hands, and a few weeks at most will, I think, bring it out. This, however, is only the first volume, and we estimate about *four*, so I dare-say we may be at work on it all our lives. A second volume, however, to include Vibrations, Waves, Elastic Solids and Fluids, Tides, and Fluid Motion generally, will, I hope, come out in a much shorter time than we have spent on the first volume. I have worked out a great deal for it about waves with friction, tides (on which I spent many a day on board the *Great Eastern*, when we were waiting for weather or making passages), and general dynamics. A great deal is to be done about the tides, and I believe observations may be reduced so as to give results that have not yet been worked out. Just now, however, *Wirbelbewegungen* have displaced everything else, since a few days ago Tait showed me in Edinburgh a magnificent way of producing them. Take one side (or the lid) off a box (any old packing-box will serve) and cut a large hole in the opposite side. Stop the open side AB loosely with a piece of cloth, and strike the middle of the cloth with your hand. If you leave anything smoking in the box, you will see a magnificent ring shot out by every blow. A piece of burning phosphorus gives very good smoke for the purpose; but I think nitric acid with pieces of zinc thrown into it, in the bottom of the box, and cloth wet with ammonia, or a large open dish of ammonia beside it, will answer better. The nitrite of ammonia makes fine white clouds in the air, which, I think, will be less pungent and disagreeable than the smoke from the phosphorus. We sometimes can make one ring shoot through another, illustrating perfectly your description; when one ring passes near another, each is much disturbed, and is seen to be in a state of violent vibration for a few seconds, till it settles again into its circular form. The accuracy of the circular form of the whole ring, and the fineness and roundness of the section, are beautifully seen. If you try it, you will easily make rings of a foot in diameter and an inch or so in section, and be able to follow them and see the constituent rotary motion. The vibrations make a beautiful subject for mathematical work. The solution for the longitudinal vibration of a straight vortex column comes out easily enough. The absolute permanence of the rotation, and the unchangeable relation you have proved between it and the portion of the fluid once acquiring such motion in a perfect fluid, shows that if there is a perfect fluid all through space, constituting the substance of all matter, a vortex-ring would be as permanent as the solid hard atoms assumed by Lucretius and his followers (and predecessors) to account for the permanent properties of bodies (as gold, lead, etc.) and the differences of their characters. Thus, if two vortex-rings were once created in a perfect fluid, passing through one another like links of a chain, they never could come into collision, or break one another, they

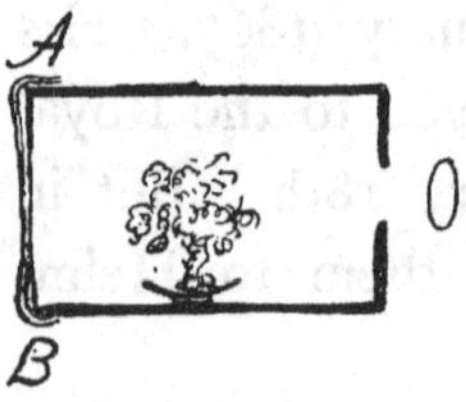

would form an indestructible atom; every variety of combinations might exist. Thus a long chain of vortex-rings, or three rings, each running through each of the other, would give each very characteristic reactions upon other such kinetic atoms. I am, as yet, a good deal puzzled as to what two vortex-rings through one another would do (how each would move, and how its shape would be influenced by the other). By experiment I find that a single vortex-ring is immediately broken up and destroyed in air by enclosing it in a ring made by one's fingers and cutting it through. But a single finger held before it as it approaches very often does not cut it and break it up, but merely causes an indentation as it passes the obstacle, and a few vibrations after it is clear.

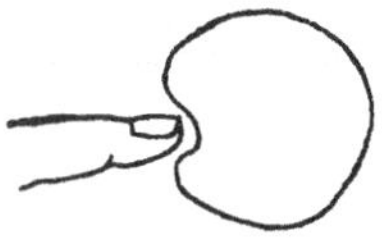

Have you seen notices of Le Roux's recent work (*Comptes rendus*) on thermo-electricity? I have had some interesting letters from him, and he has sent me proofs of a paper for the *Annales de chimie*, in which he confirms my results as to electric convection of heat, and makes a beginning of absolute measurements of the Peltier effect which promise to be very important. He tells me that hitherto no one in France would believe in the "Electric Convection of Heat".

We have been much interested in the personal matter of your letter. I hope you will have improving accounts to give of the little boy's health. We were very sorry to hear of the accident to Kirchhoff. My experience allows me to thoroughly sympathise with him, and I hope he is now much less lame than I am. I have had a letter to-day from Roscoe asking on his account about glass to hold an electric charge for electrometers, which I shall attend to immediately. The common "flint-glass" made in Glasgow answers very well. It very rarely happens that a jar made of it does not hold well enough, and many of them hold so well that they do not lose 1 per cent in a day.

I have made improvements in the divided-ring electrometer, and can now show directly (without condenser, or other multiplication of effect) differences of potential as small as $\frac{1}{500}$ of that produced by a single cell of Daniell's. A severe frost we have just had has allowed me to test (roughly) the conductive and electrolytic quality of ice, and I have found that zinc and copper separated by hard frozen ice act quite as a zinc-water-copper cell when tested by the electrometer. But the effect was quite manifest on an ordinary galvanometer, and indicated a resistance of about $120{,}000 \times 10^7 \frac{\text{metre}}{\text{seconds}}$ in a cell consisting of a few square inches of zinc separated by ice from copper, the ice being hard frozen all round the zinc. I don't know whether others have made such experiments; but electrolytic action through a solid is a new idea to me.

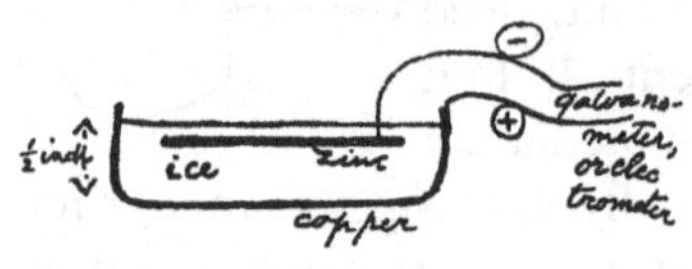

My wife has suffered so much from the cold of our Glasgow winter that she has been obliged to take refuge in a milder climate, and is now at Torquay, having found much benefit from spending part of last winter there. I am going there to-morrow, having "holidays" for three days, but except such occasional visits I am alone in the college this session. The loss of her younger sister has been, as you may conceive, a great distress.

When are we to see you again in this country? If you make a journey to see the Paris Exhibition, I hope you will extend it a little further, and let us have an opportunity of meeting. I am fixed here till the beginning of May, and after that I shall probably be somewhere in the south of England for a time, but as we have given up our house in Arran, I don't know where we shall spend the summer, but wherever we are it would always be a great pleasure to meet you.

With kind regards to Mme. Helmholtz and yourself, in which my wife would join if she were here, I remain, yours always truly, WILLIAM THOMSON.

Thomson's paper on Vortex-motion is reported as follows in the *Scotsman* of February 19, 1867:—

On Monday night the sixth ordinary meeting of the Edinburgh Royal Society was held in the Royal Institution, Sir David Brewster, president, in the chair.

Professor Sir William Thomson read an interesting communication on "Vortex-atoms," of which we give the following abstract:—After noticing Helmholtz's admirable discovery of the law of vortex-motion in a perfect liquid, that is, in a fluid perfectly destitute of viscosity (or fluid friction), Sir William said that this discovery inevitably suggests the idea that Helmholtz's rings are the only true atoms. For the only pretext seeming to justify the monstrous assumption of infinitely strong and infinitely rigid pieces of matter, the existence of which is asserted as a probable hypothesis by some of the greatest modern chemists in their rashly-worded introductory statements, is that urged by Lucretius and adopted by Newton; that it seems necessary to account for the unalterable distinguishing qualities of different kinds of matter. But Helmholtz has proved an absolutely unalterable quality in the motion of any portion of a perfect liquid, in which the peculiar motion which he calls "Wirbelbewegung" has been once created. Thus, any portion of a perfect liquid which has "Wirbelbewegung" has one recommendation of Lucretius' atoms—infinitely perennial specific quality. To generate or destroy "Wirbelbewegung" in a perfect fluid can only be an act of creative power. Lucretius' atom does not explain any of the properties of matter without attributing them to the atom itself. Thus the "clash of atoms," as it has been well called, has been invoked by his modern followers to account for the elasticity of gases. Every other property of matter has similarly required an assumption of specific forces pertaining to the atom. It is as easy (and as improbable, if not more so) to assume whatever specific forces may be required in any portion of matter which possesses the "Wirbelbewegung" as in a solid indivisible piece of matter, and hence the Lucretius atom has no *prima facie* advantage over the Helmholtz atom. A magnificent display of smoke-rings, which he recently had the pleasure of witnessing in Professor Tait's lecture room, diminished by one the number of assumptions required to explain the properties of matter, on the hypothesis that all bodies are composed of vortex-atoms in a perfect homogeneous liquid. Two smoke-rings were frequently seen to bound obliquely from one another, shaking violently from the effects of the shock. The result was very similar to that observable in two large india-rubber rings striking one another in the air. The elasticity of each smoke-ring seemed no further from perfection than might be expected in a solid india-rubber ring of the same shape from what we know of the viscosity of india-rubber. Of course, this kinetic elasticity of form is perfect elasticity for vortex-rings in a perfect liquid. This is at least as good a beginning as the "clash of atoms" to account for the elasticity of gases. It seems most probable that the beautiful investigations of D. Bernouilli, Herapath, Joule, Krönig, Clausius, and Maxwell, on the various thermodynamic properties of gases, may have all the positive assumptions they have been obliged to make as to mutual forces between two atoms, and kinetic energy acquired by individual atoms or molecules satisfied by vortex-rings, without requiring any other property in the matter whose motion composes them than inertia and incompressible occupation of space. A full mathematical investigation[1] of the mutual action between two vortex-rings of any given magnitudes and velocities, passing one another in any two lines, so directed that they never come nearer one another than a large multiple of the diameter of either, is a perfectly solvable mathematical problem; and the novelty of the circumstances contemplated presents difficulties of an exciting character. Its solution will become the foundation

[1] The complete mathematical memoir, which has never been reprinted in any of Lord Kelvin's volumes of collected works, is given in *Roy. Soc. Edin. Trans.* vol. xxv. pp. 217-260, 1869.

of a proposed new kinetic theory of gases. Another interesting problem is presented by the mutual action between closely packed vortex-atoms. For the case of cubically packed vortices he had succeeded in finding the solution expressing the motion of every particle of the fluid. By considering the variation of kinetic energy due to any variation in the sides of one of the rectangular boundaries, its area remaining constant, he found the corresponding modulus of rigidity thus constituted. It was quite certain that closely packed vortex-atoms, even of different dimensions and configurations, must produce in the aggregate an elasticity agreeing with the elasticity of real solids. Diagrams were shown to illustrate the knotted or knitted vortex-atoms, the endless variety of which is infinitely more than sufficient to explain the varieties and allotropies of known simple bodies and their mutual affinities. It is to be remarked that two ring-atoms linked together, or one knotted in any manner with its ends meeting, constitute a system which, however it may be altered in shape, can never deviate from its own peculiarity of multiple continuity, it being impossible for the matter in any line of vortex-motion to go through the line of any other matter in such motion, or any other part of its own line. In fact, a closed line of vortex matter is literally indivisible.

Professor Thomson's paper was illustrated by a series of experiments of smoke-rings, which were very successful.

Two other pieces of work were engaging Thomson's attention this spring. In February he brought before the Institution of Engineers in Scotland the subject of the rate of a clock or chronometer as influenced by the mode of suspension. This discourse (reprinted in *Popular Lectures*, vol. ii. p. 360) contains a masterly exposition of the principles involved, and an account of some curious experiments made by himself when on board the

Great Eastern in 1866, in making a chronometer watch gain or lose by special modes of hanging it up. He also pointed out that a watch regulated to go correctly when hanging on a nail cannot be expected to go at the same rate when carried about in ordinary use. The other was a dynamic method of measuring the magnetic dip.

In spite of his absorption in the subject of vortex-motion, Thomson was keeping his laboratory corps busy with experimental work on electrostatics. Form after form of electrometer had been designed and tried, and a whole series of these instruments had been completed. He now sought for some means of accumulating small electric charges, partly with the idea of rendering them measurable, partly in the endeavour to find a means of charging up to any desired degree of electrification the needles of his electrometers, or (later) the ink-siphon of his recorder. To this end he modified his water-dropping collector (p. 399), and eventually hit upon a mode of duplicating it so as to make it into a kind of reciprocal electrophorus. The drops of water which fell from one jet, being subjected to the influence of a surrounding metallic conductor charged with a small positive charge, acquired negative charges which they imparted to an insulated cup below; and this cup, connected to a metallic conductor surrounding the second jet, caused the drops falling here to acquire by influence positive charges which, in turn, fell into a cup cross-connected to the metallic conductor surrounding the first jet. In this way mutual accumulation took place, the conductors becoming highly charged on the principle now familiar in the modern influence machine. This led him to design mechanical machines to carry out the same idea of reciprocal accumulation by influence and convection. To construct the machine he employed two of his students (one of them the late Professor W. E. Ayrton). This accomplished, they were sent to search the file of patent specifications to see whether any one had previously invented a machine on the same principle. They discovered several anticipations,[1] notably one by Varley in the year 1860; but all the prior patentees had supposed that it was necessary to give an initial finite electric charge to one of the inductors of the machine, whereas Thomson's machine, designed on the compound interest principle, started with an indefinitely small initial charge. From the date, June 1867, at which Thomson announced this invention it may be inferred that he was led to the idea of a self-exciting and accumulative electrostatic machine by the controversy which had broken out at the beginning of the year as to the self-exciting magneto-electric or dynamo-electric machines. Indeed Thomson himself stated that he was thus led by the analogy. On this principle were constructed the "replenisher,"

[1] The Toepler influence machine dates from January 1865, that of Holtz from April 1865, the more modern machine of Wimshurst from 1878. All were anticipated in the use of the regenerative or "compound-interest" principle by Belli of Milan, 1831; but the most primitive sort of machine working by the principle of influence and convection was Nicholson's "Doubler" of 1788.

for use with the quadrant electrometer, and the "mouse-mill" for use with the siphon recorder.

The British Association meeting of 1867 was held at Dundee, and Sir William Thomson was chosen to preside over the section of Mathematics and Physics. It met on September 4th. Faraday had died on August 25th, and the shadow of his death touched the address with which Thomson opened the proceedings of the section :—

. . . .

It was my intention not to detain you from the interesting subjects and abundant matter for discussion which will so fully occupy our time during the meeting by an introductory address; but I must ask you to bear with me if I modify somewhat this resolution, in consequence of a recent event which, I am sure, must touch very nearly the hearts of all present, and of very many in all parts of the world, to whom the name of Faraday has become a household word for all that is admirable in scientific genius. Having had so short time for preparation, I shall not attempt at present any account of Faraday's discoveries and philosophy. But, indeed, it is very unnecessary that I should speak of what he has done for science. All that lives for us still, and parts of it we shall meet at every turn through our work in this section. I wish I could put in words something of the image which the name of Faraday always suggests to my mind. Kindliness and unselfishness of disposition; clearness and singleness of purpose; brevity, simplicity, and directness; sympathy with his audience or his friend; perfect natural tact and good taste; thorough cultivation—all these he had, each to a rare degree; and their influence pervaded his language and manner, whether in conversation or lecture. But all these combined made only a part of Faraday's charm. He had an indescribable quality of quickness and life. Something of the light of his genius irradiated his presence with a certain bright intelligence, and gave a singular charm to his manner which was felt by every one, surely, from the deepest philosopher to the simplest child who ever had the privilege of seeing him in his home—the Royal Institution. That light is now gone from us. While thankful for having seen and felt it, we cannot but mourn our loss, and feel that whatever good things, whatever brightness, may be yet in store for us, *that* light we can never see again.

Great activity prevailed at the meetings of the section. One incident was the entrance of the veteran Sir David Brewster, aged 87, who tottered into the room to read a paper on the colours of soap bubbles, which he maintained were not due to differences in thickness of the film, but were produced by the secretion from it of a new substance which flowed over the film and expanded into coloured patches. At the conclusion Sir William Thomson quietly averted criticism by observing that the mechanical questions involved in the seemingly simple operation of blowing bubbles were amongst the greatest enigmas to scientific men. After the reading of another paper, in which it was assumed that the magnetism of a steel bar resided on its surface, Thomson remarked on the fact that a very large portion of the statements made in books on Natural Philosophy were false and in direct opposition to facts published years before.

Thomson himself brought before the section the Report of the Committee on Electrical Standards, to which he contributed an important chapter on

Electrometers, a number of which he also exhibited. He also showed his new influence machine, which he colloquially described as an easy-going electrical machine built on the principle of the Successful Merchant, who, commencing life with a capital of ½d., had, after a month's persevering industry, accumulated the handsome sum of £1, and continued to go on increasing it at compound interest. He also described his uniform voltaic accumulator, a disk-dynamo machine without iron. He took part also in discussions on a large variety of topics, ranging from lunar volcanoes and aneroid barometers to the luminosity of ice, the winds of the Indian Ocean, the internal heat of the earth, and the formation of dew, which he pronounced "an admirable and wonderful exemplification of design" in the protection it afforded to leaves and flowers against destruction by cold.

In the session of 1867-8 he set two of his pupils, King[1] and Dickson,[2] to make a new determination of the value of "v," the number of electrostatic units in the electromagnetic unit. The importance of this ratio lay in the circumstances that Weber had discovered that it was not a mere numeric, but was physically of the nature of a velocity, and that Maxwell, in 1864, noticing how nearly it was equal to the velocity of light, had made the brilliant suggestion that they were the same, and that light-waves

[1] W. F. King, afterwards of King, Brown and Co., electrical engineers, Edinburgh.

[2] J. D. Hamilton Dickson, afterwards Tutor of Peterhouse, Cambridge.

were electromagnetically propagated. Thomson, who was not at all convinced of this, desired newer and more exact measurements. The method employed was to compare the values of the difference of potential between the terminals of an electrodynamometer, as measured (1) electromagnetically, by the product of the current passing through the instrument and the resistance of its coils, and (2) electrostatically, by the absolute electrometer. The whole work of that winter was preparatory, and consisted in the setting up of the electrodynamometer, and the accurate measurements of the various parts of it required for the calculation of the necessary constants. It might seem a long time to expend on merely preliminary work,[1] but Thomson insisted on the utmost accuracy; and it must be remembered that keys, batteries, and resistance boxes were commercially non-existent in those days, and it was part of the laboratory training of the students to make these instruments and to adjust them. Records of such work are still existing, showing comparison of electrometers and correction of resistance boxes for changes of temperature. In the three following years the work was continued; a new absolute electrometer was employed, and the results obtained with it by D. M'Kichan (together with those of King and Dickson) were presented[2] to the Royal Society in 1873.

[1] A summary of this research was presented to the B. A. meeting at Exeter in 1869 by W. F. King; see *Reports of Electrical Standards*, by Fleeming Jenkin, p. 186.

[2] *Philos. Transactions*, vol. clxiii. p. 409, 1873.

Lady Thomson's health compelled her to winter in Mentone, and Sir William followed her there at the end of the autumn term. A paper by him, "On Geological Time," was read on Feb. 27, 1868, to the Glasgow Geological Society. It is not known whether he was present. He read no other scientific papers for many months. A letter to his sister, from Kissingen, on July 3rd, tells that he and Lady Thomson remained in Italy, chiefly at Bellagio, on the Lake of Como, till near the end of June, and came thence northward over the Splügen. They stopped in Germany on their way home, Kissingen having been recommended to Lady Thomson for its waters. The doctor to whom they had been recommended had told him that the baths would do his leg good: "so I have commenced trying them, but my faith is not great in their efficacy."

He wrote to Helmholtz :—

KISSINGEN, *July* 24, 1868.

MY DEAR HELMHOLTZ—Many thanks for your kind letter of the 20th, which reached me here on the afternoon of the 22nd. I waited all yesterday expecting a telegram to summon me to London, but it did not come till this morning, and is to the effect that I am to wait until I hear again. Ever since we came here my plans have been very uncertain, on account of important business by which I may be called home at any moment, or else I should have written sooner to you. If I am called away just now I shall try to see you at Heidelberg either on my way to London or returning here. If I am not called to London (which I hope will be the case) we shall remain probably a fortnight longer in Germany, and in any case I shall make a point of seeing you in Heidelberg.

My wife has been feeling much better and able to walk more since she came here, and it seems as if she has derived real benefit from the waters. It is very kind of you to propose coming here to see us, and I am sorry you are engaged these next two Sundays, as I am afraid we shall have left before the 12th.

All my spare time is now spent on "Wirbelbewegung," and there will be a great deal to say on that and other matters which I must keep till we meet.

My wife joins me in kind regards, and I remain, yours always truly, WILLIAM THOMSON.

The business which Thomson alludes to was in connection with the French Atlantic Cable, from Brest to St. Pierre, which was then projected, and of which he was one of the consulting electricians. The Company was floated at the end of August.

On his return to Scotland he wrote:—

LARGS, BY GREENOCK, *Sep.* 3, 1868.

MY DEAR HELMHOLTZ—I had intended to write to you regarding your paper on a discontinuity of fluid motion, and suggest that friction at the place of rapid motion or tendency to rapid motion close to the edge may, rather than avoidance of negative pressure, be the determining cause of the slipping. Consider instead of a perfect edge an infinitely thin plate (which does not exist in nature) or tube of finite thickness with a regularly rounded lip. If the slipping which you investigate depends on avoidance of negative pressure, it should commence at A or A′ [at edge or bottom of the lip], according to lesser or greater pressure in the undisturbed fluid. But would it be so? I suspect that if the experiment were tried in a liquid under greater or less pressure, little or no difference would be found in the locality where slipping commences. Is it not possible that the real cause of the formation of a vortex-sheet may be viscosity which exists in every real liquid, and that the ideal case of a perfect liquid, perfect edge, and *infinitely thin* vortex-sheet, may be looked upon as a limiting case of more and more perfect fluid, finer and finer edge of solid, and consequently thinner and thinner vortex-sheet. I am afraid I have not made this very clear, but some time when you may be inclined to think of these subjects I should be glad to hear from you about it or to explain, if I can, better what I mean. I am very busy now writing a paper on "Vortex-motion" for the R.S.E., wh. will soon be printed. I hope you have been getting much benefit from the mountain air. My wife desires to be most kindly remembered.—Yours always truly, W. THOMSON.

On September 24 he wrote his sister, from Largs, that he would be in London at the end of the week, holding examinations for the Indian Telegraph Service, and that he was going to take a run to Clifton to see Principal Forbes, who had been all summer there in critical health. He was again in London on October 13, on cable business.

In March and April 1869 Thomson gave to the Glasgow Geological Society two papers, one on the Meteoric Theory of the Sun's Heat, the other on Geological Dynamics (see p. 547). On April 22 the University of Edinburgh conferred on Sir William Thomson the honorary degree of LL.D.

In June the French Atlantic Cable was successfully laid, and Thomson attended at Brest to superintend the tests during the laying (see p. 552).

About the end of October Thomson wrote to his sister a hurried letter from Largs telling her that Lady Thomson had been very poorly for six weeks, having taken a chill, and was not able to rise from her bed nor take part in conversation. He went on to advise her about her son, David King, junior, his nephew, a young man of great ability.

I think it will be a great thing (as I told David himself) to look forward to taking a degree. The position and influence of the Universities look more promising, and I daresay ten or twenty years hence there will be advantages in various professions (engineering included) to University graduates of which hitherto we have no experience. The Scotch Universities give a more important position to graduates than they had formerly by the Parliamentary representation; the English Universities promise (as I am sure David will agree) to take a much more important position than hitherto by becoming truly national.

The persistent ignoring of Joule's scientific writings caused Thomson to write a protest to Deville. In it he remarked:—

How necessary is a republication of Joule's papers in a conveniently accessible form is curiously illustrated by the fact that the Secretaries of the Royal Society have in the last Number of their Proceedings attributed to Foucault Joule's discovery of the heating of a piece of copper by motion in the neighbourhood of a magnet.

To Helmholtz, on January 23, 1870, he sent a long letter with a hydrodynamical discussion, and a distressing account of Lady Thomson's health. It included the following paragraphs:—

I hope you have now received a second instalment of my paper on "Vortex Motion." Some rough proofs which preceded it may be put in the fire as soon as you please. I am trying to go on with it, and have a good deal of matter ready to be written out, but I am able (as you readily conceive) for but little work of that kind now, even if I had more time for it than the rare snatches which alone are available. I am doing, besides, what I can to finish a reprint of all my electrical papers, which has stuck about p. 250 (or near the middle) since last March. . . .

I wish your conclusion had been for Bonn and Natural Philosophy, because this would have brought you nearer, both geographically and in community of pursuits, than Heidelberg and Physiology allow. I hope, however, that you are yourself satisfied in all respects. We were very glad to see that you have been at last elected to the French Institut. Has Bertrand ever had the grace to confess his error? It is clear that the verdict of his own Academy is against him. I cannot approve at all of the other election which I see reported about the same time: Mayer *above Joule*. Joule did and published more for the establishment and extension of thermodynamics in his two or three papers published before Mayer came on the field than all that Mayer has done put together. I hope to have or to make an opportunity of publicly proving this some time; but my feeling is not *against* Mayer, but *for* Joule: also for strict and judicious scientific reasoning, and against what is so much the reverse of it as Mayer's speech at Innsprück, reported in the *Revue des Cours Scientifiques*. . . .

Lady Thomson's health was so precarious that it was no longer possible for her to travel to Mentone or Torquay to winter. Dr. Matthews Duncan, who took the gravest view of her case, had on her return from Torquay in May 1869 pronounced her to be beyond hope of recovery. In spite of her continued suffering they were able, before the cold weather came on, to make a little tour in the Rob Roy country by Loch Lomond and the Trossachs. For her sake Sir William decided to winter at the coast, and was prepared rather to resign his chair than be compelled to live in the smoky purlieus of the College. They tenanted a very comfortable house, Brooksby, at Largs, within reach of other relations. He was, however, granted leave of absence from his University duties, a substitute being found for the spring term in the person of Mr. R. Kalley Miller.[1]

On April 8 Sir William wrote from Largs to Dr. King:—

> Margaret has been very poorly. There were four or five terribly suffering nights and days of scarcely less suffering. . . . I see no chance of continuing my article on the Rigidity of the Earth.

Early in May her sister, Mrs. Watson, wrote from Brooksby to Mrs. King saying that Lady Thomson was very ill, and in great suffering, but that there was a rally.

> We almost tremble to see more strength with so much suffering, but Wm. is quite delighted, and seems to have a hope that nothing can crush. He has gone to bed upstairs, having been up since three, quite bright

[1] Robert Kalley Miller (Peterhouse, Cambridge), B.A., Math. Tripos and First Smith's Prizeman, 1867. He had been a student under Thomson and Blackburn. Later, while a Fellow of Peterhouse, he was Professor of Applied Mathematics at the Royal Naval College, Greenwich, 1873-1884; and died in 1888.

Concerning the deputy lectureship a legendary story is current. The usual version is as follows:—Sir William being away laying the Atlantic cable, his class was taken by a gentleman named Day, a particularly lucid lecturer. On its being announced that the Professor was about to return, the following inscription was found one morning chalked in a student hand upon the blackboard: "Work while it is Day, for the Knight cometh when no man can work." I once asked Lord Kelvin about this story; when he replied: "I never had an assistant called Day, and the story is fabulous." Nevertheless there is a germ of truth behind it; for it was toward the close of Mr. Kalley Miller's course in the spring of 1870 that one of the students circulated round the benches a scrap of paper with the words: "Behold the knight cometh, when no man can work."

about her. . . . I think Wm. is in health wonderfully well considering all. The power he has of sleeping *at any time* is a great blessing. He goes to bed in the large drawing-room at 10 o'clock generally. Then gets up at 3 and takes Jessie's or Mrs. Tod's place. Then he goes to bed after breakfast and sleeps as long as he can, generally till about 1 o'clock.

Another letter, of May 7, from the Rev. Dr. Watson takes up the narrative :—

Things are going on to the end at Brooksby more slowly than we thought . . . less pain, more weakness . . . sleep is often rather a torpor . . . mind not clear . . . cannot find the word. She continues as sweet and patient as ever, and she has been wonderfully patient all through this last severe stage of her life. No more doubts or questionings, so that the pain to William and to the rest consists entirely in seeing the sad decay of a mind so vivid and strong. But this pain is very great—her husband is sometimes quite overwhelmed and they all feel it terribly. No one can tell how long this may last, but surely not very long now, for she is certainly weaker.

Lady Thomson died on Sunday, June 17, 1870. Mrs. Watson wrote the same day to Mrs. King :—

I had your very kind letter last evening, and sent it over to Wm. to read. This is just to tell you about him. He was in a terrible state of distress all Thursday and yesterday. I intended to remain a few days if I thought it wd. be any comfort to him, but I saw that he really liked being alone and understood the feeling so well. Mr. Watson and I came here last evg. He was much calmer before I left. . . . Wm. goes with my husband to Barra on Tuesday night, and on Wed. the friends meet at the College. . . . We have persuaded Wm. to give up Brooksby and servants when he goes to London in July, and make the Rouken his home till his College house is ready.

Lady Thomson was a highly accomplished woman. She wrote at various times a number of poems, including translations from German poets. Some of these were collected together in 1866, and printed privately as a surprise for her in a thin green-backed octavo volume. Writing to Lady Minto in 1869, Dr. John Brown says:—[1]

> I had a letter the other day from Lady Thomson, wife of Sir William Thomson, the great natural philosopher and electrician. She is one of the finest creatures I ever knew, sincere and bright and very keen-minded, and with a true insight into the real worth of things and thoughts.

And later he wrote[2] to Miss Jessie Crum:—

> I wish William to reprint the green book without change, except addition,—to put in "Bellagio" and some others; and you must have some not unfit. I often think of her and her life, and yet how much of the deepest and best delight she got out of it; and she now understands the why, or does not care to understand it, in the far more exceeding, the eternal.

The little book bears the title "Verses by M. T.; Privately printed 1866 and 1874." It was printed by MacLehose and MacDougall at the Glasgow University Press. The poems range from 1849 to 1869. All are beautiful and some very sad: a note of sadness and death runs through them. One of the more joyous, "O! Think of Me," is dated 1850. There is a fine sonnet of later date, "The Night arose and girt herself with stars."

To Mrs. King, in August, Sir William wrote a

[1] *Letters of Dr. John Brown*, p. 197. [2] *Ibid.* p. 211.

letter of condolence on the death of her daughter, the wife of Dr. J. Hall Gladstone :—

> I know well that he [Dr. Gladstone], as well as you and David, cannot want the one great consolation. Yet I know too well how insufficient even the surest conviction that the loved one is happy, and the firmest hope of a transcendently happy meeting to come, are to mitigate the bitterness of the present grief.

A photograph taken at this date of Sir William Thomson shows a face of infinite sadness. The shadow of his great sorrow lay for long across his life.

SIR WILLIAM THOMSON, 1870.

From a photograph by Fergus, Largs.

CHAPTER XIII

THE GEOLOGICAL CONTROVERSY

In attempting to preserve continuity in the preceding chapters, little mention has been made of Sir William Thomson's contribution to the science of Geology. When a mere boy he had written a masterly essay on the figure of the earth; and in his undergraduate days he had imbibed the teaching of Hopkins that the earth, instead of being a mere shell filled with an igneous molten mass, as was the once-popular belief, is a solid globe. Though Hopkins's argument was found to be of doubtful validity, Thomson demonstrated from other bases that the earth as a whole is at least as rigid as a solid globe of steel. By his studies in heat-conductivity according to the equations of Fourier, and by his investigations in Thermodynamics, he was led to assign to the age of the earth's heat, and therefore to the age of the globe itself as a habitable world, a much shorter duration than had been demanded by the geologists who were then dominated by the uniformitarian school of thought.

Professor J. W. Gregory, F.R.S., after reference

to Lord Kelvin's claim to be a naturalist in accordance with his definition of that term, says:[1] Lord Kelvin had an indisputable claim to rank as a naturalist owing to his contributions to geology, which may be grouped into four divisions:—

1. Contributions to physics which explain geological and geographical problems.

2. His calculations as to the age of the earth.

3. His disproof of the once prevalent and crude idea that the earth consists of a solid shell round a fiery or molten interior.

4. His contributions to the study of former variations of climate.[2]

Geology is deeply, although indirectly, indebted to him for his explanation of various problems in meteorology; for his investigation on tides, and descriptions of those in the Straits of Dover and the Mediterranean; his studies on wave-action; his contribution to our knowledge of the physical properties of ice; and his calculations as to the degree of stability of the sea level.

So far back as 1855 he had written[3] on the use of observations of Terrestrial Temperature for

[1] "Lord Kelvin's Contributions to Geology," *Transactions of the Geological Society of Glasgow*, vol. xiii. part ii. pp. 170-186, 1908.

[2] In 1877 Sir William Thomson discussed the cause of former local variations of climate, and showed that a different distribution of land and water in the Arctic and Sub-Arctic regions would amply account for the existence in former time of a temperate climate and of temperate vegetation in those latitudes. The persistence of polar ice was mainly due to the present Arctic Sea being land-locked, and therefore ice-bound. A thousand feet of depression of the Northern parts of Europe, Asia, and America, would give an open iceless ocean with only a few small steep islands to obstruct the free circulation of water from equatorial regions. In an open iceless circumpolar sea the climate of a small island would be quite temperate.

[3] *Rep. Brit. Assoc.*, 1855, pp. 1819. Reprinted in *Math. and Phys. Papers*, vol. ii. pp. 175-177.

the Investigation of Absolute Dates in Geology, and had followed this up with others in 1860 and 1861. In 1862 came his papers[1] on the Age of the Sun's Heat, and one[2] on the Secular Cooling of the Earth.

In the first of these Thomson pointed out that the dissipation of energy forbade the idea that the sun's heat was inexhaustible, and that the persistent loss of heat by radiation must lower the temperature of the sun unless new energy was supplied continually by the falling in of meteors, or by gravitational shrinkage of his mass. He calculated that a shrinkage by as little as one-tenth of one per cent in his diameter would account for all the heat emitted in 20,000 years, and concluded that his temperature sinks by 100° Cent. in some time greater than 700 years, but less than 700,000 years. The meteoric theory would account easily for a duration of his heat for 20,000,000 years. A million years ago the sun was sensibly hotter than he now is. Hence geological speculations assuming greater extremes of heat, more violent storms and floods, more luxuriant vegetation, and hardier and more coarse-grained plants and animals in remote antiquity, are more probable than those of the extreme quietist or "uniformitarian" school, who suppose things to have gone on as we see them now for millions of millions of years. The conclusion drawn

[1] *Rep. Brit. Assoc.*, 1862, p. 27, and article in *Macmillan's Magazine*, vol. v., March 1862, pp. 288-393.

[2] *Trans. Roy. Soc. Edinb.* vol. xxiii., 1862, pp. 157-169. Reprinted in *Math. and Phys. Papers*, vol. iii. pp. 295-311.

was that it seems, on the whole, most probable that the sun has not illuminated the earth for 100,000,000 years, and almost certain that he has not done so for 500,000,000 years.

In the second paper he stated that for eighteen years it had pressed on his mind that those geologists who oppose all paroxysmal hypotheses, and who deny[1] that the actions by which the earth's crust has been modified in geological history have been more violent in past time than they are at present, have overlooked the essential principles of thermodynamics. Considerations of underground temperatures led to the conclusion that the consolidation of the earth's globe cannot have taken place less than 20,000,000 years ago (or we should have more underground heat than we actually have), nor more than 400,000,000 years (for then the underground increment of temperature would be less than it is observed to be).

It had been the habit of the geologists[2] to make much more extravagant demands on time. For instance, Darwin, from the rate of erosion of the chalk of Kent, estimated that the excavation of the great valley between the North and South Downs "required 206,662,400 years, or say, 300,000,000

[1] With reference to the present views of geologists, Professor Gregory informs me that he thinks there is no doubt that, even before the oldest traces of life on earth, sedimentary rocks then laid down show that wind and rain and climate were practically the same as they are to-day. In the oldest sedimentary rocks of the north-west Highlands of Scotland, the prevalent wind was south-westerly, and its power of carrying sand was much the same as it is now. Still they would not deny that in the interval there may have been times of more rapid change.

[2] I take this statement, and much more of the present chapter, by permission from the previously mentioned article by Professor J. W. Gregory, F.R.S.

years, and that in all probability a far longer period than 300,000,000 years has elapsed since the latter part of the secondary period." But Jukes remarked that though Darwin's estimate might be excessive, it was just as likely that the real period was one hundred times as great, and that the erosion of the Wealden Valleys may have occupied thirty thousand million years! Confident in their views, the geologists[1] paid no heed to Thomson's calculations.

No doubt the ultra-uniformitarian view then prevalent, that for endless years the climate of the earth, and the physical causes tending to denudation and to the rise and fall of land, had remained of the same average intensity as we observe them[2] to-day, was a reaction against the cataclysmic views of Werner and others, who had ascribed all geological

[1] Exception must be made in favour of Professor John Phillips, who, when appealed to by Thomson, wrote as follows :—

OXFORD, 12*th June* 1861.

MY DEAR PROFESSOR,—. . . Darwin's computations are something *absurd* as to the wasting of the sea. *For* the calculation includes as a co-efficient the *height*! of the cliff. This astonishing error is not the only one. No one who ever does calculate (among our geologists) attaches any weight to the *result*! In my *Life on Earth* I have given some calculations which, for the period of stratification, rise to 96 millions only—and have added not unlikely estimate as to the period of coal deposition and other things of the sort. What you mention of the sun's limited power of heat is very interesting, and I much regret that your wishes as to the subterranean temperature have not been more attended to. *Topical cold* seems possible, from other causes than currents of cold water downwards, but I would trust that view a good way, I don't understand the Greenwich case. *φ*.

[2] A charming comment on the uniformitarian doctrine was given by Sir William Thomson at the close of his subsequent address on "Geological Dynamics" (see *Popular Lectures*, vol. ii. p. 131), where he introduced from Kant's *Collected Works* an apposite parable :—

"A large proportion of British popular geologists of the present day have been longer contented than other scientific men to look upon the sun as Fontenelle's roses looked upon their gardener. 'Our gardener,' say they, 'must be a very old man; within the memory of roses he is the same as he has always been; it is impossible he can ever die, or be other than he is.'"

changes to violent upheavals and abnormal paroxysms of climate.

Determined to be heard, Thomson, in 1865, threw down the gauntlet by reading a paper[1] bearing the incisive title: "The Doctrine of Uniformity in Geology briefly Refuted." The following is the entire text:—

The "Doctrine of Uniformity" in Geology, as held by many of the most eminent of British geologists, assumes that the earth's surface and upper crust have been nearly as they are at present in temperature, and other physical qualities, during millions and millions of years. But the heat which we know, by observation, to be now conducted out of the earth yearly is so great, that if *this* action had been going on with any approach to uniformity for 20,000 million years, the amount of heat lost out of the earth would have been about as much as would heat, by 100° Cent., a quantity of ordinary surface rock of 100 times the earth's bulk. (See calculation appended.) This would be more than enough to melt a mass of surface rock equal in bulk to the *whole earth.* No hypothesis as to chemical action, internal fluidity, effects of pressure at great depth, or possible character of substances in the interior of the earth, possessing the smallest vestige of probability, can justify the supposition that the earth's upper crust has remained nearly as it is, while from the whole, or from any part of the earth, so great a quantity of heat has been lost.

Thomson followed up this pronouncement with an address to the Geological Society of Glasgow, on February 1868, on "Geological Time." His opening words struck the dominant note: "A great reform in geological speculation seems now to have become

[1] *Proc. Roy. Soc. Edinb.*, Dec. 18, 1865, vol. v. p. 512. Reprinted in *Popular Lectures and Addresses*, vol. ii. pp. 6, 7.

necessary." He quoted as the orthodox opinion of geologists of the day a striking passage from Playfair's *Illustrations of the Huttonian Theory* :—

How often these vicissitudes of decay and renovation have been repeated it is not for us to determine; they constitute a series of which, as the author of this theory has remarked, we neither see the beginning nor the end, a circumstance that accords well with what is known concerning other parts of the economy of the world. In the continuation of the different species of animals and vegetables that inhabit the earth we discern neither a beginning nor an end; in the planetary motions where geometry has carried the eye so far both into the future and the past, we discover no mark either of the commencement or the termination of the present order. It is unreasonable, indeed, to suppose that such marks should exist anywhere. The Author of nature has not given laws to the universe which, like the institutions of men, carry in themselves the elements of their own destruction. He has not permitted in His works any symptom of infancy or of old age, or any sign by which we may estimate either their future or their past duration. He may put an end, as He, no doubt, gave a beginning to the present system, at some determinate time; but we may safely conclude that this great *catastrophe* will not be brought about by any of the laws now existing, and that it is not indicated by anything which we perceive.

Thomson challenged the entire doctrine. "Nothing," he said, "could possibly be further from the truth than that statement. It is pervaded by a confusion between 'present order,' or 'present system,' and 'laws now existing'—between destruction of the earth as a place habitable to beings such as now live on it, and a decline or failure of law and order in the universe

. . . and the statement that the phenomena presented by the earth's crust contain no evidence of a beginning, and no indication of progress towards an end, is founded, I think, upon what is very clearly a complete misrepresentation of the physical laws under which all are agreed that these actions take place." He pointed out that there must be a resistance to the motions of the planets, real, even if small. The tides act on the revolving earth as a friction-brake, and if there were friction there must be loss of energy; and modern theory must account for what becomes of that energy. The friction of the waters against the bottom of the sea and against one another, must, on Joule's principle, generate an equivalent of heat, and this heat-energy was in the end radiated off into space. This tidal energy being taken from the earth produced a retardation in its spinning, as James Thomson had shown; and the earth, as a time-keeper, lost about 22 seconds in each century. A thousand million years ago the earth was spinning one-seventh faster than it does now, and the centrifugal forces tending to break or distort its crust would then be about 30 per cent greater than they now are, giving rise to more violent earthquakes. Moreover, not many million years ago the earth was in a fluid state. "But," he said, "if you go back to ten thousand million years ago—which does not satisfy some great geologists—the earth must have been rotating more than twice as fast as at present,—and if it had been solid then, it must be now something totally different from what it is. Now, here is direct opposition between physical astronomy and modern geology as represented by a very large, very influential, and, I may also add, in many respects, philosophical and sound body of geological investigators, constituting perhaps a majority of British geologists. It is quite certain that a great mistake has been made—that British popular geology at the present time is in direct opposition to the principles of natural philosophy. . . . There cannot be uniformity. The earth is filled with evidences that it has not been going on for ever in the present state, and that there is a progress of events towards a state infinitely different from the present." He then recapitulated the arguments for a limit. From the slowing of the earth's rotation it followed that things have not gone on as at present a thousand million years. The sun cannot have continued to illumine the earth many times ten million years. And, "when finally we consider underground temperature, we find ourselves driven to the conclusion in every way, that the existing state of things on the earth, life on the earth, all geological history showing continuity of life, must be limited within some such period of past time as one hundred million years."

Such a challenge the geologists could not refuse to face. Told in such authoritative terms that they could no longer presume on unlimited time, but must "hurry up the phenomena," they resented the intervention. The champion who came forth to answer the challenge was no other than Professor Huxley, a keen controversialist and master of dialectic, but imbued with a profound passion for veracity. On the 19th February 1869, Huxley, as President of the Geological Society of London, gave an address,[1] in which he sought to combat the necessity for the reform that Thomson had urged, and to repel the notion that the current geological doctrine was counter to physical principles. As he prefaced his address by likening himself to a special pleader in the courts who gets up his case and gains his cause mainly by the force of mother-art and common sense, he seems to have been conscious that his attitude was not judicial. On behalf of the geologists he appealed to "that higher court of educated scientific opinion to which we are all amenable" for a verdict of "not guilty." He began by defining *Catastrophism*, the doctrine which, in order to account for geological phenomena, supposes the operation of forces different in their nature, or immeasurably different in power, from those we at present see in operation. The doctrine of violent upheavals, *débâcles*, and cataclysms in general is catastrophic, so far as it assumes that these were brought about by causes which have now no parallel. By *Uniformitarianism*, he designated the doctrine pre-eminently of Hutton and Lyell. Hutton had persistently, in practice, shut his eyes to the existence

[1] *Quarterly Journal of the Geological Society of London*, vol. xxv. pt. i. pp. xxxviii. to lii., 1869.

of that prior and different state of things before the formation of the stratified rocks which in theory he admitted. Both Hutton and Lyell agreed in their indisposition to carry their speculations a step beyond the period recorded in the most ancient strata now open to observation in the crust of the earth. Catastrophism had insisted upon the existence of a practically unlimited bank of force. Uniformitarianism had equally insisted on a practically unlimited bank of time. It had kept before our eyes the power of the infinitely little, time being granted, and had compelled us to exhaust known causes before flying to the unknown. To these two doctrines Huxley now added a third—*Evolutionism*—in which geology is conceived as the history of the earth in the same sense as biology is the history of living things. For this school he claimed Kant, and cited a passage from his *Cosmogony* as anticipating some of the "great principles" taught by Hutton. "On the one hand," he declared, "Kant is true to science. He knows no bounds to geological speculation but those of intellect. He reasons back to a beginning of the present state of things; he admits the possibility of an end." Evolution embraced in one stupendous analogy the growth of a solar system from molecular chaos, "the shaping of the earth from the nebulous cub-hood of its youth," through innumerable changes and immeasurable ages to its present form, and the development of a living being from the shapeless mass of protoplasm which we term a germ.

"I do not suppose," he pursued, "that, at the present day, any geologist would be found to maintain absolute Uniformitarianism, to deny that the rapidity of the rotation of the earth *may* be diminishing, that the sun *may* be waxing dim, or that the earth itself *may* be cooling. Most of us, I suspect, are Gallios, 'who care for none of these things,' being of opinion that, true or fictitious, they have made no practical difference to the earth during the period of which a record is preserved in stratified rocks. If Hutton and Playfair declare the course of the world to have been always the same, point out the fallacy by all means, but in so doing, do not imagine that you are proving modern geology to be in opposition to natural philosophy." But it was said that it is biology, not geology, which asks for so much time, that the succession of life demands vast intervals. Biology takes her time from geology. If the geological clock is wrong, all the naturalist will have to do is to modify his notions of the rapidity of change accordingly. Then dealing with the three lines of mathematical inference that seemed to limit the age of the earth to about 100 million years, Huxley pronounced the following well-known passage:—

But I desire to point out that this seems to be one of the many cases in which the admitted accuracy of mathematical processes is allowed to throw a wholly inadmissible appearance of authority over the results obtained by them. Mathematics may be compared to a mill of exquisite workmanship, which grinds you stuff of any degree of fineness; but, nevertheless, what you get out depends on what you put in ; and as the grandest mill in the world will not extract wheat-flour from peascods, so pages of formulae will not get a definite result out of loose data.

This last allusion shows that Huxley did not appreciate the fact that the most precise and rigid mathematics may yield definite and sure values for the higher and lower limits of a numerical result, without assigning any *necessary* value between those limits. He ended thus :—

My functions as your advocate are at end. I speak with more than the sincerity of a mere advocate when I express the belief that the case against us has entirely broken down. The cry for reform which had been raised without is superfluous, inasmuch as we have long been reforming from within with all needful speed. And the critical examination of the grounds upon which the very grave charge of opposition to the principles of natural philosophy has been brought against us rather shows that we have exercised a wise discrimination in declining to meddle with our foundations at the bidding of the first passer-by who fancies our house is not so well built as it might be.

Thomson's reply took the form of a masterly address, *Of Geological Dynamics*, to the Glasgow Geological Society on April 5, 1869. Though he subsequently returned to the subject in 1876 (see p. 675) and again in 1897, this address marks the culmination of his views. He asked first to state that the very root of the evil to which he objected was that so many geologists were content to regard the general principles of physics as matters foreign to their ordinary pursuits. He commented on Huxley's attitude as an advocate :—

Though a clever counsel may, by force of mother-wit and common sense, aided by his very peculiar intellectual training, readily carry a jury with him to either side when a scientific question is before the Court, or may even succeed in perplexing the mind of a judge, I do not think that the high court of educated scientific opinion will ever be satisfied by pleadings conducted on such precedents. But jury and judge may be somewhat perplexed as to what it is on which they are asked to give verdict and sentence, when they learn that Professor Huxley himself makes the gravest of the accusations against Hutton and uniformity, which he repels as made to me.

Huxley had cited the *Cosmogony* of Kant as anticipating, and had pronounced Kant to be true to science, because he knows no bounds to geological speculation but those of intellect, and reasons back to a beginning of the present. "Professor Huxley does not use words without a meaning," said Thomson, "and these mean that Hutton was *not* true to science, when he said, 'The result, therefore, of this physical inquiry is, that we find no vestige of a beginning, no prospect of an end.'" It was precisely because so many geologists had been Gallios who cared for none of these things that they had brought so much of British popular geology into direct opposition to the principles of natural philosophy. He then pointed out that already in 1860 Phillips had estimated, from stratigraphical evidence, the antiquity of life on the earth as possibly being between thirty-six millions and ninety-six millions of years. "How many orthodox geologists accepted this estimate fourteen months ago?" He excepted from this reflection Mr. Geikie,[1] who had twelve months before declared his secession from the prevailing orthodoxy. He then criticized Huxley's commentary on the numerical data as to retardation of the earth's rotation and the cooling of the sun ; and turning to Huxley's comparison of mathematics to a mill, he assented to the dictum that pages of formulae will not bring a definite result out of loose data. " I have not presented definite results," he replied, " I have amply indicated how 'loose' my data are ; and I have taken care to make my results looser."

Then, returning to the charge, Thomson added new paragraphs in support of the view that because of the former greater heat of the sun and earth metamorphic causes must have been more active in ancient times than now ; and he cited from the 1868 edition of Lyell's *Principles* a statement full of the same evil tendency to overlook essential principles of thermodynamics of which he had complained. The quotations justified him in maintaining that his call for reform was far from superfluous.

In conclusion he turned to Huxley's rather scornful reference to the passer-by who had bid them meddle with the foundations of their house, and ended with these dignified words :—

I cannot pass from Professor Huxley's last sentence without asking, Who are the occupants of " our house," and who is the " passer-by " ? Is geology not a branch of physical science ? Are investigations, experimental and mathematical, of underground temperature, not to be

[1] Now Sir Archibald Geikie, P.R.S.

regarded as an integral part of geology? Are suggestions from astronomy and thermodynamics, when adverse to a tendency in geological speculation recently become extensively popular in England through the brilliancy and eloquence of its chief promoters, to be treated by geologists as an invitation to meddle with their foundations, which a "wise discrimination" declines? For myself, I am anxious to be regarded by geologists, not as a mere passer-by, but as one constantly interested in their grand subject, and anxious, in any way, however slight, to assist them in their search for truth.

The unflinching tenacity with which Thomson stuck to his propositions was not less remarkable than the unvarying courtesy with which he treated his opponent. Two years later came Huxley's turn. As the retiring President of the British Association of 1871, at Edinburgh, it fell to him to introduce his successor in the chair, which he did in these memorable words:—

Permit me, finally, to congratulate the Association that its deliberations will be presided over by a gentleman, who in spite of what I must call the trifling and impertinent accident of birth, is to all intents and purposes a Scotchman. For it is in Glasgow that Sir William Thomson has carried out, now for five-and-twenty years, that series of remarkable researches which, as I am told by the most competent authority, places him at the head of those who apply mathematics to physical science, and in the front rank of physical philosophers themselves—an achievement, which, in this age of the cultivation of science, and in the pressing rivalry of able and accomplished men in all directions, confers upon him who realises it the title of an intellectual giant. On the one hand Sir William Thomson has followed out to the utmost limit of even that which has been dealt with at a previous meeting of the Association—I mean the scientific imagination—those speculations which carry men further forward in the pursuit of truth; and, upon the other hand, he has, by the versatility which is conferred only on the ablest men, been able to turn his vast knowledge and his remarkable ingenuity to the perfection and the carrying out of one of those great feats of engineering which may be looked upon as among the great practical triumphs of this age and generation. Those are the public, notorious, and obvious feats of your President-elect. What is less known and less obvious—his personal qualities—are such as I dare not, and will not, here dwell upon; but upon one matter which lies within my own personal knowledge I may be permitted to say of him, as the old poet says of Lancelot, that

"Gentler knight
There never broke a lance."

CHAPTER XIV

LATER TELEGRAPHIC WORK—THE SIPHON RECORDER

AFTER the triumphant success of the Atlantic telegraph, Sir William Thomson was associated for many years with various projects for other submarine cables. One of the first of these was the French Atlantic cable from Brest to S. Pierre, a stretch of 2580 miles. Though the concessions were negotiated by French financiers, the Company was essentially an English one, and was floated in 1868, Thomson and Varley being the consulting electricians, and Latimer Clark, Henry C. Forde, and Fleeming Jenkin[1] the engineers. The cable was manufactured by the Telegraph Construction Company, at Blackwall, and laid in June 1869 by the *Great Eastern*, under the superintendence of Mr. Willoughby Smith. Thomson was stationed at Brest to superintend the tests from the land end during laying. All went smoothly from first to last.

Thomson, Varley, and Jenkin had in 1865 joined partnership as to their patents; and Varley's

[1] Jenkin, writing home from the *Great Eastern*, off Ushant, June 20, says: . . . "Thomson shook hands and wished me well. I *do* like Thomson" (Memoir of Fleeming Jenkin, p. cx.).

patent for the interposition of condensers to accelerate signalling proved almost as profitable as Thomson's patent for the mirror galvanometer. The partnership continued in force till all the patents lapsed. Further, Thomson and Jenkin formed a partnership as consulting engineers; and they continued in association until Jenkin's death in 1885. In this way, and also in connection with the new inventions presently to be mentioned, Thomson was associated with other enterprises, amongst which may be mentioned the Falmouth, Gibraltar, and Malta Company, and the British Indian Company, both later merged in the Eastern Telegraph Company; and the Great Western Telegraph Company.

In April 1869 Lord Stanley (later Earl of Derby), then Secretary of State for Foreign Affairs, was installed as Lord Rector of Glasgow University. He became a firm friend of Thomson, who for many successive years used to spend a few days at Christmas in visits to Knowsley Park. At the installation banquet of April 2, 1869, in proposing the toast of the Progress of Science and Literature, and coupling it with the name of Sir William Thomson, the Chairman, Sir William Stirling-Maxwell, humorously rallied him on having encroached on the province of the professor of literature:—

The electric telegraph, with which Sir William Thomson's name is so intimately associated, has taught our writers many wholesome lessons. It has imposed on the expression of thought in this country, and all over the world, new and very admirable conditions. I cannot imagine any two writers working under circumstances and under laws more different than that writer who contributes to periodical literature, as it is sometimes supposed, for the humble contribution of a penny a line, and the writer who writes for the electric telegraph, having to pay ten guineas a line for the expression of his thoughts. The one writer may be said to be a man of words, whereas the other is essentially a man of letters. This lesson has been taught in many parts of the world by Sir William Thomson. His electric cable is preaching already all over the world that doctrine which was so ably enforced by my noble friend the guest of this evening, that "brevity is the soul of wit."

In replying, Sir William alluded to the benefit of free trade, to the promotion of liberal education, and to the maintenance of religious instruction in the national schools of Scotland.

Sir William Thomson now began to reap substantial pecuniary benefit from his telegraphic inventions. His first act was to devote a share of his gains to the University in which he laboured. He wrote to Principal Barclay:—

LARGS, *Sep.* 1, 1869.

MY DEAR PRINCIPAL—A few weeks ago I received a sum of £3000 for the use of my instruments on the Atlantic cables. This is the first payment which has been made to me in recompense for what has cost me a great deal of thought, labour, and money from 1856 to the present time, and I have now a prospect of further return.

I have always felt grateful to yourself and my colleagues for the liberal and friendly spirit which has been shown me in respect to my connection with telegraphic enterprise. I also feel that whatever success may now come I owe in a great measure to the facilities for experimenting which the College has afforded me. I am anxious to mark my sense of these benefits by setting aside something of what I have received for promoting the cultivation of experimental science in Glasgow College.

I am writing now to offer £1000 for the acceptance of the Senate for this purpose. I shall be much obliged to you for any suggestions as to the best way of arranging to make it useful that may occur to you.

A similar letter was written to his colleague Professor Allen Thomson, who immediately replied as follows:—

MORLAND, SKELMORLIE, *1st Septr.* 1869.

MY DEAR THOMSON—I received your letter last evening on my return from a pleasant sail in his steam yacht with John Burns.[1] I am sorry you cannot come to us this week, but I hope that Sharpey will stay at least next week with us, and that you will meet him in any way that may be most agreeable or convenient for yourself, so I beg you to let me know as soon as you return.

The latter part of your letter has given me much gratification. First of all, that you should begin to receive a part of the material reward of so much ingenuity, thought, and labour as you have bestowed on electrical and magnetic researches, and next that you should so generously and appropriately devote a part of your receipts to the purpose of enabling others to follow in your own distinguished career. As to the manner in which this object may be best accomplished, I confess that I think it a very difficult matter to decide. So far as I have any opinion at present, I am much more in favour of bursaries than scholarships, meaning by the first assistance *during* and by the second *after* College study. But there may be other ways, and I will consider the whole matter, and we may have the benefit among others of Sharpey's opinion, as he has now had very great experience in such matters.

[1] The late Lord Inverclyde.

I hope the weather may continue favourable to your trip, and that you may find it easy and beneficial to Lady Thomson.—Ever sincerely yours,

ALLEN THOMSON.

Principal Barclay's reply was equally cordial :—

THE COLLEGE, *Sep.* 11, 1869.

MY DEAR SIR WILLIAM—As I presume you have by this time returned from the north, I can delay no longer to congratulate you on the receipt of the first fruits of your important discoveries and inventions in connection with telegraphic science. In common with every one of your colleagues, I have always felt that it was our duty and our interest to afford every facility in our power for experiments by one whose talents and researches reflect so much honour on our University. Even your verbal acknowledgment of our desire to aid you in your investigations would of itself have been sufficiently gratifying—and you have really over-estimated that aid; but you have added a more substantial acknowledgment in devoting a large proportion of your pecuniary reward to promote in the University the cultivation of that science of which you are so distinguished a professor. Your very liberal donation of £1000 for that purpose shall be announced to the Senate at its next meeting, when it will be officially acknowledged.—Believe me ever, yours most truly,

T. BARCLAY.

Sir Wm. Thomson.

The formal acceptance by the Senate was duly recorded in the following minute of November 4th, 1869 :—

The Senate resolve to record their thanks to Sir William Thomson for the donation of £1000 which he has placed at their disposal, and which they have much pleasure in accepting; and they appoint Dr. Allen Thomson, Dr. Rankine, and Mr. Blackburn (Mr. Blackburn convener) a Committee to consult with Sir William

Thomson, and to report to the Senate in which way the sum may be best employed so as to promote the object contemplated.

The money thus given provided means for establishing the "Thomson Experimental Scholarships" to assist deserving students in the Natural Philosophy Laboratory. They proved for many years a great boon, and most attractive to students of the best sort, and still continue. On several occasions in after years Thomson himself contributed additional sums to increase the available annual dividend. Nor was this his only benefaction of the kind, for in 1874, when the Neil Arnott Demonstratorship was created with a sum of £1000 given by Dr. and Mrs. Neil Arnott, Thomson quietly added another sum of £2000, thus raising the total capital to £3000.

In September 1869 the Parliamentary representation of the Universities of Glasgow and Aberdeen became vacant by the appointment of Mr. Moncreiff to the office of Lord Justice-Clerk, and Sir William Thomson was asked to stand in the Liberal interest. At first he declined, and urged the claims of his friend Archibald Smith, Q.C. Smith agreed tentatively, then hesitated, partly on account of the expense; and, thinking that Thomson's hesitation had been on his account, he urged the Liberal Committee to put forward Thomson, whose nomination would have been much more acceptable, at any rate in Glasgow. But Thomson, after consultation with Mr. William

Graham, M.P., as to the actual duties of a Parliamentarian, telegraphed on October 8 that he found even the minimum attendance permissible to a Member of Parliament inconsistent with his duties in Glasgow, and definitely declined. Archibald Smith was therefore put forward, Thomson being one of the chairmen of his committee. A speech of Thomson's in his favour was printed and circulated, eulogizing Smith's mathematical and magnetic work, pronouncing him to be "a very advanced Liberal," and a safe representative in "the Liberal triumph to which we all look forward." The election turned largely upon the then burning question of religious education in the primary schools. In the sequel Mr. Gordon, the Conservative candidate, was elected.

Late in 1869 a new prospect was disclosed. For some years a movement had been afoot in Cambridge for the establishment of a Physical Laboratory, and the appointment of a professorship or lecturership in Experimental Physics. The Royal Commission on Scientific Education had given an additional impulse, and the new statutes of some of the colleges had paved the way to its realization. As the affair ripened hopes were raised that Sir William Thomson might be induced to accept the proposed post at Cambridge; and *pourparlers* were opened by the Master of Trinity in the following unofficial letter, written on the suggestion of the Senior Fellows, offering him a Prælectorship in Science:—

TRINITY LODGE, CAMBRIDGE,
10th Nov. 1869.

DEAR SIR W. THOMSON—A rumour having reached the ears of some of our Senior Fellows that on sanitary grounds, and possibly for other reasons, you will not be altogether indisposed, if occasion should offer, to exchange your post at Glasgow for another elsewhere, it has occurred to me and others that probably your own University of Cambridge might not have lost its hold on your regard. The reasons for troubling you with a remark which might otherwise have seemed impertinent is briefly this. By the statutes of our College we are enjoined, so soon as certain income is ensured to every Fellow, to elect three Prælectors, each of whom shall give lectures to the students of this and other colleges in some branch of Science or Literature, and for a certain fixed remuneration. By an amended statute, which has passed the Queen in Council, we are further enabled to offer any such Prælector a place on the roll of Fellows of the College at the annual examination for Fellowships in October. The amended statute removes certain practical difficulties which produced delay in acting upon that which it replaces, and we are desirous that another academical year should not pass without the appointment of at least one Prælector. A Fellowship is now worth in money value, irrespective of commons, etc., about £275, with a fair prospect of increase as the property of the College improves. A Prælector, in addition to his Fellowship, is entitled to a stipend of £250. The appointment is, therefore, worth £525, in addition to fees which the Prælector is entitled to receive from the students of other colleges. In considering the mode of filling up this important office, we had regard—1st, to the scientific requirements of the University; 2nd, to the probability of obtaining the services of a first-rate representative of the particular branch of science we might fix upon. It has been suggested to us by those who ought to know that the higher parts of experimental Physics constitute a department as yet insufficiently represented among us, and this information, joined to the rumour to which I have adverted at the beginning of this letter, has induced the Seniors unanimously to request me to ask you whether it would be agreeable to you that I should propose you for election at our next examination in October in the capacity of Prælector-Fellow. Of course we do not expect an immediate answer to a proposition for which you are unprepared; but if the thing is not altogether out of the question I shall be happy, if you wish it, to give you the most detailed information bearing on the office, its duties, and conditions. Our desire is to make such an appointment as may add to the reputation of the College, while it confers a benefit upon the University at large.—I remain, dear Sir William, yours very faithfully, W. H. THOMPSON.

P.S.—I forgot to add that the office I have described is statutably tenable with any Professorship or Public Lectureship in the University.

Thomson's reply has not been found, but its content is shown by the following minute of the Seniority, of November 16:—

The Master read a letter from Sir William Thomson, in which he regrets, that owing to Lady Thomson's serious illness, he is unable at present to accept the office of Prælector, or to form any new plans. He asks for more particulars of the office, and also when a decided answer must be given.

TRINITY LODGE, CAMBRIDGE,
19 *Nov.* 1869.

DEAR SIR WILLIAM THOMSON—We are all exceedingly sorry for the cause of your unwillingness, for the present, to entertain the proposal I made to you in my last; and none more so than our good Professor Sedgwick, who would have hailed your acceptance with enthusiasm.

There is, however, *no hurry*. We should like, if you found that the change did not suit you, to have a few months before next October in which to make inquiries in other quarters; but we need not think of this before the Lent Term—practically, that is, before the 1st February 1870.

From the provisions in our amended statutes I select the following:—

1. The Prælectors are "to give lectures in such subjects as the Master and Seniors shall from time to time determine."

2. Such lectures shall be open, if the M. and Srs. so determine, to students of other colleges.

3. The Prælectors hold office during the pleasure of the M. and Srs., and, those who are Fellows elected in virtue of their Prælectorship hold their Fellowship on the same tenure. After ten years' tenure of the Prælectorship they can, by a vote of the M. and Srs., retain their Fellowship for life, whether married or otherwise, *after resigning* the office of Prælector.

These statutes obviously give the M. and Srs. the right to determine the minimum residence of the Prælectors. This would be, probably, the major part or possibly $\frac{2}{3}$ds of each term. Lectures, too, would be required—a certain number every week during such major part of the term. If the Prælector were elected to a Professorship in the University, account would doubtless be taken of this in fixing the minimum N° of lectures, *e.g.* his University lectures might be accepted in lieu of College lectures on certain conditions favourable to students of Trinity. Here, however, I am speaking *de meo*; but in any case I venture to pledge the College to be reasonable in its demands on the time of a philosopher.

I made allusion to a "Professorship" in my previous letter. It is perhaps not unknown to you that our scientific men have long perceived our want of an adequate public representative of the department described as that of the Higher Physics, and—I am again only going by an impression—the prospect of having this want supplied does *not* seem very distant.

I have to add that I gave a low, not to say minimum

estimate of the emoluments of the office in my last. The M. and Srs. would have the power of making an additional allowance out of the so-called Tuition Fund, proportional, let us say, to the number of Trinity men attending the P.'s lectures. This, I think, would not be less than £100. All these points, however, could be definitely settled in case you thought fit to proceed further in this matter.

With earnest hopes for the abatement of your domestic anxieties.—I am, dear Sir William, yours very truly,

W. H. THOMPSON.

That Sir William had cherished a possibility is shown by the third letter of the correspondence :—

TRINITY LODGE, CAMBRIDGE,
11th Mar. 1870.

MY DEAR SIR—The Seniors would desire to see their way, if possible, to the appointment of a Prælector before the end of this term, and I should be obliged if you would inform me whether you have our proposal still under consideration, and whether, before the end of this month, you will be able to give us a final answer.—Yours very truly,

W. H. THOMPSON.

Professor Sir W. Thomson,
etc., etc.

Again Lady Thomson's extremely precarious state must have been the overt reason for Thomson's hesitant refusal, and the matter dropped at least until after her death in June 1870. Later in the year the matter progressed a stage. The Duke of Devonshire, who was Chancellor of the University of Cambridge, signified his desire to build and equip a Physical Laboratory for Cambridge. This magnificent offer resulted in the foundation of the Cavendish Laboratory, which was opened in 1874. The building was thus assured: the man to create the teaching was still to seek. On November 28, 1870, the Vice-Chancellor gave notice that there would be holden a congregation on Thursday the 9th day of February 1871, when the following grace will be offered to the Senate: That there be established in the University a Professorship of Experimental Physics. At once Cookson, now Master of Peterhouse, took the initiative and invited Sir William to offer himself for the proposed chair. His reply has happily been preserved:—

GLASGOW COLLEGE, *Dec.* 1, 1870.

MY DEAR COOKSON—I thank you very much for your letter of the 29th Nov.

Every decision that I can make in such a matter is equally painful to me. In other circumstances the comparative mildness of the climate and the congeniality of Cambridge society would have decided me without hesitation to accept. But now I feel that there is nothing for me in life but to make the most of what time remains for doing everything I can do in scientific work. The great advantages I have here with the new College, the apparatus and the assistance provided, the convenience of Glasgow for getting mechanical work done, give me means of action which I could not have in any other place. It would be impossible for me to feel as free from anxiety regarding the duties of any new position, and especially one of so great importance and responsibility as the proposed professorship in Cambridge, as I feel here after 24 years' habit, and without that freedom from anxiety I could do nothing in scientific work. But above all I have an invincible repugnance, in the circumstances in which I am left, to the idea of beginning a new life at all, and especially one which would have been so full of interest.—Believe me, yours very truly,

WILLIAM THOMSON.

Tait, to whom Thomson sent Cookson's letter and a copy of his reply, answered by the following characteristic note :—

O T.—Many thanks for the opportunity of perusing the enclosed. I do not quite agree with all you say to Cookson, but that does not affect my high opinion of your resolution and reasons taken as a whole.

Y^rs^, T′.

Then Thomson wrote to Helmholtz to sound him on the subject.

LONDON, *Jan.* 28, 1871,
THE ATHENÆUM.[1]

MY DEAR HELMHOLTZ — I have been asked by Stokes and by the Master and Tutor of my College at Cambridge to write to you asking if you could be induced to accept a new professorship of Experimental Physics to be established there. It is much to be desired to create in Cambridge a school of experimental science, not merely by a system of lectures with experimental illustrations, but by a physical laboratory in which students, under the professor and his assistants, would perform experiments, and the professor would have all facilities attainable for making experimental investigations. The Duke of Devonshire has already given £6000 for the building of the laboratory and making a commencement of providing it with instruments. I think it may be confidently expected that funds will not be wanting for carrying out this part of the plan well. The University proposes to give £500 per annum to the professor, whose income will also be increased somewhat by fees from the students. But if you accept the professorship it is, I may say, quite certain that you would be also appointed at least to a fellowship in St. Peter's College, which would bring an additional income of £250 or £300. Thus, I believe, the income to be relied on

[1] Sir William Thomson had been elected in 1870 a member of the *Athenæum Club*, under Rule II., for distinguished eminence in science.

would be not less than £800. But I think it probable that a prælector-fellowship of Trinity College would be offered to you, although on this point I speak merely from my own judgment. The duties of the prælector-fellowship, a new institution in Trinity College (of which one in physiology has been given to Dr. Michael Foster, and it was intended there should be also one in experimental physics), would be fulfilled by giving facilities to members of Trinity College for attending the professorial lectures: that is to say, if the University *professor* is appointed to the prælectorship. Otherwise the prælector would lecture on physics in Trinity College simply. The income of the prælectorship would be about £600, so that the professor if appointed to the prælectorship would have £1100 per annum.

I know that the question of amount of income is very far from the first in respect to the inducements which might possibly cause you to think of accepting such a proposal, and I am quite prepared to learn that you are "absolute for" Berlin. Still, it is a question that must be weighed before you could decide to accept, and therefore I have told you what I can regarding it. If you could at all entertain the idea of coming to this country, I think that the situation at Cambridge would present many advantages. It would certainly be a most interesting field in respect to our profession, as the desire for physical science is growing stronger and stronger in the University, and the force of public opinion is steadily advancing in support of it, and to stimulate it when stimulus is needed. The duties of *lecturing* would occupy only twenty weeks out of a year, and all the remainder of the time would be available for experimental or mathematical investigation and scientific writing. Will you give me a line at the earliest to say if you think you could be induced to accept, or if on the contrary you decide at once against it. I need not say that it would be a great gratification and advantage to English scientific men to have you among us instead of merely having very rare opportunities of seeing you, and that I myself would consider the difference of distance from Glasgow to Cambridge and Berlin a great gain.

I write necessarily in haste, and can only add that irrespective of this question I wished to write to you asking for news of yourself and of your family, and particularly how you have been affected by this terrible war. I hope very much that you have not now cause for anxiety in respect to your son or other relations.

I am in London on account of a committee invited to advise the Admiralty on Designs for Ships of War, especially in respect to stability, in consequence of the loss of the *Captain*, which was upset with 500 men in the Bay of Biscay last September. When you write address The College, Glasgow, as I am in London only for a few days each fortnight.—Believe me, yours always truly,

WILLIAM THOMSON.

Helmholtz, however, was unable to leave Berlin, where he had but just been appointed to take charge of the new Institute of Physics.

Clerk Maxwell was then invited to take the chair. He was elected in March 1871, and held it till his death in 1879.

When in 1870 the weekly journal *Nature* was founded as an organ for scientific news, Thomson was keenly interested. To the first volume he contributed an important original article "On the Size of Atoms," which appeared[1] on March 31, 1870. Estimates were afforded by four independent lines of arguments, which may be thus summarised:—

[1] This article was reprinted in 1883 as an Appendix to the second edition of Thomson and Tait's *Treatise*. In the same year the same subject was the topic of a discourse at the Royal Institution, reprinted in *Popular Lectures and Addresses*, vol. i. p. 147.

(1) Cauchy's theory of the dispersion of light by the prism indicates that the atoms must be much smaller than the size of a wave-length of light. In short, optical dynamics leaves no alternative but to admit that the diameter of a molecule, or the distance from the centre of a molecule to the centre of a contiguous molecule in glass, water, or any other of our transparent liquids and solids, exceeds a ten-thousandth of the wave-length, or a two-hundred-millionth of a centimetre.

(2) Comparing the forces observed in experiments on the contact-electricity of metals, with the work-equivalent of the heat evolved in chemical combinations of the same metals, it would appear that a close approximation to a chemical compound might be produced if plates of zinc and copper of a thickness of one-three-hundred-millionth part of a centimetre were placed close together alternately, if indeed such thin plates could be made without splitting atoms.

(3) From observations on the thickness of the thinnest films of soapy water, and on the work required to overcome the forces of surface-tension in drawing them out, and from the work-value of the heat required to convert water into vapour, it is probable that there are not several molecules in the thickness of a water-film that is a twenty-millionth of a millimetre thick.

(4) From the length of the mean free path of the molecules of a gas, according to the kinetic theory, the number of molecules in a given volume of gas may be inferred; and as the densities of gases, liquids, and solids are known, it may be deduced that the distance from centre to nearest centre in liquids and solids is not greater than one one-hundred-and-forty-millionth part, nor less than one four-hundred-and-sixty-millionth part of a centimetre.

The result is thus summed up. To form some conception of the degree of coarse-grainedness indicated, imagine a rain-drop or a globe of glass as large as a pea[1]

[1] In the later discourse at the Royal Institution, in 1883, this conclusion was modified by substituting for a *pea* a *football*, "or say a globe of 16 centimetres diameter."

to be magnified up to the size of the earth, each constituent molecule being magnified in the same proportion. Then the magnified structure would be more coarse-grained than a heap of small shot, but probably less coarse-grained than a heap of cricket-balls.

Glasgow University underwent a great transformation in the year 1870. Founded in 1451 by a Bull of Pope Alexander V., its once handsome quadrangles had grown dingy, its buildings dilapidated and inadequate, and its surroundings altogether squalid, insanitary, and even dangerous. For many years an agitation had been on foot for its removal to more suitable quarters, when the development of the North British Railway, by demanding the site for the purposes of a goods station, brought matters to a crisis. If the city would but assign a new site, the sum payable by the Railway for compensation would go far towards the cost of rebuilding. Government aid was also sought for funds that the new buildings might be adequate. A commanding and worthy site was, after protracted negotiations, assigned on the top of Gilmore Hill, overlooking the Kelvin burn half a mile from the point where it empties itself into the Clyde. Here, with the aid of a grant of some £120,000 from the Imperial Treasury, the noble pile of the present University was erected. For some three or four years the buildings had been rising toward completion.

From a speech delivered by Sir William Thomson at Greenock on April 18, 1870, in aid of the

NEW UNIVERSITY BUILDINGS, GILMORE HILL, GLASGOW.

building fund, it appears that for the purchase of the site, the erection of the building including the hospital, medical school, and the official residences of the professors, the estimated cost was £360,000. The railway company paid £100,000 for the site of the old College, and £127,000 was raised by public subscription. The fine Bute Hall, between the two quadrangles of the new building, was added a few years later at the cost of the Marquis of Bute.

On May-Day 1870 the last degree-giving in the old College was celebrated. Sir William Thomson was not present; and the severe domestic bereavement which fell upon him six weeks later prevented him from taking any active part in the removal of his laboratory to its new quarters in the western end of the chief block on Gilmore Hill. The work of superintending the removal fell upon his nephew, Dr. James T. Bottomley, who had resided with him for several years, and upon his faithful old assistant Donald MacFarlane. The official residence assigned to Sir William Thomson was at the end of the Professors' Court, only a few paces from his lecture-theatre and laboratory. In the new theatre were installed many of the time-honoured appliances used in the old College; and to the well-lighted new rooms provided for laboratories were brought the treasured instruments and paraphernalia which had been accumulating in the dark cellar and the deserted Blackstone lecture-room of the earlier time. A capital account of the new Physical

Laboratory of the University was given in *Nature* of May 9, 1872.

This is the appropriate place to narrate the new and important development in submarine telegraphy with which at this stage Sir William Thomson enriched the world, when he devised his celebrated instrument called the *Siphon Recorder.*

For nearly two years he had been very closely occupied in the realization of an idea which had long before presented itself to him of constructing an apparatus which, while as sensitive to minute impulses as his earlier mirror galvanometer, should leave an actual record of the message received through the cable from the distant sending station. Already in 1860, in his *Encyclopædia Britannica* article (see p. 386 *supra*), he had written—

> To ink a pen so that it may mark a dash on the moving paper as long as it is held on by the electro-magnetic force of the feeblest signal-current that it is in other respects desirable to use, is the invention wanting to supersede relays in recording apparatus for the Morse system.

He had also realized—as his *Good Words* article (p. 508) shows—the gain in substituting for Morse's method of dot and dash the Steinheil plan of signals, in which, by the use of currents of different polarities, movements to right and left are employed. A dash takes more time than a dot, but a movement to the left may be as rapid as one to the right; moreover, the use of alternating polarity helps to reduce the cable retardations. He had used the Steinheil plan for years with his mirror galvanometer; but to *record* the movements automatically was a new problem.

The friction of a pen or pencil on a moving tape of paper is far too great to permit the recording point to be thrust to right or left by the delicate electric impulses that suffice to deflect the little mirror of the galvanometer. To record cable-signals, therefore, some means must be found of abolishing friction and of making the recording pen light as gossamer.

One of the earliest attempts at such a receiving instrument was the "Spark Recorder" which, though never introduced commercially, is of interest on account of the principles used in its construction, and because it to some degree foreshadowed the more successful instruments of later date. Its action was as follows:—An indicator, actuated by the currents constituting the signals, was caused to take a to-and-fro movement. This indicator was connected to a Ruhmkorff coil capable of sending a continual succession of sparks to a horizontal metal plate laid below the moving part of the indicator. Over the surface of this plate there was passed mechanically, by clockwork, at a uniform speed, a ribbon of prepared paper or other material capable of being acted on by the sparks, either by their chemical action, their heat, or their perforating force. The record of the signals presented an undulating line of spots or fine perforations; and from the character and succession of these undulations the signals could be interpreted.

After many months of thought and labour in building experimental instruments, success came with a different apparatus of the following nature. Between the poles of a powerful electro-magnet (or later a very strong compound horseshoe magnet) there was hung a lightly-suspended coil of fine wire forming part of the circuit. This coil, when any small electric current traversed it, experienced forces tending to turn it to right or left in its place between the poles, just as in the mirror galvanometer the tiny magnet and mirror turn within the surrounding coil. A hair-like glass tube

Compensated 9.1 Words per minute

can you read this yes

(vaccine tubing, in fact) about two inches long, bent sharply to form a siphon, was suspended by silk threads and attached to the suspended coil, so that the end of the siphon would participate in any movement of the coil. It was filled with ink, and the ink electrified by a small special electrifying machine, causing it to spurt forward from the end of the siphon in a jet of exceeding fineness. Opposite this jet glides the paper ribbon receiving the fine ink line — straight when no signals are being sent, jagged with indentations to one side or the other as the signals drive the siphon hither or thither. The accompanying facsimile is from a sample of the records so made. Sir William gave the name of *Siphon Recorder* to this ingenious and beautiful apparatus.

The accompanying figure shows the instrument in one of its later simplified forms. Two upright bundles of steel plates constitute the permanent magnet, between the upper ends of which swings the signalling coil that moves the delicate ink siphon. The paper ribbon travels slowly

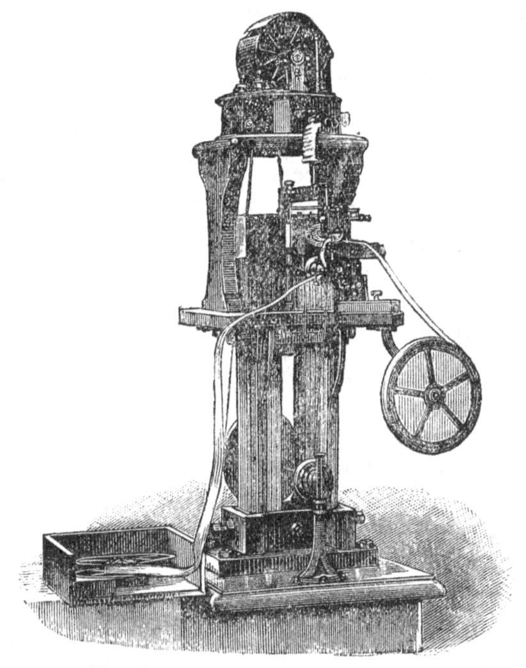

THE SIPHON RECORDER (later pattern).

from one reel to the other beneath the recording tip of the siphon. At the top is situated the "mouse-mill," a little influence-machine of compact form which electrifies the ink.

It was patented in 1867, but required nearly three years of incessant work to bring it into satisfactory shape. On March 9, 1869, Sir William was able to speak of it to Sir James Anderson in the following terms:—

March 9, 1869.

My dear Sir James—I enclose some specimens of my new recorder's performance. I have succeeded in giving it much increased sensibility, and, indeed, I can increase it farther almost without limit. Thus, through any length of line, however great, it will, I believe, be able to do all that the mirror can do, and as quickly, with the advantage of leaving a permanent record. It may be read off by eye, I think, probably with as much ease as the mirror during the actual arrival of the signals. But I would advise (however confident we may come to feel as to the new instrument) that the mirror should be trusted to in the first place. The new instrument may be put in circuit simultaneously with the mirror without either disturbing the indications of the other injuriously; and thus, without loss of any practical work, we may have a competitive trial of the two systems.

As the mirror is to be trusted to for opening work, I do not urge you to go to the expense of having recorders made just now for the three stations. But as it cannot be very great (certainly, I should think, less than £150), you may perhaps think it right to order them at once. If so, please let me know as soon as possible, as there is not much time to spare. If you are satisfied to try the new instrument at one station, you may have my present one, which, though a trial instrument, is available for actual work, as long as you require it, in testing its usefulness.—Yours very truly,

William Thomson.

One advantage of such a recorder is that its operation involves far less strain upon the nerves of the clerk in charge of the receiving station, who, when watching the quivering spot of light from a mirror galvanometer, fears to take his eyes from the instrument lest he miss a movement of the spot. There is no such anxiety with the recorder, as the signals on the ribbon can be read off at leisure.

The instruments, both experimental and commercial, were constructed in the workshops of Mr. James White of Glasgow, whose name was associated for forty years with Sir William Thomson's inventions in the capacity of instrument maker.

A few months later the recorder was installed for trial on the French Atlantic cable, one instrument being sent to Brest under the charge of J. D. H. Dickson, another to St. Pierre under that of W. F. King, another pupil. By April 1870 the Anglo-Mediterranean and British-Indian Telegraph Companies were prepared to entertain proposals to adopt the recorder, and Sir William wrote to say that he and his partners would expect a royalty of £1000 per annum from each of these companies for the duration of the patent.

The first exhibition of the Siphon Recorder in England took place on June 23, 1870, when, on the occasion of the completion of the British-Indian submarine telegraph by cables stretching from Falmouth to Bombay, a demonstration was given at the London residence, Arlington Street, of Mr. (afterwards Sir John) Pender, and was witnessed by the Prince of Wales (now our King) and the Duke of Cambridge. *Punch* made the exhibition of the new instrument the occasion of a mock-heroic poem, in the issue of July 9th, under the title of *Tempus Fugit*:—

When Piccadilly's ablaze, in the height and the heat of the season—
Rises a gaily hung tent in the yard of the mansion of Pender—
Mansion belit and bepictured and crowded with stateliest swelldom,
Swelldom that, down from blood royal, in Wales and in Cambridge embodied,
Flows through the pipes of the Peerage—Diplomacy—Ministers—Members—
Thence to the Magnates of Money and so to the syndics of Science.
Ceaseless the buzz and the bowing, the flashing of stars and garters,
Ceaseless the mopping of brows and imbibing of cooling refreshments,
Endless the glare and the glitter and gossip—the wealth and the wittles,
What have they met to accomplish, these leaders of fashion and science?
What is it brings them together, before the small syphon that, waving,
Scatters its fine jet of ink in accord with the pulses electric,
So making plain to the eye what the spark through the wire is conveying;
What is transacting to-night in the tent of the mansion of Pender?
Lo, 'tis Britannia stretching invisible hands under ocean,
Bringing the furthermost East and the uttermost West into contact;

. . . .

So does the spark of our wires outpace e'en the fleet foot of Chronos!
Miracle-workers are we—sitting here in the mansion of Pender,
Gossiping thus at our ease, over Continents, Hemispheres, Oceans,
Saying to space "Be no more," and to baffled Time, "Get thou behind me!"

As so usual in journalism, this account omits the name of the man of science whose achievement was thus being celebrated.

A month later Sir William Thomson was busily occupied with the installation of the recorder on the Falmouth-Malta cable. Thence he wrote to his sister-in-law, Miss Jessie Crum, to tell of his work:—

PORTHCURNO, *July* 21.

We found Leitch here with the instrument on Saturday, and by Monday we had it running fairly well. On Tuesday we could offer it as quite ready to do the Company's work, if their clerks could read it. One or two of them in the course of a few odd half-hours have learned to decipher it, but a little practice (a few days) will be necessary to allow them to read it easily enough for practical purpose. Meantime the superintendent here (Bull, one of my old *Agamemnon* A.T.C. staff) and his clerks seem quite convinced that it will do their work with more ease to themselves, more speed, and more accuracy than the mirror. So the days of signalling by the "spot of light" are numbered, and a luminous electrified pen will succeed the mirror. I enclose a small but very clear specimen of its work. . . . This evening we expect extra clerks from London to learn it, or take the place of some of those here and allow them to learn it. As soon as we see them fairly in train I shall leave, but that cannot be before Monday, as there is a great block on the line owing to this wicked war, stopping European telegraphs, and causing numerous commercial messages to "cancel orders," which will likely last till Saturday night. Sunday may possibly be a quiet day, and if so *must* be used for trials which cannot be done when messages are passing. I am afraid it will be a week from now before I can be with you at the Rouken, as I must be two days in London. I am very glad poor Fairy is better.—Your affectionate brother, W. T.

A kindly letter from Helmholtz elicited the following reply :—

PORTHCURNO, PENZANCE, *July* 29, 1870.

MY DEAR HELMHOLTZ—I thank you most warmly for your kind letter. It is indeed, as you say, an unspeakable great loss which I have suffered. My sense of it goes on increasing every day and through all my occupations, after the first shock. That the end was certainly a happy relief for my dear wife from incessant suffering has done nothing yet to diminish the desolation in which it has left me. I am very sorry to think that there is now no prospect of meeting you soon as I had hoped. I could have spoken to you of these things but I feel it impossible to write. Meantime what I suppose is the best medicine for me has been forced on me—sheer hard work. The completion of the cables between England and India two months ago has led to an urgent demand for my recording instrument [an electrified pen (a very fine capillary siphon) shooting ink at the paper] of which, I think, I told you when I saw you last in Heidelberg. I have been here, at the English terminus of the Falmouth,[1] Lisbon, Gibraltar, and Malta cable, for nearly a fortnight, and have got the recorder into full action, writing down every signal that passes through the cable either way, between this and Lisbon. I enclose a specimen of its performance, on which I have written the interpretation in ordinary (but less clear!) lettering and figures. The power used at Lisbon is only eight small sawdust (Daniell's) cells. At this end the cable is connected with one plate of a condenser equal in electrostatic capacity to about 100 nautical miles of cable, the other plate of which is connected with the earth (that is in this case the iron sheath of the cable), a mirror galvanometer (of very dull construction) with 2500×10^9 cms. per second of resistance[2] reckoned in absolute electro-magnetic units, and my new instrument, the coil of which has only 100×10^9 of resistance, are both in circuit between the cable and one plate of the condenser, or the other plate and the earth. The resistance of the cable itself is 8200×10^9, and its length 823 nautical miles. The messages are as yet received practically on the mirror, but operators are daily learning the new instrument. Even in the specimen I send you you may see an advantage of the new instrument. The repetition of the word "Anglais" would, had the practical operators been using it, have been unnecessary; and half a minute of time would have been saved. The cable is full of work throughout the twenty-four hours.

I hope to be able to leave for Glasgow (where the removal to the new College is going on energetically) in two days, all being nearly in train here.

I would feel much obliged by a single line from you to-day if this reaches you. I address Berlin because I read in *Nature* a fortnight ago that you *had gone there*

[1] So called, no doubt, because the first intention was to have the terminus at or near Falmouth.

[2] The signalling will be much clearer in the received messages when the galvanometer is thrown out.

from Heidelberg; and judge at all events that it will run less risk so of finding communications stopped than if it were addressed to Heidelberg. Tell me also if possibly after all you will be in England this autumn. I shall not go to the British Association, but I hoped to persuade you to come to Scotland.—Yours always truly,

W. THOMSON.

He was detained at Porthcurno some time longer, as the following letters to Miss Crum show:—

PORTHCURNO, *Aug.* 1.

Day after day passes, and every one has up to now brought some fresh matter of importance requiring attention. Even yesterday (after church, evening service, and my late bath in the sea, which I generally take just before going away for the night) there came an opportunity for trying the recorder alone (till then the "mirror" having been always in circuit with it, to be used for the practical work, while a clerk is learning the recorder in a separate room). The result of putting out the mirror, although I expected an improvement, was surprising. Bull (the superintendent) was much struck with it. . . . So he has set to-day to make arrangements for doing practical work on the recorder, and discarding the mirror as soon as possible. . . . The business is really of great importance. As soon as Leitch can leave the instrument here (or rather a new and more perfect one to be sent from White's) he will have to commence a progress eastward with seven recorders to be installed at Lisbon, Gibraltar, Malta, Alexandria, Suez, Aden, and Bombay. So the more thoroughly I do the thing now, the less trouble there will be to myself and others henceforth. For instance, I hope to escape having to go to Lisbon, Gibraltar, and Malta myself in Sept. and October by getting all well in train here. . . .

I have been to Land's End this morning, which is a thing not to be missed. The scene is very grand—wild precipitous headlands of granite—cyclopean architecture, and a raging tide below, though the sea was very calm.

Aug. 3, PORTHCURNO.

You would scarcely believe, and perhaps can scarcely understand, what a pain the intelligence of your letter, which I have just received, has given me. I had thought of the poor little creature,[1] though not likely to be very well again, still giving me a kind welcome and showing some joy when I return. But I feel something quite different from the regret for a faithful little friend like that. It seems to me as if a link were gone, and the bitterness of my great grief comes strongly on me. This is unspeakable and not to be written about, but it is always present, and worse after the first shock, when what I lost was nearer.

He had now purchased the yacht *Lalla Rookh*, which was destined to play so large a part in the next few years of his life (see p. 585), and on leaving Cornwall sailed in her to the Hebrides. Returning he wrote to Helmholtz.

GLASGOW COLLEGE, *Sept.* 8, 1870.

MY DEAR HELMHOLTZ—I received your letter only last night, on returning from a yacht cruise of a fortnight among the Western Islands. I have read it with the liveliest sympathy, and I thank you for it very much. I hope you will not be displeased that I have given an extract from it to be published (without names) in the *Glasgow Herald* (newspaper).

We are all deeply interested in your case,[2] and all (with scarcely an exception) agree in abhorrence of the whole action of the French from the declaration of war till now. I believe France itself will be better and happier ten years hence for the bitter lesson you are now teaching it. But it is a terrible price you are paying for what is only your

[1] [This refers to the little black-and-tan terrier "Fairy," which had belonged to Lady Thomson.]

[2] The reference here is to the circumstance of Helmholtz's son then serving in the campaign of the Franco-Prussian War. See p. 588.

right. There is only one point on which we can feel neutral here, and that is sorrow for the wounded and those who are bereaved. We are trying to raise funds to assist in some slight degree, but all the subscriptions that can possibly be raised cannot be more than a mite towards the vast amount required. Before I left the cruise I had sent a subscription to a fund that was instituted for aid to the German wounded, and intended to send an equal sum for the French as soon as a subscription for them should be commenced. But during my absence it was arranged that the fund first instituted should be for *German and French* wounded. Instead of sending the other sum as a second subscription to this fund, if you will allow me I will send it through you for any purpose in connection with the hospitals or with bereaved families that you may judge right. I therefore send you halves of Bank of England notes for £25, and shall be obliged by a line from you to say that you have received them. I keep the other halves meantime, and shall send you them instantly when I hear that you have the first halves. I hope that this is not troublesome to you, and wish I could do anything to assist in the smallest way to relieve the vast mass of suffering in the middle of which you are now working.—Yours always truly, WILLIAM THOMSON.

On September 11 he wrote from Thornliebank to Mrs. King that he hoped to go to Arran to see her, but that he "must let the British Association be either finished or nearly so before I can leave, as I have a good deal of writing to do for it."

He did not go to the Association at Liverpool, but helped to prepare the reports of two committees—on Harmonic Analysis, as applied to Tidal Observations, and on Electrical Standards—and soon returned south. From Cowes, on September 20, he wrote to Miss Crum that matters had advanced about agreement for the use of the recorder on the Indian telegraphs. The message, pencilled on a couple of pages torn from his "green book," continued:—

At Cowes the *Lalla Rookh* exceeded my expectations. I came back to Southampton intending to get as far towards Swanage (unknown region) same night. I got as far as Bournemouth Sat. night late. Sunday forenoon exquisite church, pitiful service (would have been less fatiguing in Latin, as then one would not have had the fatigue of straining the ear to try to hear). Boat afternoon to Swanage (10 miles) where very kind reception from A. S. and his wife. I shall probably sail to Belfast in *Lalla Rookh*, Frid. to Monday. Arran, Tuesday. Rouken, Thursday or Frid. No time for more as plank ashore.—Yours, W

It was early in this month of September, just when all Europe was watching the Prussian armies closing in upon Louis Napoleon at Sedan, that England was horrified by the news that H.M.S. *Captain* had gone down with five hundred hands while on her trial voyage in Vigo Bay. Sir William Thomson's labours as a member of the Admiralty Committee that resulted from the inquiry into the loss are recounted elsewhere (see p. 731).

The formal transfer of the University to its new quarters took place on Monday, November 7, 1870, the Principal and professors marching in procession from the old College in High Street to the new buildings on Gilmore Hill. A civic banquet celebrated the occasion, but Sir William Thomson, being in mourning, did not attend it.

With the new year came the labours of the Admiralty Committee on the Design of Ships of War, under Lord Dufferin as chairman, which met in London, fortnightly, for several months. Sir William was compelled to depute much of his duties to his nephew, James T. Bottomley, until the end of the session. The following letter to James Thomson shows the nature of the questions in which Sir William Thomson interested himself.

COLLEGE, GLASGOW, *Jan.* 31, '71.

MY DEAR JAMES—I have just got back from London this morning, where I have been since Friday attending meetings of the Committee on Designs for Ships of War. This is the second set of meetings we have had, and I shall have to go again on Thursday night next week, and so on every second week, to attend three meetings (Frid., Sat., and Monday) each time. We have, although not yet part of the proper business of the committee, had a good deal of discussion of the "hydraulic" method of propulsion (see pamphlet sent by book post, which please return soon, not later, say, than by Tuesday's dispatch of next week). The facts regarding the *Waterwitch* seem to show that when the thing is done on better principles and with better details great results may be anticipated

I have been making some calculations, and it seems, so far as I can judge, that if water is discharged by an orifice, properly shaped both outside and inside, under the level of the sea at the stern, it might be a more economical propeller than the screw.

Sundry letters to Miss Jessie Crum reveal the busy occupations of these months.

Mar. 4 [a post card].—Rec[d.] your letter in London just as I was leaving for Plymouth on Monday afternoon. Tried the *Hotspur* (Ram) outside the breakwater on

Tuesday, to see the green water breaking over her bow and washing up to the turret. Tried also the *Waterwitch* (hydraulic propeller) with little satisfaction. Dined at Admiral Houston Stewart's. . . . Next day attacked merchant ship . . . with a torpedo . . . etc., etc. . . . I got back in time for my lecture on Thursday morning Heard Dr. Raleigh on Sunday, also saw Mrs. Faraday and Miss Barnard.

Mar. 11, ADMIRALTY, *Sat.*

I write in the middle of the meeting, during a general sea-fight, the admirals all attacking the *Cerberus*, *Hydra*, and *Cyclops* very vehemently. You must not, therefore, expect anything involving continuous thought or collected ideas, which the din of action renders impossible. I was engaged to dine with Varley (at his club) yesterday, to discuss with Lord Sackville Cecil (who has charge of my last recorder sent recently to Porthcurno) and go to the Royal Institution to hear Dr. Carpenter on deep-sea results. . . . I go this evening to Cambridge to stay till Monday. Porter (Mrs. Tait's brother, who is tutor of Peterhouse, and was a student of my father's, and one of my own earliest) had asked me to spend a Sat. till Monday along with Mr. Dickenson[1] at Peterhouse. . . . On Tuesday we go to see the *Glatton*, just launched at Chatham. And on Tuesday night I return to Glasgow. . . .

He was eager to get over the few remaining weeks of the session, that he might begin cruising on his yacht.

[1] Mr. Lowes Dickenson, who painted the portraits of Cookson and of Lord Kelvin now hanging in Peterhouse.

END OF VOL. I

Printed in the United States
By Bookmasters